OpenStat Reference Manual

William Miller

OpenStat Reference Manual

Springer

William Miller
Iowa State University
Ames, IA, USA

ISBN 978-1-4614-5739-8   ISBN 978-1-4614-5740-4 (eBook)
DOI 10.1007/978-1-4614-5740-4
Springer New York Heidelberg Dordrecht London

Library of Congress Control Number: 2012951039

© Springer Science+Business Media New York 2013
This work is subject to copyright. All rights are reserved by the Publisher, whether the whole or part of the material is concerned, specifically the rights of translation, reprinting, reuse of illustrations, recitation, broadcasting, reproduction on microfilms or in any other physical way, and transmission or information storage and retrieval, electronic adaptation, computer software, or by similar or dissimilar methodology now known or hereafter developed. Exempted from this legal reservation are brief excerpts in connection with reviews or scholarly analysis or material supplied specifically for the purpose of being entered and executed on a computer system, for exclusive use by the purchaser of the work. Duplication of this publication or parts thereof is permitted only under the provisions of the Copyright Law of the Publisher's location, in its current version, and permission for use must always be obtained from Springer. Permissions for use may be obtained through RightsLink at the Copyright Clearance Center. Violations are liable to prosecution under the respective Copyright Law.
The use of general descriptive names, registered names, trademarks, service marks, etc. in this publication does not imply, even in the absence of a specific statement, that such names are exempt from the relevant protective laws and regulations and therefore free for general use.
While the advice and information in this book are believed to be true and accurate at the date of publication, neither the authors nor the editors nor the publisher can accept any legal responsibility for any errors or omissions that may be made. The publisher makes no warranty, express or implied, with respect to the material contained herein.

Printed on acid-free paper

Springer is part of Springer Science+Business Media (www.springer.com)

*To the hundreds of graduate students and users of my statistics programs. Your encouragement, suggestions and patience have kept me motivated to maintain my interest in statistics
and measurement.*

To my wife who has endured my hours of time on the computer and wonders why I would want to create free material.

Contents

1	Introduction	1
2	Installing OpenStat	3
3	Starting OpenStat	5
4	**Files**	7
	Creating a File	7
	Entering Data	10
	Saving a File	10
	Help	11
	The Variables Menu	12
	The Edit Menu	15
	The Analyses Menu	19
	The Simulation Menu	20
	Some Common Errors!	20
	Empty Cells	20
	Incorrect Format for Floating Point Values	20
	String Labels for Groups	21
	Floating Point Errors	21
	Values Too Large (or Small)	21
5	**Distributions**	23
	Using the Distribution Parameter Estimates Procedure	23
	Using the Breakdown Procedure	23
	Using the Distribution Plots and Critical Values Procedure	25
6	**Descriptive Analyses**	27
	Frequencies	27
	Cross-Tabulation	30
	Breakdown	32
	Distribution Parameters	35
	Box Plots	35

	Three Variable Rotation	38
	X Versus Y Plots	39
	Histogram/Pie Chart of Group Frequencies	41
	Stem and Leaf Plot	43
	Compare Observed and Theoretical Distributions	44
	QQ and PP Plots	44
	Normality Tests	46
	Resistant Line	47
	Repeated Measures Bubble Plot	49
	Smooth Data by Averaging	51
	X Versus Multiple Y Plot	52
	Compare Observed to a Theoretical Distribution	56
	Multiple Groups X versus Y Plot	57
7	**Correlation**	**61**
	The Product Moment Correlation	61
	Testing Hypotheses for Relationships Among Variables: Correlation	62
	Simple Linear Regression	63
	Testing Equality of Correlations in Two Populations	65
	Differences Between Correlations in Dependent Samples	65
	Binary Receiver Operating Characteristics	66
	Partial and Semi_Partial Correlations	70
	Partial Correlation	70
	Autocorrelation	72
8	**Comparisons**	**79**
	One Sample Tests	79
	Proportion Differences	82
	t-Tests	85
	One, Two or Three Way Analysis of Variance	87
	Analysis of Variance: Treatments by Subjects Design	90
	One Between, One Repeated Design	92
	Two Factor Repeated Measures Analysis	95
	Nested Factors Analysis of Variance Design	99
	A, B and C Factors with B Nested in A	100
	Latin and Greco-Latin Square Designs	103
	Plan 2	105
	Plan 3 Latin Squares Design	108
	Analysis of Greco-Latin Squares	111
	Plan 5 Latin Square Design	113
	Plan 6 Latin Squares Design	116
	Plan 7 for Latin Squares	118
	Plan 9 Latin Squares	121
	2 or 3 Way Fixed ANOVA with 1 Case Per Cell	126
	Two Within Subjects ANOVA	129
	Analysis of Variance Using Multiple Regression Methods	132
	An Example of an Analysis of Covariance	132

	Sums of Squares by Regression	136
	The General Linear Model	141
	Using OpenStat to Obtain Canonical Correlations	141
	Binary Logistic Regression	145
	Cox Proportional Hazards Survival Regression	147
	Weighted Least-Squares Regression	148
	2-Stage Least-Squares Regression	153
	Non-linear Regression	158
9	**Multivariate**	163
	Discriminant Function / MANOVA	163
	An Example	163
	Cluster Analyses	172
	Hierarchical Cluster Analysis	172
	K-Means Clustering Analysis	177
	Average Linkage Hierarchical Cluster Analysis	178
	Path Analysis	181
	Example of a Path Analysis	181
	Factor Analysis	189
	General Linear Model (Sums of Squares by Regression)	194
	Example 1	195
	Example Two	199
	Median Polish Analysis	203
	Bartlett Test of Sphericity	204
	Correspondence Analysis	206
	Log Linear Screening, AxB and AxBxC Analyses	210
	The Screening Procedure	212
	The A × B Log Linear Analysis	214
	The A × B × C Log Linear Analysis	217
10	**Non-parametric**	237
	Contingency Chi-Square	237
	Example Contingency Chi Square	237
	Spearman Rank Correlation	240
	Example Spearman Rank Correlation	240
	Mann-Whitney U Test	241
	Fisher's Exact Test	243
	Kendall's Coefficient of Concordance	245
	Kruskal-Wallis One-Way ANOVA	246
	Wilcoxon Matched-Pairs Signed Ranks Test	248
	Cochran Q Test	249
	Sign Test	250
	Friedman Two Way ANOVA	251
	Probability of a Binomial Event	252
	Runs Test	253
	Kendall's Tau and Partial Tau	255

	Kaplan-Meier Survival Test	256
	The Kolmogorov-Smirnov Test	265
11	**Measurement**	269
	The Item Analysis Program	269
	Analysis of Variance: Treatment by Subject and Hoyt Reliability	275
	Kuder-Richardson #21 Reliability	277
	Weighted Composite Test Reliability	278
	Rasch One Parameter Item Analysis	280
	Guttman Scalogram Analysis	285
	Successive Interval Scaling	287
	Differential Item Functioning	289
	Adjustment of Reliability For Variance Change	307
	Polytomous DIF Analysis	308
	Generate Test Data	312
	Spearman-Brown Reliability Prophecy	315
12	**Statistical Process Control**	317
	XBAR Chart	317
	An Example	317
	Range Chart	320
	S Control Chart	322
	CUSUM Chart	324
	p Chart	326
	Defect (Non-conformity) c Chart	328
	Defects Per Unit U Chart	330
13	**Linear Programming**	333
	The Linear Programming Procedure	333
14	**Using MatMan**	337
	Purpose of MatMan	337
	Using MatMan	337
	Using the Combination Boxes	338
	Files Loaded at the Start of MatMan	338
	Clicking the Matrix List Items	339
	Clicking the Vector List Items	339
	Clicking the Scalar List Items	339
	The Grids	339
	Operations and Operands	340
	Menus	340
	Combo Boxes	340
	The Operations Script	341
	Getting Help on a Topic	341
	Scripts	341
	Print	342
	Clear Script List	342

Contents   xi

    Edit the Script 342
    Load a Script 343
    Save a Script 343
    Executing a Script 344
    Script Options 344
Files 344
    Keyboard Input 345
    File Open 346
    File Save 346
    Import a File 346
    Export a File 347
    Open a Script File 347
    Save the Script 347
    Reset All 348
Entering Grid Data 348
    Clearing a Grid 349
    Inserting a Column 349
    Inserting a Row 349
    Deleting a Column 349
    Deleting a Row 349
    Using the Tab Key 350
    Using the Enter Key 350
    Editing a Cell Value 350
    Loading a File 350
Matrix Operations 351
    Printing 351
    Row Augment 352
    Column Augmentation 352
    Extract Col. Vector from Matrix 352
    SVDInverse 352
    Tridiagonalize 354
    Upper-Lower Decomposition 354
    Diagonal to Vector 354
    Determinant 354
    Normalize Rows or Columns 355
    Pre-multiply By 355
    Post-multiply By 356
    Eigenvalues and Vectors 356
    Transpose 357
    Trace 357
    Matrix A+Matrix B 357
    Matrix A−Matrix B 358
    Print 358
Vector Operations 358
    Vector Transpose 359

        Multiply a Vector by a Scalar 359
        Square Root of Vector Elements 359
        Reciprocal of Vector Elements 359
        Print a Vector 360
        Row Vector Times a Column Vector 360
        Column Vector Times Row Vector 360
    Scalar Operations 360
        Square Root of a Scalar 360
        Reciprocal of a Scalar 361
        Scalar Times a Scalar 361
        Print a Scalar 361

15 The GradeBook Program 363
    The GradeBook Main Form 363
    The Student Page Tab 363
    Test Result Page Tabs 364

16 The Item Banking Program 367
    Introduction 367
    Item Coding 368
    Using the Item Bank Program 369
    Specifying a Test 369
    Generate a Test 370

17 Neural Networks 373
    Using the Program 373
        The Neural Form 373
        Example Control File for Prediction 376
    Examples 379
        Regression Analysis with One Predictor 379
        Regression Analysis with Multiple Predictors 382
        Classification Analysis with Multiple Classification Predictors 386
        Pattern Recognition 389
        Exploration of Natural Groups 391
        Time Series Analysis 400

Bibliography 407

Index 411

List of Figures

Fig. 3.1	OpenStat main form	6
Fig. 4.1	The Variables Definition form	8
Fig. 4.2	The Options form	9
Fig. 4.3	The form for saving a file	11
Fig. 4.4	The Variable Transformation form	12
Fig. 4.5	The Variables Equation option	13
Fig. 4.6	Result of using the Equation option	14
Fig. 4.7	The Sort form	14
Fig. 4.8	The Select Cases form	16
Fig. 4.9	The Select If form	17
Fig. 4.10	Random selection of cases form	17
Fig. 4.11	Selection of a range of cases	18
Fig. 4.12	The Recode form	18
Fig. 4.13	Selection of an analysis from the main menu	19
Fig. 5.1	Central tendency and variability estimates	24
Fig. 6.1	Frequency analysis form	28
Fig. 6.2	Frequency interval form	28
Fig. 6.3	Frequency Distribution plot	30
Fig. 6.4	Cross-Tabulation dialog form	31
Fig. 6.5	The Breakdown form	32
Fig. 6.6	The Box Plot form	36
Fig. 6.7	Box and whiskers plot	38
Fig. 6.8	Three Dimension plot with rotation	39
Fig. 6.9	X Versus Y Plot form	40
Fig. 6.10	Plot of regression line in X versus Y	41
Fig. 6.11	Form for a pie chart	42
Fig. 6.12	Pie chart	42
Fig. 6.13	Stem and Leaf form	43
Fig. 6.14	Dialog form for examining theoretical and observed distributions	45

Fig. 6.15	The QQ / PP Plot Specification form	45
Fig. 6.16	A QQ plot	46
Fig. 6.17	Normality tests	47
Fig. 6.18	Resistant Line dialog	48
Fig. 6.19	Resistant Line plot	49
Fig. 6.20	Dialog for the repeated measures bubble plot	50
Fig. 6.21	Bubble plot	50
Fig. 6.22	Dialog for smoothing data by averaging	52
Fig. 6.23	Smoothed data frequency distribution plot	53
Fig. 6.24	Cumulative frequency of smoothed data	53
Fig. 6.25	Dialog for an X versus multiple Y plot	54
Fig. 6.26	X versus multiple Y plot	55
Fig. 6.27	Dialog for comparing observed and theoretical distributions	56
Fig. 6.28	Comparison of an observed and theoretical distribution	57
Fig. 6.29	Dialog for multiple groups X versus Y plot	58
Fig. 6.30	X versus Y plot for multiple groups	58
Fig. 7.1	Correlation regression line	62
Fig. 7.2	Simulated bivariate scatterplot	63
Fig. 7.3	Single sample tests form for correlations	64
Fig. 7.4	Comparison of two independent correlations	64
Fig. 7.5	Comparison of correlations for dependent samples	66
Fig. 7.6	Dialog for the ROC analysis	67
Fig. 7.7	ROC plot	67
Fig. 7.8	Form for calculating partial and semi-partial correlations	70
Fig. 7.9	The Autocorrelation form	72
Fig. 7.10	Moving Average form	73
Fig. 7.11	Smoothed plot using moving average	74
Fig. 7.12	Plot of residuals obtained using moving averages	74
Fig. 7.13	Polynomial regression smoothing form	74
Fig. 7.14	Plot of polynomial smoothed points	75
Fig. 7.15	Plot of residuals from polynomial smoothing	75
Fig. 7.16	Auto and partial autocorrelation plot	78
Fig. 8.1	Single Sample Tests Dialog form	80
Fig. 8.2	Single Sample Proportion test	81
Fig. 8.3	Single Sample Variance test	82
Fig. 8.4	Test of equality of two proportions	83
Fig. 8.5	Test of Equality of Two Proportions form	84
Fig. 8.6	Comparison of Two Sample Means form	85
Fig. 8.7	Comparison of two sample means	86
Fig. 8.8	One, two or three way ANOVA dialog	87
Fig. 8.9	Plot of sample means from a one-way ANOVA	89
Fig. 8.10	Specifications for a two-way ANOVA	89
Fig. 8.11	Within subjects ANOVA dialog	91
Fig. 8.12	Treatment by subjects ANOVA dialog	93

Fig. 8.13	Plot of treatment by subjects ANOVA means	94
Fig. 8.14	Dialog for the two-way repeated measures ANOVA	96
Fig. 8.15	Plot of factor A means in the two-way repeated measures analysis	97
Fig. 8.16	Plot of factor B in the two-way repeated measures analysis	97
Fig. 8.17	Plot of factor A and factor B interaction in the two-way repeated measures analysis	98
Fig. 8.18	The nested ANOVA dialog	99
Fig. 8.19	Three factor nested ANOVA dialog	101
Fig. 8.20	Latin and Greco-Latin squares dialog	103
Fig. 8.21	Latin squares analysis dialog	104
Fig. 8.22	Four factor Latin square design dialog	106
Fig. 8.23	Another Latin Square Specification form	108
Fig. 8.24	Latin Square Design form	111
Fig. 8.25	Latin Square Plan 5 Specifications form	114
Fig. 8.26	Latin square plan 6 specification	116
Fig. 8.27	Latin Squares Repeated Analysis Plan 7 form	119
Fig. 8.28	Latin Squares Repeated Analysis Plan 9 form	122
Fig. 8.29	Dialog for 2 or 3 way ANOVA with one case per cell	127
Fig. 8.30	One case ANOVA plot for factor 1	127
Fig. 8.31	Factor 2 plot for one case ANOVA	128
Fig. 8.32	Interaction plot of two factors for one case ANOVA	128
Fig. 8.33	Dialog for two within subjects ANOVA	130
Fig. 8.34	Factor one plot for two within subjects ANOVA	130
Fig. 8.35	Factor two plot for two within subjects ANOVA	131
Fig. 8.36	Two way interaction for two within subjects ANOVA	131
Fig. 8.37	Analysis of covariance dialog	133
Fig. 8.38	Sum of squares by regression	137
Fig. 8.39	Example 2 of sum of squares by regression	139
Fig. 8.40	Canonical Correlation Analysis form	142
Fig. 8.41	Logistic Regression form	145
Fig. 8.42	Cox Proportional Hazards Survival Regression form	147
Fig. 8.43	Weighted least squares regression	149
Fig. 8.44	Plot of ordinary least squares regression	149
Fig. 8.45	Plot of weighted least squares regression	150
Fig. 8.46	Two Stage Least Squares Regression form	153
Fig. 8.47	Non-linear Regression Specifications form	158
Fig. 8.48	Scores predicted by non-linear regression versus observed scores	159
Fig. 8.49	Correlation plot between scores predicted by non-linear regression and observed scores	159
Fig. 8.50	Completed non-linear regression parameter estimates of regression coefficients	161
Fig. 9.1	Specifications for a discriminant function analysis	164

Fig. 9.2	Plot of cases in the discriminant space	164
Fig. 9.3	Hierarchical Cluster Analysis form	172
Fig. 9.4	Plot of grouping errors in the discriminant analysis	173
Fig. 9.5	The K Means Clustering form	177
Fig. 9.6	Average Linkage dialog form	179
Fig. 9.7	Path Analysis dialog form	181
Fig. 9.8	Factor Analysis dialog form	190
Fig. 9.9	Screen plot of eigenvalues	190
Fig. 9.10	The GLM dialog form	195
Fig. 9.11	GLM Specifications for a repeated measures ANOVA	199
Fig. 9.12	A × B × R ANOVA dialog form	201
Fig. 9.13	Dialog for the Median Polish analysis	203
Fig. 9.14	Dialog for the Bartlett Test of Sphericity	205
Fig. 9.15	Dialog for Correspondence Analysis	206
Fig. 9.16	Correspondence Analysis plot 1	207
Fig. 9.17	Correspondence Analysis plot 2	207
Fig. 9.18	Correspondence Analysis plot 3	208
Fig. 9.19	Dialog for Log Linear Screening	211
Fig. 9.20	Dialog for the A × B Log Linear Analysis	211
Fig. 9.21	Dialog for the A × B × C Log Linear Analysis	212
Fig. 10.1	Contingency Chi-Square Dialog form	238
Fig. 10.2	The Spearman rank correlation dialog	240
Fig. 10.3	The Mann-Whitney U Test dialog form	242
Fig. 10.4	Fisher's Exact Test dialog form	244
Fig. 10.5	Kendal's coefficient of concordance	245
Fig. 10.6	Kruskal-Wallis one way ANOVA on ranks dialog	247
Fig. 10.7	Wilcoxon matched pairs signed ranks test dialog	248
Fig. 10.8	Cochran Q Test Dialog form	249
Fig. 10.9	The matched pairs sign test dialog	250
Fig. 10.10	The Friedman Two-Way ANOVA dialog	251
Fig. 10.11	The binomial probability dialog	253
Fig. 10.12	A sample file for the runs test	254
Fig. 10.13	The Runs Dialog form	254
Fig. 10.14	Kendal's Tau and Partial Tau dialog	256
Fig. 10.15	The Kaplan-Meier dialog	259
Fig. 10.16	Experimental and control curves	264
Fig. 10.17	A sample file for the Kolmogorov-Smirnov test	265
Fig. 10.18	Dialog for the Kolmogorov-Smirnov test	266
Fig. 10.19	Frequency distribution plot for the Kolmogorov-Smirnov test	266
Fig. 11.1	Classical item analysis dialog	270
Fig. 11.2	Distribution of test scores (classical analysis)	270
Fig. 11.3	Item means plot	271
Fig. 11.4	Hoyt reliability by ANOVA	275
Fig. 11.5	Within subjects ANOVA plot	276

List of Figures

Fig. 11.6	Kuder-Richardson Formula 20 Reliability form	278
Fig. 11.7	Composite test reliability dialog	279
Fig. 11.8	The Rasch item analysis dialog	281
Fig. 11.9	Rasch item log difficulty estimate plot	281
Fig. 11.10	Rasch log score estimates	282
Fig. 11.11	A Rasch item characteristic curve	282
Fig. 11.12	A Rasch test information curve	283
Fig. 11.13	Guttman scalogram analysis dialog	286
Fig. 11.14	Successive interval scaling dialog	287
Fig. 11.15	Differential item functioning dialog	290
Fig. 11.16	Differential item function curves	291
Fig. 11.17	Another item differential functioning curve	291
Fig. 11.18	Reliability adjustment for variability dialog	308
Fig. 11.19	Polytomous item differential functioning dialog	309
Fig. 11.20	Level means for polytomous item	309
Fig. 11.21	The item generation dialog	312
Fig. 11.22	Generated item data in the main grid	313
Fig. 11.23	Plot of generated test data	314
Fig. 11.24	Test of normality for generated data	314
Fig. 11.25	Spearman-Brown Prophecy dialog	315
Fig. 12.1	XBAR chart dialog	318
Fig. 12.2	XBAR chart for boltsize	319
Fig. 12.3	XBAR chart plot with target specifications	320
Fig. 12.4	Range chart dialog	321
Fig. 12.5	Range chart plot	321
Fig. 12.6	Sigma chart dialog	323
Fig. 12.7	Sigma chart plot	323
Fig. 12.8	CUMSUM chart dialog	324
Fig. 12.9	CUMSUM chart plot	325
Fig. 12.10	p control chart dialog	326
Fig. 12.11	p control chart plot	327
Fig. 12.12	Defect c chart dialog	329
Fig. 12.13	Defect control chart plot	329
Fig. 12.14	Defects U chart dialog	331
Fig. 12.15	Defect control chart plot	331
Fig. 13.1	Linear programming dialog	334
Fig. 13.2	Example specifications for a linear programming problem	335
Fig. 14.1	The MatMan dialog	338
Fig. 14.2	Using the MatMan files menu	345
Fig. 15.1	The GradeBook dialog	364
Fig. 15.2	The GradeBook summary	365
Fig. 15.3	The GradeBook Measurement Specifications form	366

Fig. 16.1	The Item Bank form	369
Fig. 16.2	The item banking Test Specification form	370
Fig. 16.3	The form to generate a test	371
Fig. 16.4	Student verification form for a test administration	371
Fig. 16.5	A test displayed on the computer	371
Fig. 17.1	The Neural form	374
Fig. 17.2	The neural file menu	374
Fig. 17.3	The neural control file generation options	375
Fig. 17.4	The control file generation form for prediction problems	375
Fig. 17.5	The form for generating a classification control file	377
Fig. 17.6	Form for specifying a Kohonen network control file	378
Fig. 17.7	Groups versus between group error	393
Fig. 17.8	Plot of subjects in three groups, each subject measured on two variables	400
Fig. 17.9	Original daily sales of creamed chicken with smoothed averages (3 values in each average)	401
Fig. 17.10	Auto and partial correlations for lags from Sunday (lag 1 = Saturday, etc.)	402

Chapter 1
Introduction

OpenStat, among others, are ongoing projects that I have created for use by students, teachers, researchers, practitioners and others. The software is a result of an "overactive" hobby of a retired professor (Iowa State University.) I make no claim or warranty as to the accuracy, completeness, reliability or other characteristics desirable in commercial packages (as if they can meet these requirement also.) They are designed to provide a means for analysis by individuals with very limited financial resources. The typical user is a student in a required social science or education course in beginning or intermediate statistics, measurement, psychology, etc. Some users may be individuals in developing nations that have very limited resources for purchase of commercial products.

Because I do not warrant them in any manner, you should insure yourself that the routines you use are adequate for your purposes. I strongly suggest analyses of text book examples and comparisons to other statistical packages where available. You should also be aware that I revise the program from time to time, correcting and updating OpenStat. For that reason, some of the images and descriptions in this book may not be exactly as you see when you execute the program. I update this book from time to time to try and keep the program and text coordinated.

Chapter 2
Installing OpenStat

OpenStat has been successfully installed on Windows 95, 98, ME, XT, NT, VISTA and Windows 7 systems. A free setup package (INNO) has been used to distribute and install OpenStat. Included in the setup file (OpenStatSetup.exe) is the executable file and Windows Help files. Sample data files that can be used to test the analysis programs are also available. Several Linux system users have also found that the free WINE software will allow OpenStat to run on a Linux platform.

To install OpenStat for Windows, follow these steps:

1. Connect to the internet address: http://statprograms4U.com
2. Click the download link for the OpenStatSetup.exe file
3. After the file has been downloaded, double click that program to initiate the installation of OpenStat. At the same website in 1 above, you will also find a link to a zip file containing sample data files that are useful for acquainting yourself with OpenStat. In addition, there are multiple tutorial files in Windows Media Video (.WMV) format as well as Power Point slide presentations.

Chapter 3
Starting OpenStat

To begin using a Windows version of OpenStat simply click the Windows "Start" button in the lower left portion of your screen, move the cursor to the "Programs" menu and click on the OpenStat entry. The following form should appear (Fig. 3.1):

The form contains several important areas. The "grid" is where data values are entered. Each column represents a "variable" and each row represents an "observation" or case. A default label is given for the first variable and each case of data you enter will have a case number. At the top of this "main" form there is a series of "drop-down" menu items. When you click on one of these, a series of options (and sometimes sub-options) that you can click to select. Before you begin to enter case values, you probably should "define" each variable to be entered in the data grid. Select the "VARIABLES" menu item and click the "Define" option. More will be said about this in the following pages.

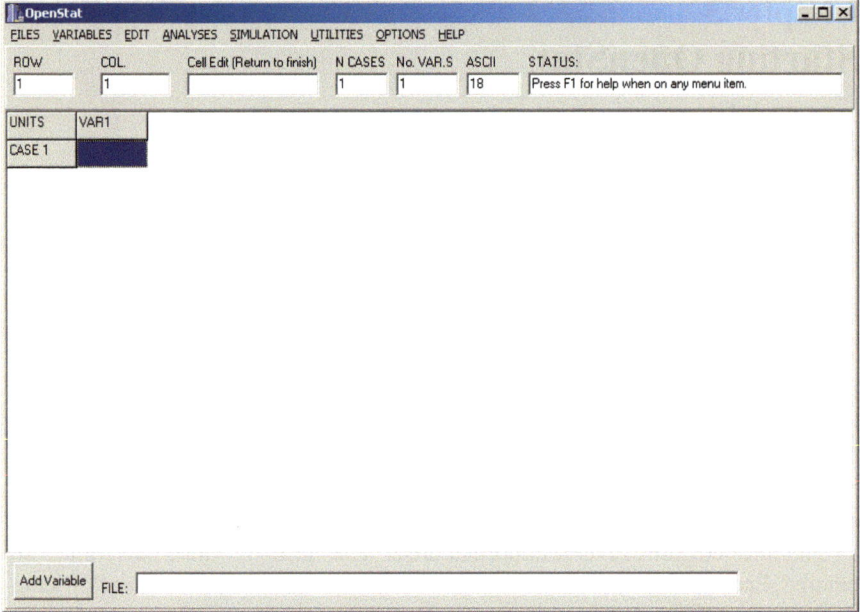

Fig. 3.1 OpenStat main form

Chapter 4
Files

The "heart" of OpenStat or any other statistics package is the data file to be created, saved, retrieved and analyzed. Unfortunately, there is no one "best" way to store data and each data analysis package has its own method for storing data. Many packages do, however, provide options for importing and exporting files in a variety of formats. For example, with Microsoft's Excel package, you can save a file as a file of "tab" separated fields. Other program packages such as SPSS can import "tab" files. Here are the types of file formats supported by OpenStat:

1. OPENSTAT binary files (with the file extension of .BIN .)
2. Tab separated field files (with the file extension of .TAB.)
3. Comma separated field files (with the file extension of .CSV.)
4. Space separated field files (with the file extension of .SSV.)
5. Text files (with the extension .TEX) NOTE: the file format in this text file is unique to OpenStat!
6. Epidata files (this is a format used by Epidemiologists)
7. Matrix files previously saved by OpenStat
8. Fixed Format files in which the user specifies the record format

My preference is to save files as .TEX files. Alternatively, tab separated field files are often used. This gives you the opportunity to analyze the same data using a variety of packages. For relatively small files (say, for example, a file with 20 variables and 1,000 cases), the speed of loading the different formats is similar and quite adequate. The default for OPENSTAT is to save as a binary file with the extension .TEX to differentiate it from other types of files.

Creating a File

When OPENSTAT begins, you will see a "grid" of two rows and two columns. The left-most column will automatically contain the word "Case" followed by a number (1 for the first case.) The top row will contain the names of the variables that you

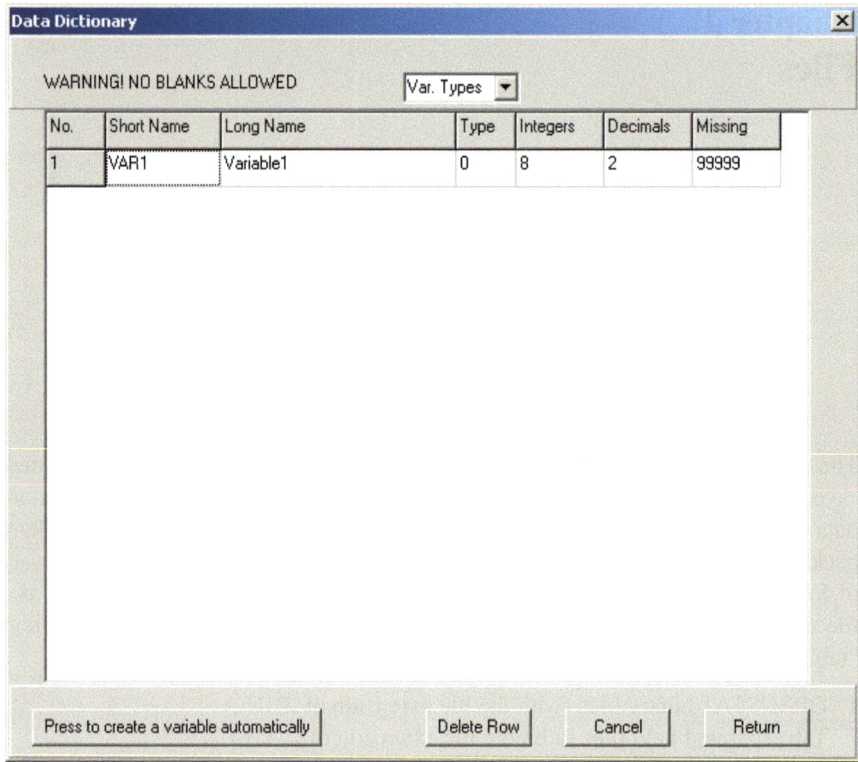

Fig. 4.1 The Variables Definition form

assign when you start entering data for the first variable. If you click your mouse on the "Variables" menu item, a drop-down list will appear that contains the word "define". If you click on this label, the above form appears:

In the above figure (Fig. 4.1) you will notice that a variable name has automatically been generated for the first variable. To change the default name, click the box with the default name and enter the variable name that you desire. It is suggested that you keep the length of the name to eight characters or less. Do NOT have any blanks in the variable name. An underscore (_) character may be used. You may also enter a long label for the variable. If you save your file as an OPENSTAT file, this long name (as well as other descriptive information) will be saved in the file (the use of the long label has not yet been implemented for printing output but may be in future versions.) To proceed, simply click the Return button in the lower right of this form. The default type of variable is a "floating point" value, that is, a number which may contain a decimal fraction. If a data field (grid cell) is left blank, the program will usually assume a missing value for the data. The default format of a data value is eight positions with two positions allocated to fractional decimal values (format 8.2.) By clicking on any of the specification fields you can modify

Creating a File

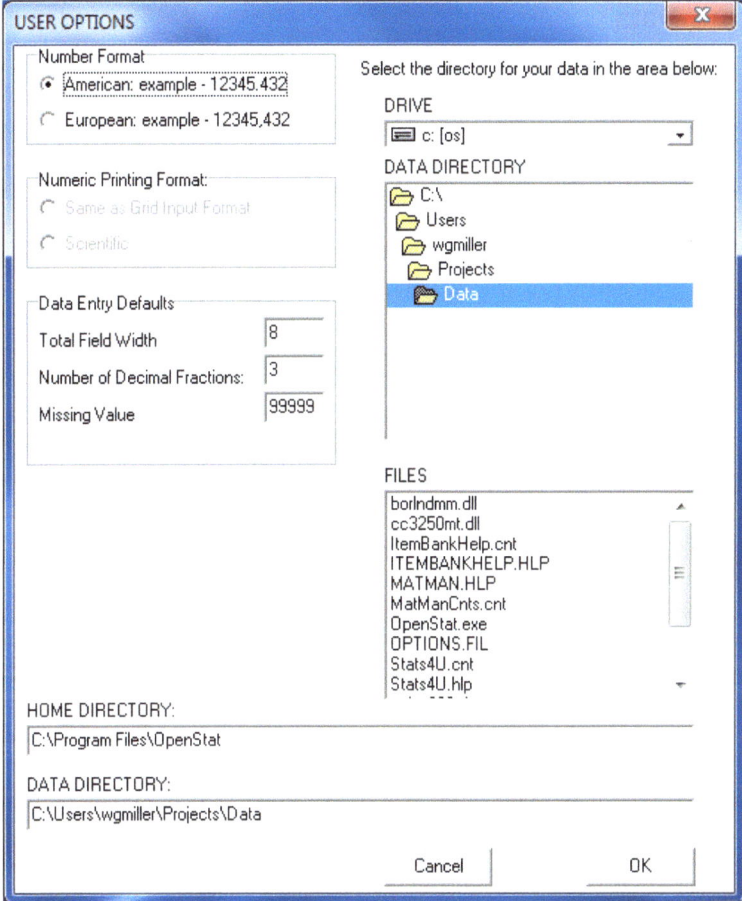

Fig. 4.2 The Options form

these defaults to your own preferences. You can change the width of your field, the number of decimal places (0 for integers.) Another way to specify the default format and missing values is by modifying the "Options" file. When you click on the Options menu item and select the change options, the above form appears (Fig. 4.2):

In the options form you can specify the Data Entry Defaults as well as whether you will be using American or European formatting of your data (American's use a period (.) and Europeans use a comma (,) to separate the integer portion of a number from its fractional part.) The Printer Spacing section is currently ignored but may be implemented in a future version of OpenStat. You can also specify the directory in which to find the data files you want to process. I recommend that you save data in the same directory that contains the OpenStat program (the default directory).

Entering Data

When you enter data in the grid of the main form there are several ways to navigate from cell to cell. You can, of course, simply click on the cell where you wish to enter data and type the data values. If you press the "enter" key following the typing of a value, the program will automatically move you to the next cell to the right of the current one or down to the next cell if you are at the last variable. You may also press the keyboard "down" arrow to move to the cell below the current one. If it is a new row for the grid, a new row will automatically be added and the "Case" label added to the first column. You may use the arrow keys to navigate left, right, up and down. You may also press the "Page Up" button to move up a screen at a time, the "Home" button to move to the beginning of a row, etc. Try the various keys to learn how they behave. You may click on the main form's Edit menu and use the delete column or delete row options. Be sure the cursor is sitting in a cell of the row or column you wish to delete when you use this method. A common problem for the beginner is pressing the "enter" key when in the last column of their variables. If you do accidentally add a case or variable you do not wish to have in your file, use the edit menu and delete the unused row or variable. If you have made a mistake in the entry of a cell value, you can change it in the "Cell Edit" box just below the menu. In this box you can use the delete key, backspace key, enter characters, etc. to make the corrections for a cell value. When you press your "Enter" key, the new value will be placed in the corresponding cell. Notice that as you make grid entries and move to another cell, the previous value is automatically formatted according to the definition for that variable. If you try to enter an alphabetic character in an integer or floating point variable, you will get an error message when you move from that cell. To correct the error, click on the cell that is incorrect and make the changes needed in the Cell Edit box.

Saving a File

Once you have entered a number of values in the grid, it is a good idea to save your work (power outages do occur!) Go to the main form's File menu and click it. You will see there are several ways to save your data. The first time you save your data you should click the "Save a Text Type of File" option. A "dialog box" will then appear as shown below (Fig. 4.3):

Simply type the name of the file you wish to create in the File name box and click the Save button. After this initial save as operation, you may continue to enter data and save with the Save button on the file menu. Before you exit the program, be sure to save your file if you have made additions to it.

If you do not need to save specifications other than the short name of each variable, you may prefer to "export" the file in a format compatible to other programs. The "Export Tab File option under the File menu will save your data in a text file in

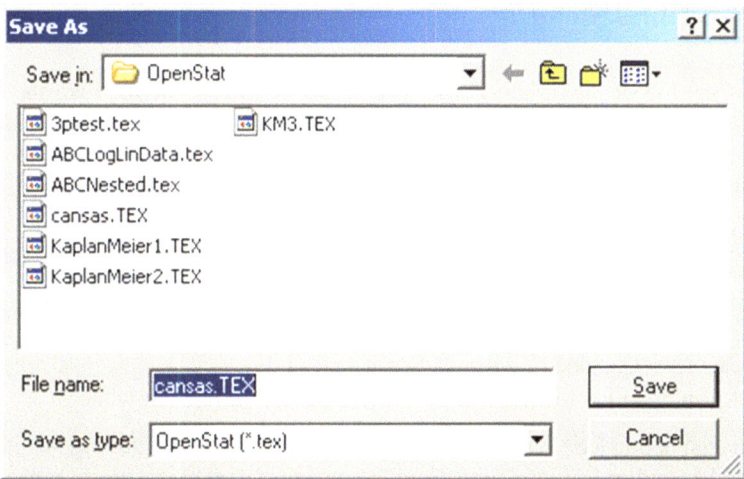

Fig. 4.3 The form for saving a file

which the cell values in each row are separated by a tab key character. A file with the extension .TAB will be created. The list of variables from the first row of the grid are saved first, then the first row of the data, etc. until all grid rows have been saved.

Alternatively, you may export your data with a comma or a space separating the cell values. Basic language programs frequently read files in which values are separated by commas or spaces. If you are using the European format of fractional numbers, DO NOT USE the comma separated files format since commas will appear both for the fractions and the separation of values - clearly a design for disaster!

Help

Users of Microsoft Windows are used to having a "help" system available to them for instant assistance when using a program. Most of these systems provide the user the ability to press the "F1" key for assistance on a particular topic or by placing their cursor on a particular program item and pressing the right mouse button to get help. OpenStat for the Microsoft Windows does have a help file. Place the cursor on a menu topic and press the F1 key to see what happens! You can use the help system to learn more about OpenStat procedures. Again, as the program is revised, there may not yet be help topics for all procedures and some help topics may vary slightly from the actual procedure's operation. Vista and Windows 7 users may have to download a file from MicroSoft to provide the option for reading ".hlp" files.

The Variables Menu

Across the top of the "Main Form" is a series of "menu" items. Like the "File" menu, each of these menu items "drops-down" a series of options and these options may have sub-options. The "Variables" menu contains a variety of options to assist you in working with the variables (columns of data). These options include:

1. Define
2. Transform
3. Print Dictionary
4. Sort
5. Create An Expanded File from a Frequencies File
6. Enter an Equation to Combine Variables to Create a New Variable

The first option lets you enter or change a variable definition (see Fig. 4.1 above.)

Another option lets you "transform" an existing variable to create a new variable. A variety of transformations are possible. If you elect this option, you will see the following dialogue form (Fig. 4.4):

You will note that you can transform a variable by adding, subtracting, multiplying, dividing or raising a value to a power. To do this you select a variable to transform by clicking on the variable in the list of available variables and then clicking the right arrow. You then enter a constant by clicking on the box for the constant and entering a value. You select the transformation with a constant from among the first 10 possible transformations by clicking on the desired transformation (you will see

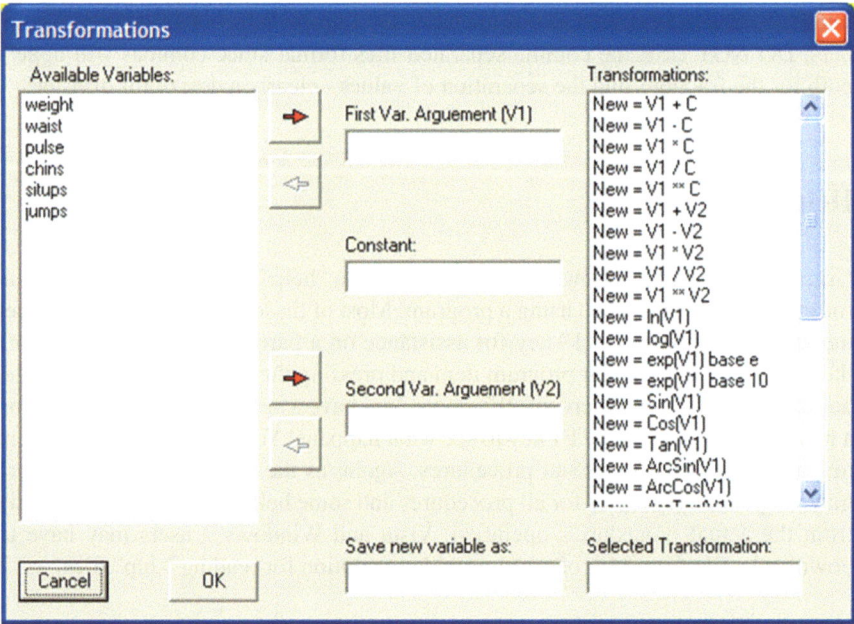

Fig. 4.4 The Variable Transformation form

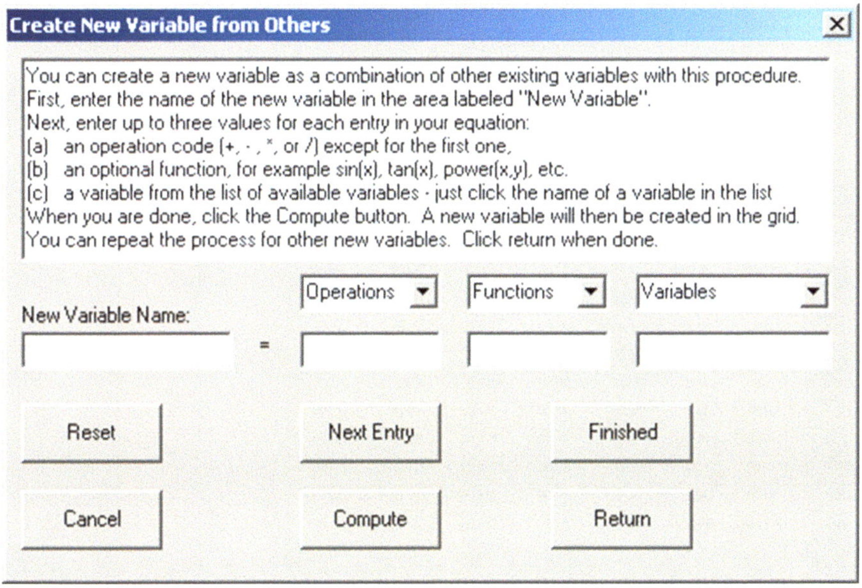

Fig. 4.5 The Variables Equation option

it entered automatically in the lower right box.) Next you enter a name for the new variable in the box labeled "Save new variable as:" and click the OK button.

Sometimes you will want to transform a variable using one of the common exponentiation or trigonometric functions. In this case you do not need to enter a constant - just select the variable, the desired transformation and enter the variable name before clicking the OK button.

You can also select a transformation that involves two variables. For example, you may want a new variable that represents the sum, product, difference, etc. of two variables. In this case you select the two variables for the first and second arguments using the appropriate right-arrow key after clicking one and then the other in the available variables list.

The "Print Dictionary" option simply creates a list of variable definitions on an "output" form which may be printed on your printer for future reference.

The option to create a new variable by means of an equation can be useful in a variety of situations. For example, you may want to create a new variable that is simply the sum of several other variables (or products of, etc.) We have selected a file labeled "cansas.tab" from our sample files and will create a new variable labeled "physical" that adds the first three variables. When we click the equation option, the above form appears (Fig. 4.5):

To use the above, enter the name of your new variable in the box provided. Following this box are three additional "edit" boxes with "drop-down" boxes above each one. For the first variable to be added, click the drop-down box labeled "Variables" and select the name of your first variable. It will be automatically placed in the third box. Next, click the "Next Entry" button. Now click the "Operations" drop-down arrow and select the desired operation (plus in our example) and again

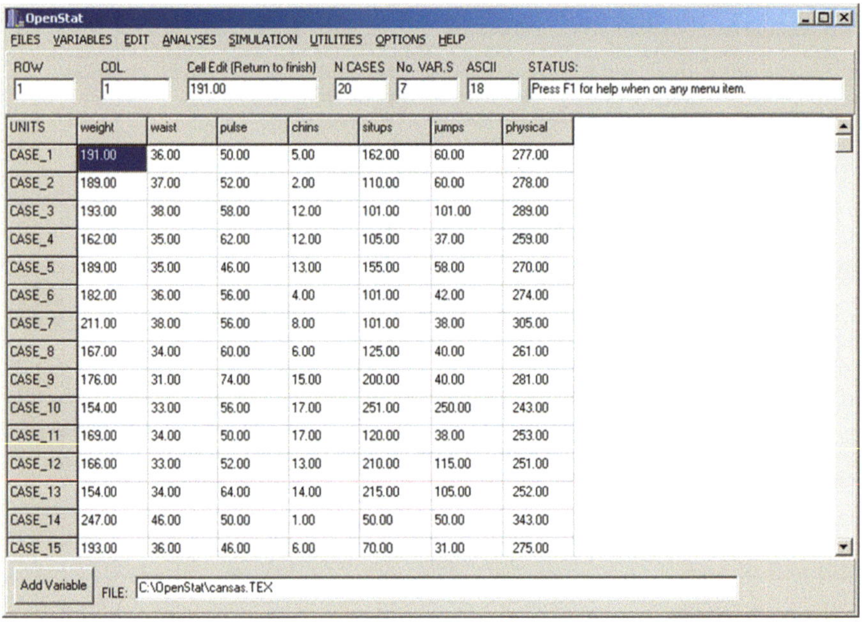

![OpenStat screenshot showing data grid with columns UNITS, weight, waist, pulse, chins, situps, jumps, physical for CASE_1 through CASE_15]

Fig. 4.6 Result of using the Equation option

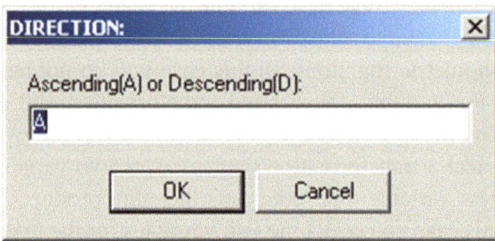

Fig. 4.7 The Sort form

select a variable from the Variables drop-down box. Again click the "Next Entry" button. Repeat the Operations and Variables for the last variable to be added. Click the "Finished" button to end the creation of the equation. Click the Compute button and then the Return button. An output of your equation will be shown first as below:

```
Equation Used for the New Variable
physical = weight + waist + pulse
```

You will see the new variable in the grid (Fig. 4.6):

The "Sort" option involves clicking on a cell in the column on which the cases are to be sorted and then selecting the Variables/Sort option. You then indicate whether you want to sort the cases in an ascending order or a descending order. The form above demonstrates the sort dialogue form (Fig. 4.7):

The Edit Menu

The Edit menu is provided primarily for deleting, cutting and pasting of cells, rows or columns of data. It also provides the ability to insert a new column or row at a desired position in the data grid. There is one special "paste" operation provided for users that also have the Microsoft Excel program and wish to copy cells from an Excel spreadsheet into the OpenStat grid. These operations involve clicking on a cell in a given row and column and the selecting the edit operation desired. The user is encourage to experiment with these operations in order to become familiar with them. The following options are available:

1. Copy
2. Delete
3. Paste
4. Insert a New Column
5. Delete a Column
6. Copy a Column
7. Paste a Column
8. Insert a New Row
9. Delete a Row
10. Copy a Row
11. Paste a row
12. Format Grid Values
13. Select Cases
14. Recode
15. Switch USA to Euro or Vice Versa
16. Swap Rows and Columns
17. Open Output Form / Word Processor

The first 11 of these options involve copying, deleting, pasting a row, column or block of grid cells or inserting a new row or column. You can also "force" grid values to be reformatted by selecting option 12. This can be useful if you have changed the definition of a variable (floating point to integer, number of decimal places, etc.)

In some cases you may need to swap the cell values in the rows and columns so that what was previously a row is now a column. If you receive files from an individual using a different standard than yourself, you can switch between European and USA standards for formatting decimal fraction values in the grid. Another useful option lets you "re-code" values in a selected variable. For example, you may need to recode values that are currently 0 to a 1 for all cases in your file.

The "Select Cases" option lets you analyze only those cases (rows) which you select. When you press this option you will see the following dialogue form (Fig. 4.8):

Notice that you may select a random number of cases, cases the exhibit a specific range of values or cases if a specific condition exists. Once selection has been made, a new variable is added to the grid called the "Filter" variable. You can subsequently

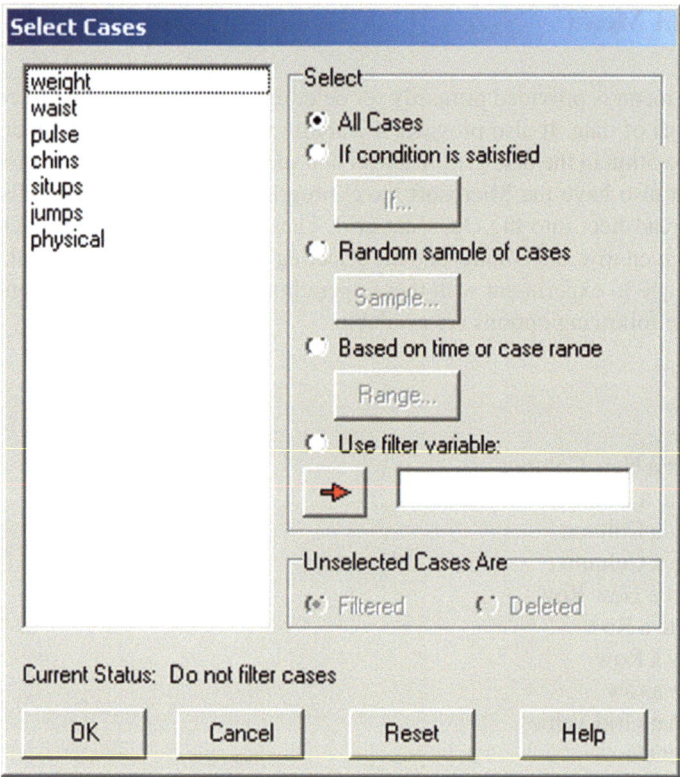

Fig. 4.8 The Select Cases form

use this filter variable to delete unneeded cases from your file if desired. Each of the selection procedures invokes a dialogue form that is specific to the type of selection chosen. For example, if you select the "if condition is satisfied" button, you will see the following dialogue form (Fig. 4.9):

An example has been entered on this form to demonstrate a typical selection criteria. Notice that compound statements involve the use of opening and closing parentheses around each expression You can directly enter values in the "if" box or use the buttons provided on the pad.

Should you select the "random" option in Fig. 4.8 you would see the following form (Fig. 4.10):

The user may select a percentage of cases or select a specific number from a specified number of cases.

Finally, the user may select a specified range of cases. This option produces the following dialogue form (Fig. 4.11):

The Variables/Recode option is used to change the value of cases in a given variable. For example, you may have imported a file which originally coded gender as

The Edit Menu

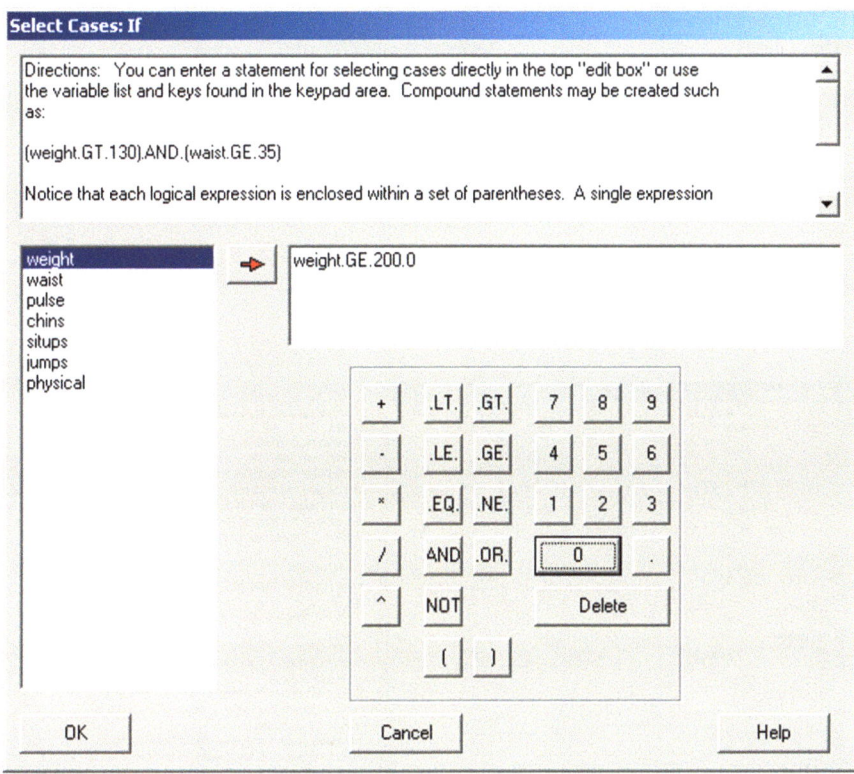

Fig. 4.9 The Select If form

Fig. 4.10 Random selection of cases form

Fig. 4.11 Selection of a range of cases

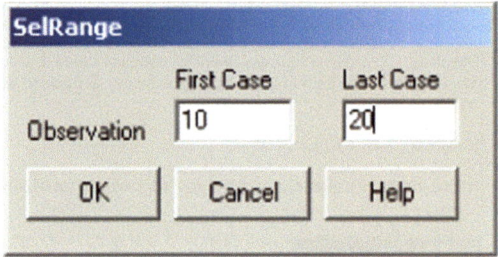

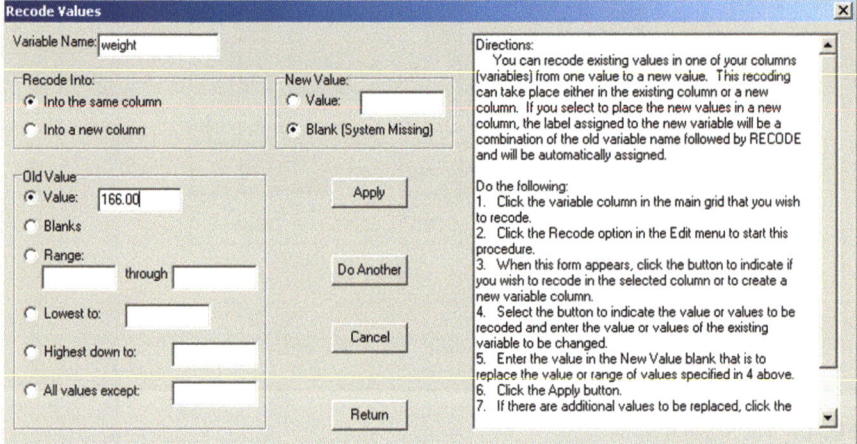

Fig. 4.12 The Recode form

"M" or "F" but the analysis you want requires a coding of 0 and 1. You can select the recode option and get the above form to complete (Fig. 4.12):

Notice that you first click on the column of the variable to recode, enter the old value (or value range) and also enter the new value before clicking the Apply button. You can repeat the process for multiple old values before returning to the Main Form.

Some files may require the user to change all column values to row values and row values to column values. For example, a user may have created a file with rows that represent subjects measured on 10 variables. One of the desired analysis however requires the calculation of correlations among subjects, not variables. To obtain a matrix of this form the user can swap rows and columns. Clicking on this option will switch the rows and columns. In doing this, the original variable labels are lost. The previous cases are now labeled Var1, Var2, etc. and the original variables are labeled CASE 1, CASE 2, etc. Clearly, one should save the original file before completing this operation! Once the swap has occurred, you can save the new file under a different name.

The last option under the variables menu lets you switch between the American and European format for decimal fractions. This may be useful when you have imported a file from another country that uses the other format. OpenStat will attempt to convert commas to periods or vice-versa as required.

The Analyses Menu

The heart of any statistics package is the ability to perform a variety of statistical analyses. Many of the typical analyses are included in the options and sub-options of the Analyses menu. The figure below (Fig. 4.13) shows the options and the sub-options under the descriptive option. No attempt will be made at this point in the text to describe each analysis - these are described further in the text.

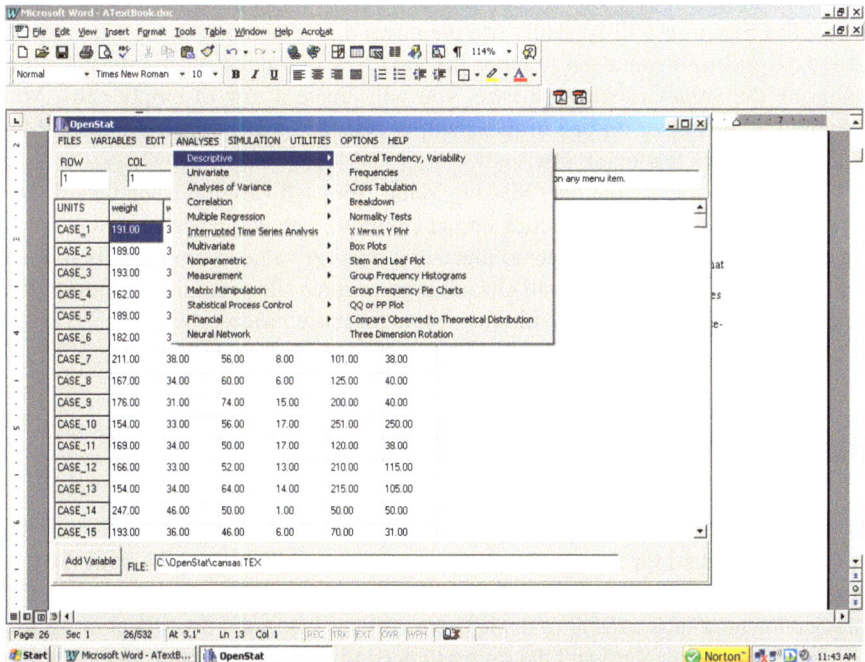

Fig. 4.13 Selection of an analysis from the main menu

The Simulation Menu

As you read about and learn statistics, it is helpful to be able to simulate data for an analysis and see what the distribution of the values looks like. In addition, the concepts of "type I error", "type II error", "Power", correlation, etc. may be more readily grasped if the student can "play" with distributions and the effects of choices they might make in a real study. Under the simulation menu the user may generate a sequence of numbers, may generate multivariate data, may generate data that are a sample from a theoretical population or generate bivariate-normal data for a correlation. One can even generate data for a two-way analysis of variance!

Some Common Errors!

Empty Cells

The beginning user will often see a message something like "" is not a valid floating point value. The most common cause of this error occurs when a procedure attempts to read a blank cell, that is, a cell that has been left empty by the user. The new user will typically use the down-arrow to move to the next row in the data grid in preparation to enter the next row of values. If you do this after entering the values for the last case, you will create a row of empty cells. You should put the cursor on one of these empty cells and use the Edit->Delete Row menu to remove this blank row.

The user should define the "Missing Value" for each variable when they define the variable. One should also click on the Options menu and place a missing value in that form. OpenStat attempts to place that missing value in empty cells when a file is saved as .TEX file. Not all OpenStat procedures allow missing values so you may have to delete cases with missing values for those procedures.

Incorrect Format for Floating Point Values

A second reason you might receive a "not valid" error is because you are using the European standard for the format of values with decimal fractions. Most of the statistical procedures contain a small "edit" window that contains a confidence level or a rejection area such as 95.0 or 0.05. These will NOT be valid floating point values in the European standard and the user will need to click on the value and replace it with the correct form such as 95,0 or 0,05. This has been done for the user in some procedures but not all!

String Labels for Groups

Users of other statistics packages such as SPSS or Excel may have used strings of characters to identify different groups of cases (subjects or observations.) OpenStat uses sequential integer values only in statistical analyses such as analyses of variance or discriminant function analysis. An edit procedure has been included that permits the conversion of string labels to integer values and saves those integers in a new column of the data grid. An attempt to use a string (alphanumeric) value will cause an "not valid" type of error. Several procedures in OpenStat have been modified to let you specify a string label for a group variable and automatically create an integer value for the analysis in a few procedures but not all. It is best to do the conversion of string labels to integers and use the integer values as your group variable.

Floating Point Errors

Sometimes a procedure will report an error of the type "Floating Point Division Error". This is often the outcome of a procedure attempting to divide a quantity by zero (0.) As an example, assume you have entered data for several variables obtained on a group of subjects. Also assume that the value observed for one of those variables is the same (a constant value) for all cases. In this situation there is no variability among the cases and the variance and standard deviation will be zero! Now an attempt to use that zero variance or standard deviation in the calculation of z scores, a correlation with another variable or other usage will cause an error (division by zero is not defined.)

Values Too Large (or Small)

In some fields of study such as astronomy the values observed may be very, very large. Computers use binary numbers to represent quantities. Nearly all OpenStat procedures use "double precision" storage for floating point values. The double precision value is stored in 64 binary "bits" in the computer memory. In most computers this is a combination of 8 binary "bytes" or words. The values are stored with a characteristic and mantissa similar to a scientific notation. Of course bits are also used to represent the sign of these parts. The maximum value for the characteristic is typically something like 2 raised to the power of 55 and the mantissa is 2 to the 7th power. Now consider a situation where you are summing the product of several of very large values such as is done in obtaining a variance or correlation. You may very well exceed the 64 bit storage of this large sum of products! This causes an "overflow" condition and a subsequent error message. The same thing can be said of values too small. This can cause an "underflow" error and associated error message.

The solution for these situations of values too large or too small is to "scale" your initial values. This is typically done by dividing or multiplying the original values by a constant to move the decimal point to decrease (or increase) the value. This does, of course, affect the "precision" of your original values but it may be a sacrifice necessary to do the analysis. In addition, the results will have to be "re-scaled" to reflect the original measurement scale.

Chapter 5
Distributions

Using the Distribution Parameter Estimates Procedure

One of the procedures which may be executed in your OpenStat package is the Analyses/Statistics/Central Tendency and Variability procedure. The procedure will compute the mean, variance, standard deviation, range, skew, minimum, maximum and number of cases for each variable you have specified. To use it, you enter your data as a column of numbers in the data grid or retrieve the data of a file into the data grid. Click on the Statistics option in the main menu and click on the Mean, Variance, Std.Dev, Skew, Kurtosis option under the Descriptive sub-menu. You will see the following form (Fig. 5.1):

Select the variables to analyze by clicking the variable name in the left column followed by clicking the right arrow. You may select ALL by clicking the All button. Click on the Continue button when you have selected all of your variables. Notice that you can also convert each of the variables to standardized z scores as an option. The new variables will be placed into the data grid with variable names created by combining z with the original variable names. The results will be placed in the output form which may be printed by clicking the Print button of that form.

Using the Breakdown Procedure

The Breakdown procedure is an OpenStat program designed to produce the means and standard deviations of cases that have been classified by one or more other (categorical) variables. For example, a sample may contain subjects for which have values for interest in school, grade in school, gender, and rural/urban home environment. A researcher might be interested in reporting the mean and standard deviation of "interest in school" for persons classified by combinations of the other three (nominal scale) variables grade, gender and rural/urban.

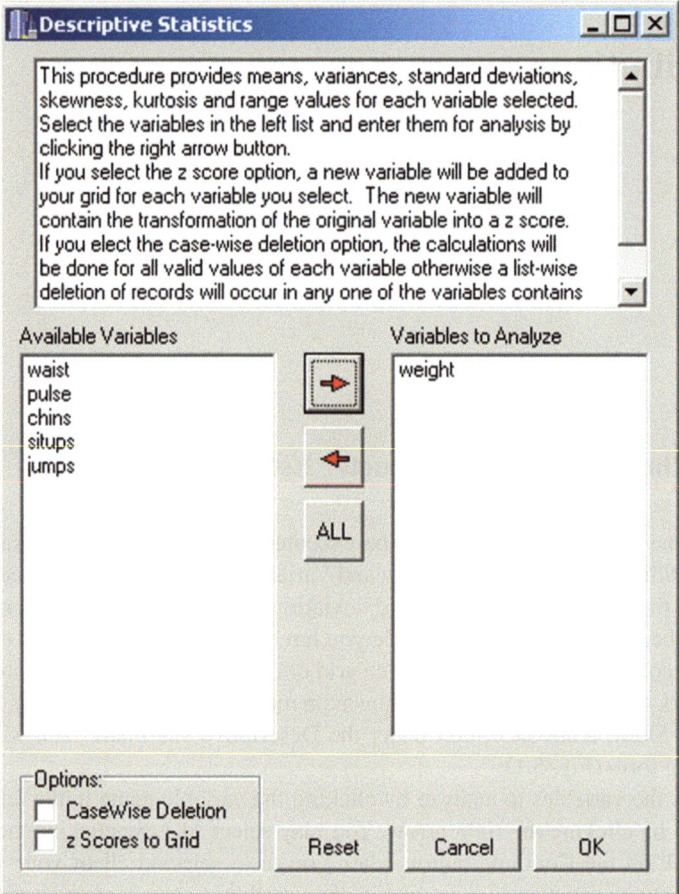

Fig. 5.1 Central tendency and variability estimates

The Breakdown program summarizes the means and standard deviations for each level of the variable entered last within levels of the next-to-last variable, etc. In our example, the statistics would be given for rural and urban codes within male and female levels first, then statistics for males and females within grade level and finally, the overall group means and standard deviations. The order of specification is therefore important. The variable receiving the finest breakdown is listed last, the next-most relevant breakdown next-to-last, etc. If the order of categorical variables for the above example were listed as 2, 4, 3 then the summary would give statistics for males and females within rural and urban codes, and rural and urban students (genders combined) within grade levels. Optionally, the user may request one-way analysis of variance results. An ANOVA table will be produced for the continuous variable for the categories of each of the nominal variables.

Using the Distribution Plots and Critical Values Procedure

This simulation procedure generates three possible distributions, i.e. (a) z scores, (b) Chi-squared statistics or (c) F ratio statistics. If you select either the Chi-squared or the F distribution, you will be asked to enter the appropriate degrees of freedom. You are also asked to enter the probability of a Type I error. The default value of 0.05 is commonly used. You may also elect to print the distribution that is created.

Chapter 6
Descriptive Analyses

Frequencies

Selecting the Descriptive/Distribution Frequencies option from the Analyses menu results in the following form being displayed. The cansas.TEX file has been loaded and the weight variable has been selected for analysis. The option to display a histogram has also been selected, the three dimensional vertical bars has been selected and the plotting of the normal distribution has been checked (Fig. 6.1).

When the OK button is clicked, each variable is analyzed in sequence. The first thing that is displayed is a form shown below (Fig. 6.2):

You will notice that the number of intervals shown for the first variable (weight) is 16. You can change the interval size (and press return) to increase or decrease the number of intervals. If we change the interval size to 10 instead of the current 1, we would end up with 11 categories.

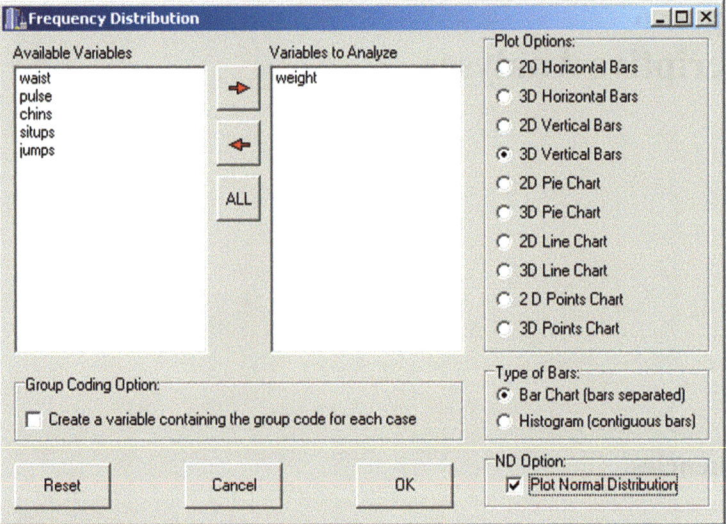

Fig. 6.1 Frequency analysis form

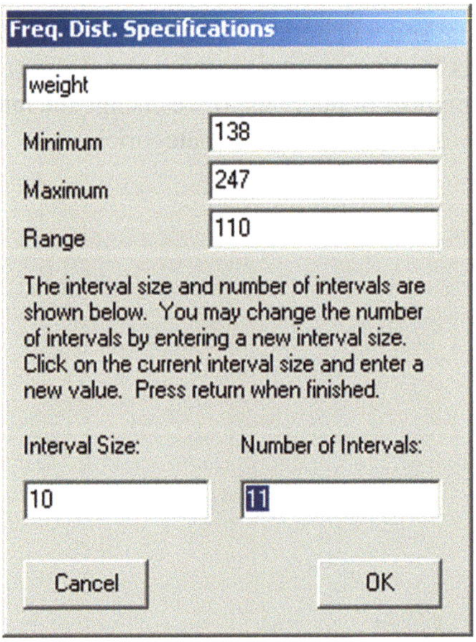

Fig. 6.2 Frequency interval form

Now when the OK button on the specifications form is clicked the following results are displayed:

```
FREQUENCY ANALYSIS BY BILL MILLER
Frequency Analysis for waist
FROM    TO      FREQ.   PCNT    CUM.FREQ.   CUM.PCNT.   %ILE RANK
31.00   32.00   1       0.05    1.00        0.05        0.03
32.00   33.00   1       0.05    2.00        0.10        0.07
33.00   34.00   4       0.20    6.00        0.30        0.20
34.00   35.00   3       0.15    9.00        0.45        0.38
35.00   36.00   2       0.10    11.00       0.55        0.50
36.00   37.00   3       0.15    14.00       0.70        0.63
37.00   38.00   3       0.15    17.00       0.85        0.78
38.00   39.00   2       0.10    19.00       0.95        0.90
39.00   40.00   0       0.00    19.00       0.95        0.95
40.00   41.00   0       0.00    19.00       0.95        0.95
41.00   42.00   0       0.00    19.00       0.95        0.95
42.00   43.00   0       0.00    19.00       0.95        0.95
43.00   44.00   0       0.00    19.00       0.95        0.95
44.00   45.00   0       0.00    19.00       0.95        0.95
45.00   46.00   0       0.00    19.00       0.95        0.95
46.00   47.00   1       0.05    20.00       1.00        0.97
```

The above results of the output form show the intervals, the frequency of scores in the intervals, the percent of scores in the intervals, the cumulative frequencies and percents and the percentile ranks. Clicking the Return button then results in the display of the frequencies expected under the normal curve for the data:

```
Interval ND Freq.
1        0.97
2        1.42
3        1.88
4        2.26
5        2.46
6        2.44
7        2.19
8        1.79
9        1.33
10       0.89
11       0.54
12       0.30
13       0.15
14       0.07
15       0.03
16       0.01
17       0.00
```

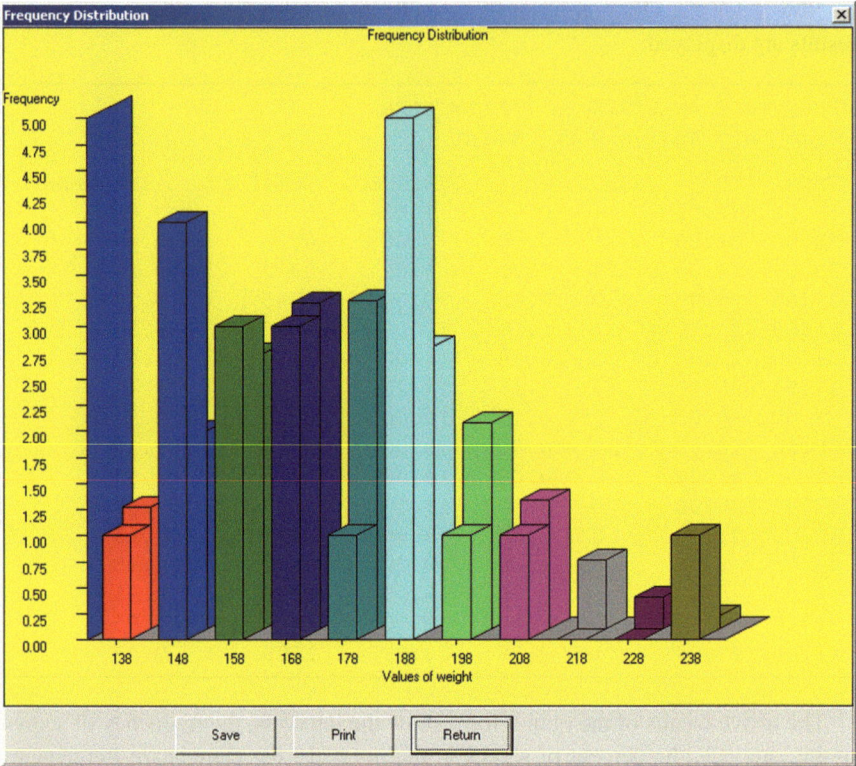

Fig. 6.3 Frequency Distribution plot

When the Return button is again pressed the histogram is produced as illustrated above (Fig. 6.3):

Cross-Tabulation

A researcher may observe objects classified into categories on one or more nominal variables. It is desirable to obtain the frequencies of the cases within each "cell" of the classifications. An example is shown in the following description of using the cross-tabulation procedure. Select the cross-tabulation option from the Descriptive option of the Statistics menu. You see a form like that below (Fig. 6.4):

Cross-Tabulation

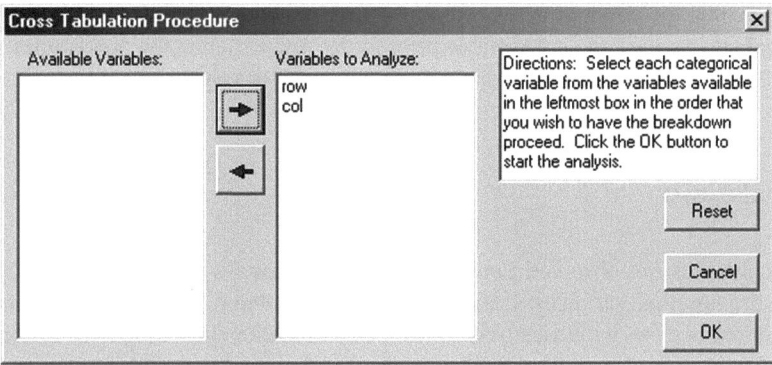

Fig. 6.4 Cross-Tabulation dialog form

In this example we have opened the chisquare.tab file to analyze. Cases are classified by "row" and "col" variables. When we click the OK button we obtain:

```
CROSSTABULATION ANALYSIS PROGRAM

VARIABLE SEQUENCE FOR THE CROSSTABS:
row (Variable 1) Lowest level = 1 Highest level = 3
col (Variable 2) Lowest level = 1 Highest level = 4

FREQUENCIES BY LEVEL:
For Cell Levels: row : 1 col: 1 Frequency = 5
For Cell Levels: row : 1 col: 2 Frequency = 5
For Cell Levels: row : 1 col: 3 Frequency = 5
For Cell Levels: row : 1 col: 4 Frequency = 5
Number of observations for Block 1 = 20
For Cell Levels: row : 2 col: 1 Frequency = 10
For Cell Levels: row : 2 col: 2 Frequency = 4
For Cell Levels: row : 2 col: 3 Frequency = 7
For Cell Levels: row : 2 col: 4 Frequency = 3
Number of observations for Block 2 = 24
For Cell Levels: row : 3 col: 1 Frequency = 5
For Cell Levels: row : 3 col: 2 Frequency = 10
For Cell Levels: row : 3 col: 3 Frequency = 10
For Cell Levels: row : 3 col: 4 Frequency = 2
Number of observations for Block 3 = 27
Cell Frequencies by Levels

col
             1        2        3        4
Block 1    5.000    5.000    5.000    5.000
Block 2   10.000    4.000    7.000    3.000
Block 3    5.000   10.000   10.000    2.000

Grand sum for all categories = 71
```

Note that the count of cases is reported for each column within rows 1, 2 and 3. If we had specified the col variable prior to the row variable, the procedure would summarize the count for each row within columns 1 through 4.

Breakdown

If a researcher has observed a continuous variable along with classifications on one or more nominal variables, it may be desirable to obtain the means and standard deviations of cases within each classification combination. In addition, the researcher may be interested in testing the hypothesis that the means are equal in the population sampled for cases in the categories of each nominal variable. We will use sample data that was originally obtained for a three-way analysis of variance (threeway. tab.) We then select the Breakdown option from within the Descriptive option on the Statistics menu and see (Fig. 6.5):

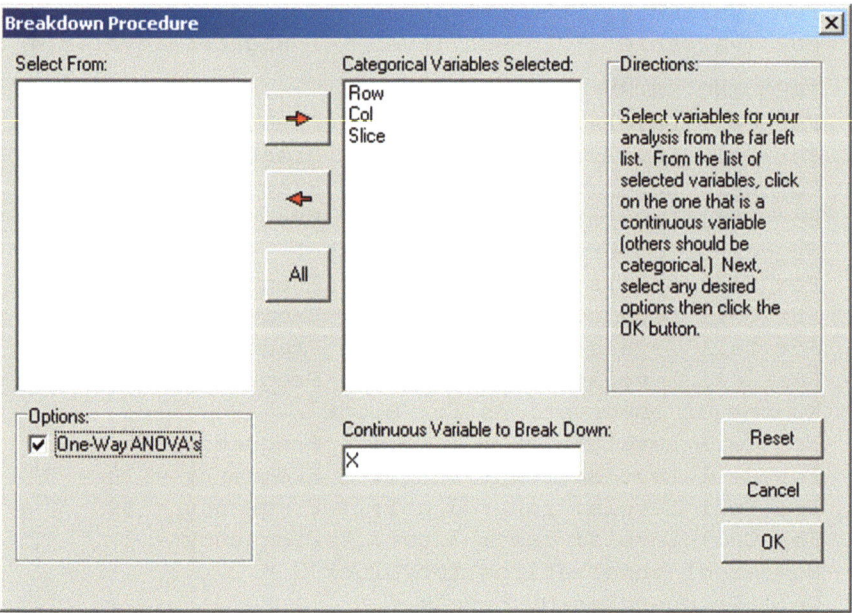

Fig. 6.5 The Breakdown form

Breakdown

We have elected to obtain a one-way analysis of variance for the means of cases classified into categories of the "Slice" variable for each level of the variable "Col." and variable "Row". When we click the Continue button we obtain the first part of the output which is:

```
BREAKDOWN ANALYSIS PROGRAM

VARIABLE SEQUENCE FOR THE BREAKDOWN:
Row    (Variable 1) Lowest level = 1 Highest level = 2
Col.   (Variable 2) Lowest level = 1 Highest level = 2
Slice  (Variable 3) Lowest level = 1 Highest level = 3

Variable levels:
Row    level = 1
Col.   level = 1
Slice  level = 1

Freq.      Mean      Std. Dev.
  3        2.000       1.000

Variable levels:
Row    level = 1
Col.   level = 1
Slice  level = 2

Freq.      Mean      Std. Dev.
  3        3.000       1.000

Variable levels:
Row    level = 1
Col.   level = 1
Slice  level = 3

Freq.      Mean      Std. Dev.
  3        4.000       1.000

Number of observations across levels = 9
Mean across levels = 3.000
Std. Dev. across levels = 1.225
```

We obtain similar output for each level of the "Col." variable within each level of the "Row" variable as well as the summary across all rows and columns. The procedure then produces the one-way ANOVA's for the breakdowns shown. For example, the first ANOVA table for the above sample is shown below:

```
Variable levels:
Row    level = 1
Col.   level = 2
Slice  level = 1

Freq.      Mean      Std. Dev.
  3        5.000       1.000
```

```
Variable levels:
Row     level = 1
Col.    level = 2
Slice   level = 2

Freq.     Mean      Std. Dev.
  3       4.000     1.000

Variable levels:
Row         level = 1
Col.        level = 2
Slice       level = 3

Freq.     Mean      Std. Dev.
  3       3.000     1.000

Number of observations across levels = 9
Mean across levels = 4.000
Std. Dev. across levels = 1.225

ANALYSES OF VARIANCE SUMMARY TABLES

Variable levels:
Row         level = 1
Col.        level = 1
Slice       level = 1

Variable levels:
Row         level = 1
Col.        level = 1
Slice       level = 2

Variable levels:
Row         level = 1
Col.        level = 1
Slice       level = 3

SOURCE    D.F.    SS        MS      F        Prob.>F
GROUPS     2      6.00      3.00    3.000    0.3041
WITHIN     6      6.00      1.00
TOTAL      8     12.00
```

The last ANOVA table is:

```
ANOVA FOR ALL CELLS

SOURCE    D.F.    SS         MS      F         Prob.>F
GROUPS    11     110.75     10.07   10.068     0.0002
WITHIN    24      24.00      1.00
TOTAL     35     134.75
FINISHED
```

You should note that the analyses of variance completed do NOT consider the interactions among the categorical variables. You may want to compare the results above with that obtained for a three-way analysis of variance completed by either the 1,2, or 3 way randomized design procedure or the Sum of Squares by Regression procedure listed under the Analyses of Variance option of the Statistics menu.

Distribution Parameters

The distribution parameters procedure was previously described.

Box Plots

Box plots are useful graphical devices for viewing both the central tendency and the variability of a continuous variable. There is no one "correct" way to draw a box plot hence various statistical packages draw them in somewhat different ways. Most box plots are drawn with a box that depicts the range of values between the 25th percentile and the 75 percentile with the median at the center of the box. In addition, "whiskers" are drawn that extend up from the top and down from the bottom to the 90th percentile and 10th percentile respectively. In addition, some packages will also place dots or circles at the end of the whiskers to represent possible "outlier" values (values at the 99th percentile or 1 percentile. Outliers are NOT shown in the box plots of OpenStat. In OpenStat, the mean is plotted in the box so one can also get a graphical representation of possible "skewness" (differences between the median and mean) for a set of values.

Now lets plot some data. In the Breakdown procedure described above, we analyzed data found in the threeway.tab file. We will obtain box plots for the continuous variable classified by the three categories of the "Slice" variable. Select Box Plots from the Descriptives option of the Statistics menu. You should see (after selecting the variables) (Fig. 6.6):

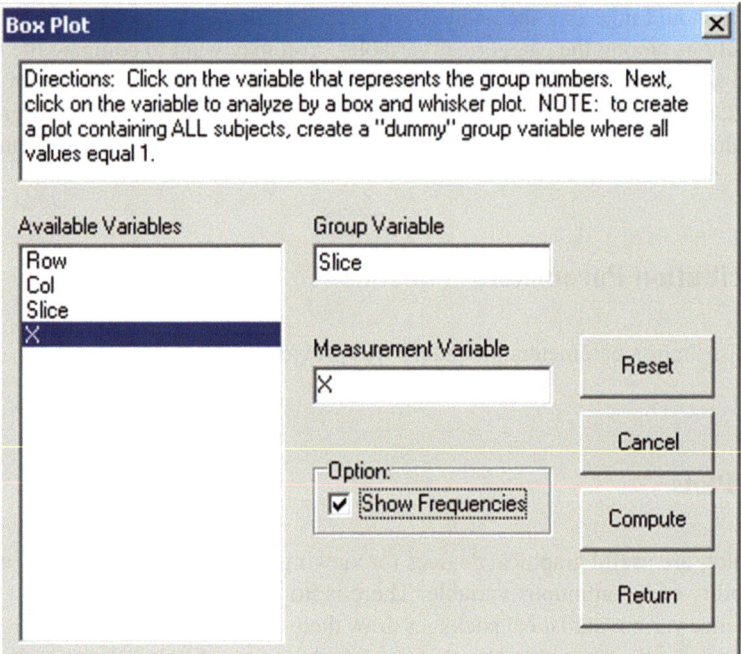

Fig. 6.6 The Box Plot form

Having selected the variables and option, click the Return button. In our example you should see (Fig. 6.7):

```
Box Plot of Groups

Results for group 1, mean = 3.500
Centile         Value
Ten             1.100
Twenty five     2.000
Median          3.500
Seventy five    5.000
Ninety          5.900
Score Range     Frequency   Cum.Freq.   Percentile Rank
───────────────────────────────────────────────────────
0.50 - 1.50     2.00        2.00         8.33
1.50 - 2.50     2.00        4.00        25.00
2.50 - 3.50     2.00        6.00        41.67
3.50 - 4.50     2.00        8.00        58.33
4.50 - 5.50     2.00       10.00        75.00
5.50 - 6.50     2.00       12.00        91.67
6.50 - 7.50     0.00       12.00       100.00
7.50 - 8.50     0.00       12.00       100.00
```

```
 8.50 -  9.50    0.00    12.00    100.00
 9.50 -10.50    0.00    12.00    100.00
10.50 -11.50    0.00    12.00    100.00
```

Results for group 2, mean = 4.500
Centile Value
Ten   2.600
Twenty five   3.500
Median   4.500
Seventy five   5.500
Ninety 6.400

```
Score Range     Frequency   Cum.Freq.   Percentile Rank
_____
 0.50 -  1.50    0.00     0.00       0.00
 1.50 -  2.50    1.00     1.00       4.17
 2.50 -  3.50    2.00     3.00      16.67
 3.50 -  4.50    3.00     6.00      37.50
 4.50 -  5.50    3.00     9.00      62.50
 5.50 -  6.50    2.00    11.00      83.33
 6.50 -  7.50    1.00    12.00      95.83
 7.50 -  8.50    0.00    12.00     100.00
 8.50 -  9.50    0.00    12.00     100.00
 9.50 -10.50    0.00    12.00     100.00
10.50 -11.50    0.00    12.00     100.00
```

Results for group 3, mean = 4.250
Centile Value
Ten 1.600
Twenty five   2.500
Median   3.500
Seventy five   6.500
Ninety   8.300
Score Range Frequency Cum.Freq. Percentile Rank

```
_____
 0.50 -  1.50    1.00     1.00       4.17
 1.50 -  2.50    2.00     3.00      16.67
 2.50 -  3.50    3.00     6.00      37.50
 3.50 -  4.50    2.00     8.00      58.33
 4.50 -  5.50    1.00     9.00      70.83
 5.50 -  6.50    0.00     9.00      75.00
 6.50 -  7.50    1.00    10.00      79.17
 7.50 -  8.50    1.00    11.00      87.50
 8.50 -  9.50    1.00    12.00      95.83
 9.50 -10.50    0.00    12.00     100.00
10.50 -11.50    0.00    12.00     100.00
```

Fig. 6.7 Box and whiskers plot

Three Variable Rotation

The option for 3D rotation of 3 variables under the Descriptive option of the Statistics menu will rotate the case values around the X, Y and Z axis! In the example below we have again used the cansas.tab data file which consists of six variables measuring weight, pulse rate, etc. of individuals and measures of their physical abilities such as pull ups, sit ups, etc. By "dragging" the X, Y or Z bars up or down with your mouse, you may rotate up to 180° around each axis (see Figs. 6.8–6.9 below (Fig. 6.8)):

X Versus Y Plots

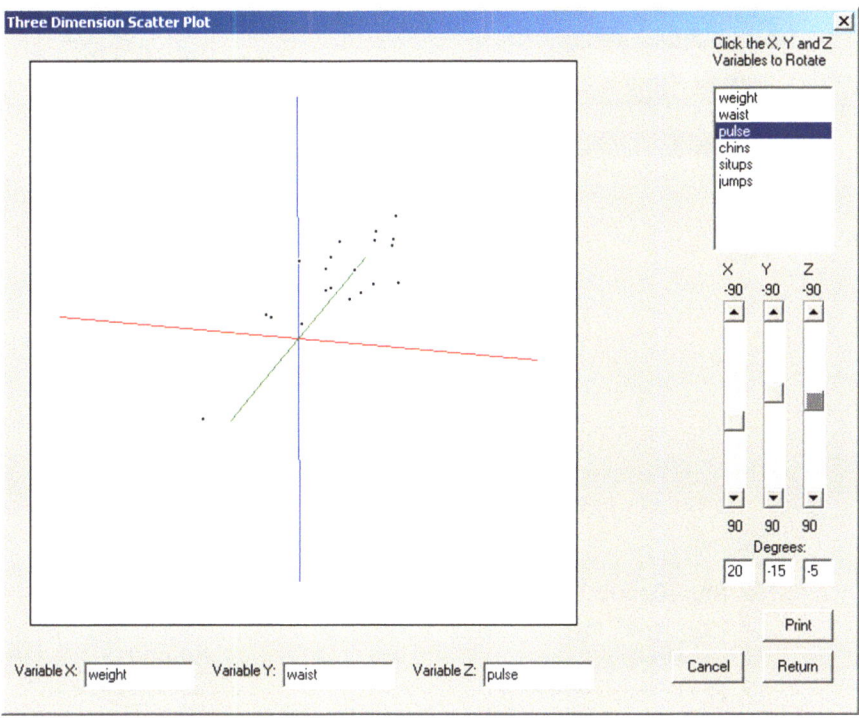

Fig. 6.8 Three Dimension plot with rotation

X Versus Y Plots

As mentioned above, plotting one variable's values against those of another variable in an X versus Y scatter plot often reveals insights into the relationships between two variables. Again we will use the same cansas.tab data file to plot the relationship between weight and waist measurements. When you select the X Versus Y Plots option from the Statistics/Descriptive menu, you see the form below (Fig. 6.9):

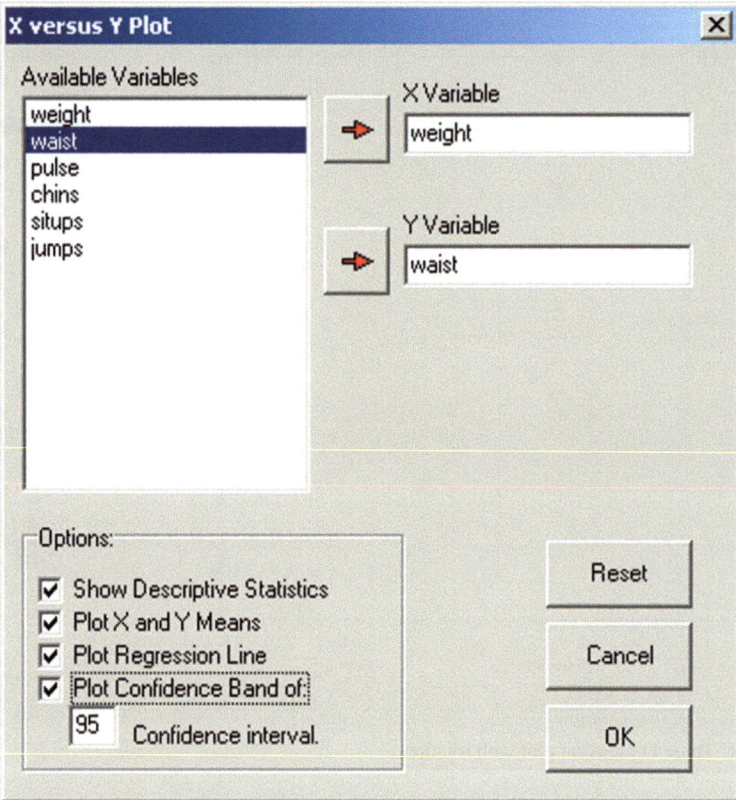

Fig. 6.9 X Versus Y Plot form

In the above form we have elected to print descriptive statistics for the two variables selected and to plot the linear regression line and confidence band for predicted scores about the regression line drawn through the scatter of data points. When you click the Compute button, the following results are obtained for the descriptive statistics in the output form:

```
X versus Y Plot

X = weight , Y = waist from file:
C:\Projects\Delphi\OpenStat\cansas.txt

Variable    Mean        Variance      Std.Dev.
weight      178.60      609.62        24.69
waist        35.40       10.25         3.20
Correlation = 0.8702, Slope = 0.11, Intercept = 15.24
Standard Error of Estimate =  1.62
```

When you press the Return button on the output form, you then obtain the desired plot (Fig. 6.10):

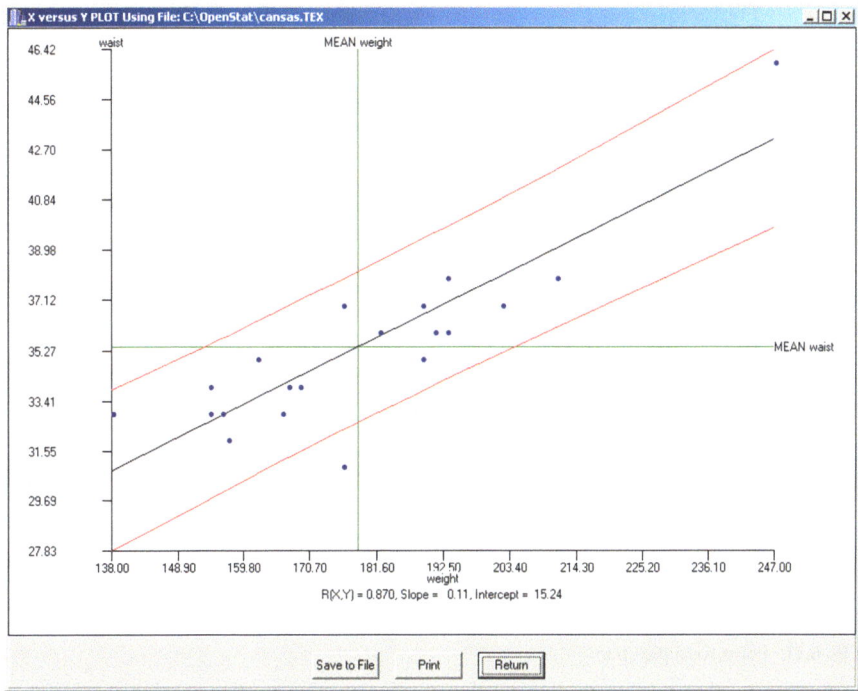

Fig. 6.10 Plot of regression line in X versus Y

Notice that the measured linear relationship between the two variables is fairly high (.870) however, you may also notice that one data point appears rather extreme on both the X and Y variables. Should you eliminate the case with those extreme scores (an outlier?), you would probably observe a reduction in the linear relationship! I would personally not eliminate this case however since it "seems reasonable" that the sample might contain a subject with both a high weight and high waist measurement.

Histogram/Pie Chart of Group Frequencies

You may obtain a histogram or pie chart plot of frequencies for a variable using the Analyses/Descriptive options of either the Histogram of Group Frequencies of Pie Chart of Group Frequencies option. Selecting either of these procedures results in the following dialogue form (Fig. 6.11):

In this example we have loaded the chisqr.TEX OpenStat file and have chosen to obtain a pie chart of the col variable. The result is shown below (Fig. 6.12):

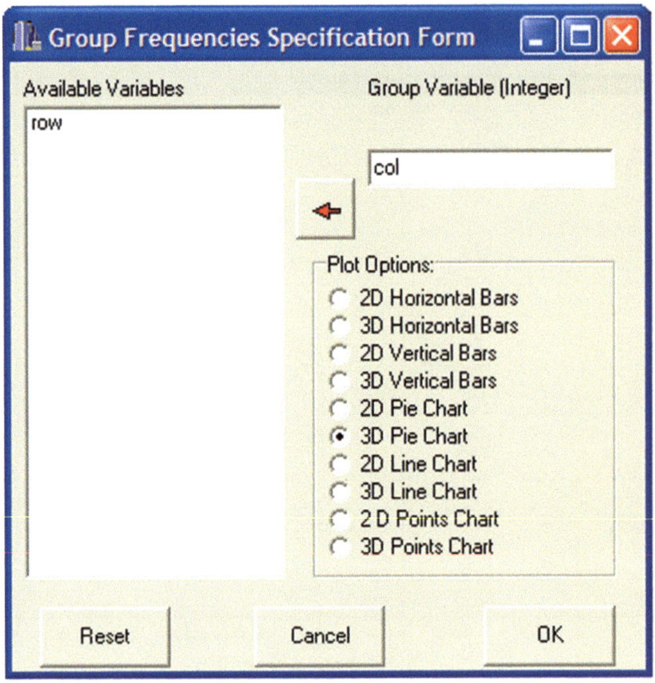

Fig. 6.11 Form for a pie chart

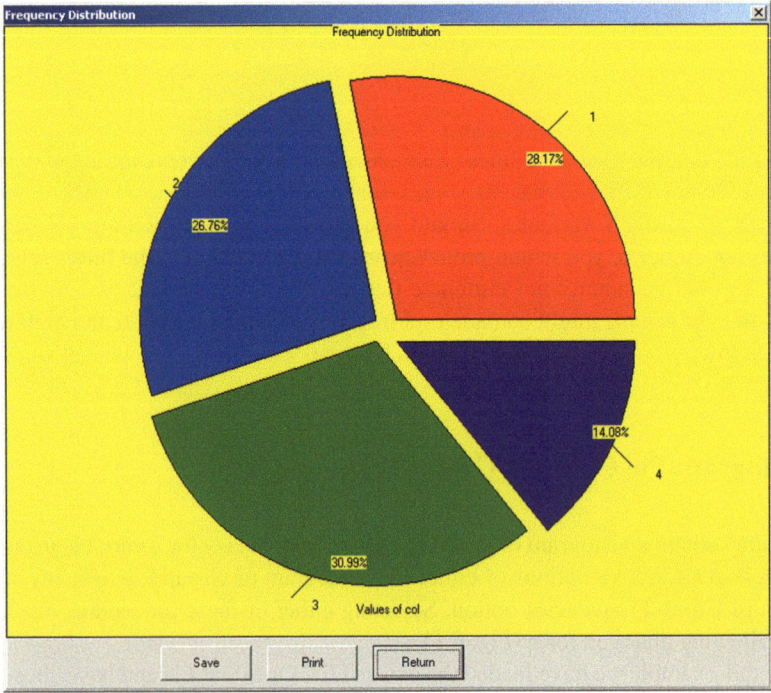

Fig. 6.12 Pie chart

Stem and Leaf Plot

One of the earliest plots in the annals of statistics was the "Stem and Leaf" plot. This plot gives the user a view of the major values found in a frequency distribution. To illustrate this plot, we will use the file labeled "StemleafTest2.TAB". If you select this option from the Descriptive option of the Analyses menu, you will see the dialogue form below (Fig. 6.13):

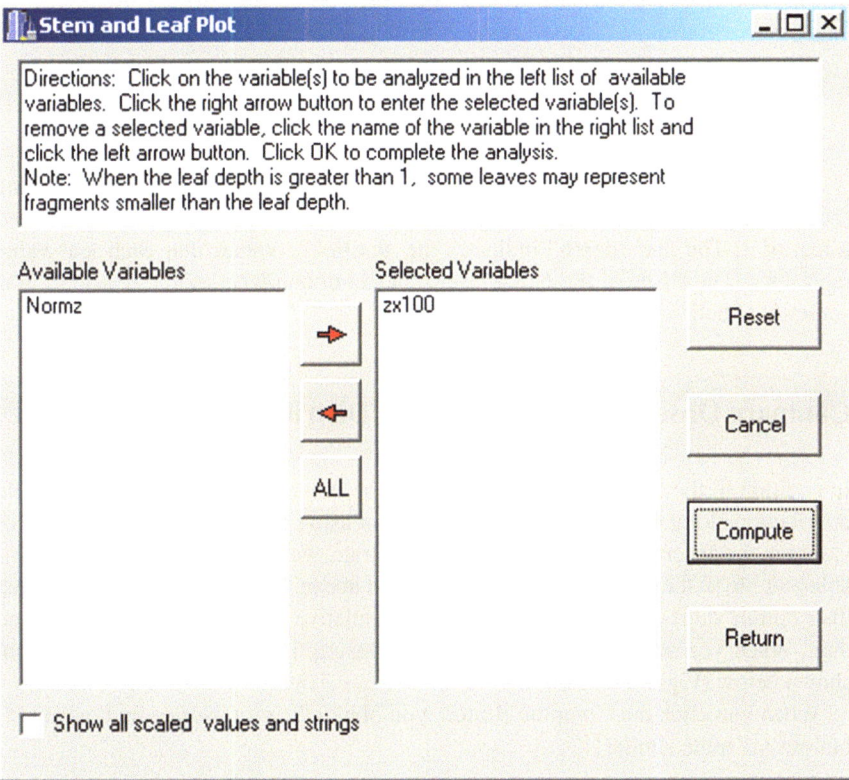

Fig. 6.13 Stem and Leaf form

We will choose to plot the zx100 variable to obtain the following results:

```
STEM AND LEAF PLOTS
Stem and Leaf Plot for variable: zx100
Frequency  Stem & Leaf
    1       -3      0
    6       -2      0034
   12       -1      0122234
    5       -1      6789
   71        0      000111111122222222233333334444444444
   78        0      555555556666666677777777788888889999999
   16        1      00011223
    7        1      56789
    2        2      03
    2        2      57
Stem width = 100.00, max. leaf depth = 2
Min. value = -299.000, Max. value = 273.600
No. of good cases = 200
```

The results indicate that the Stem has values ranging from −300 to +200 with the second digits shown as leaves. For example, the value 111.6 has a stem of 100 and a leaf of 1. The leaf "depth" indicates the number of values that each leaf value represents. The shape of the plot is useful in examining whether the distribution is somewhat "bell" shaped, flat, skewed, etc.

Compare Observed and Theoretical Distributions

In addition to the Stem and Leaf Plot described above, one can also plot a sample distribution along with a theoretical distribution using the cumulative proportion of values in the observed distribution. To demonstrate, we will again use the same variable and file in the stem and leaf plot described above. We will examine the normal distribution values expected for the same cumulative proportions of the observed data. When you select this option from the Descriptive option, you see the form shown below (Fig. 6.14):

When you click the Compute Button, you obtain the plot. Notice that our distributions are quite similar!

QQ and PP Plots

In a manner similar to that shown above, one can also obtain a plot of the theoretical versus the observed data. You may select to plot actual values observed and expected or the proportions (probabilities) observed and expected. Show below is the dialogue form and a QQ plot for the save data of the previous section (Figs. 6.15, 6.16):

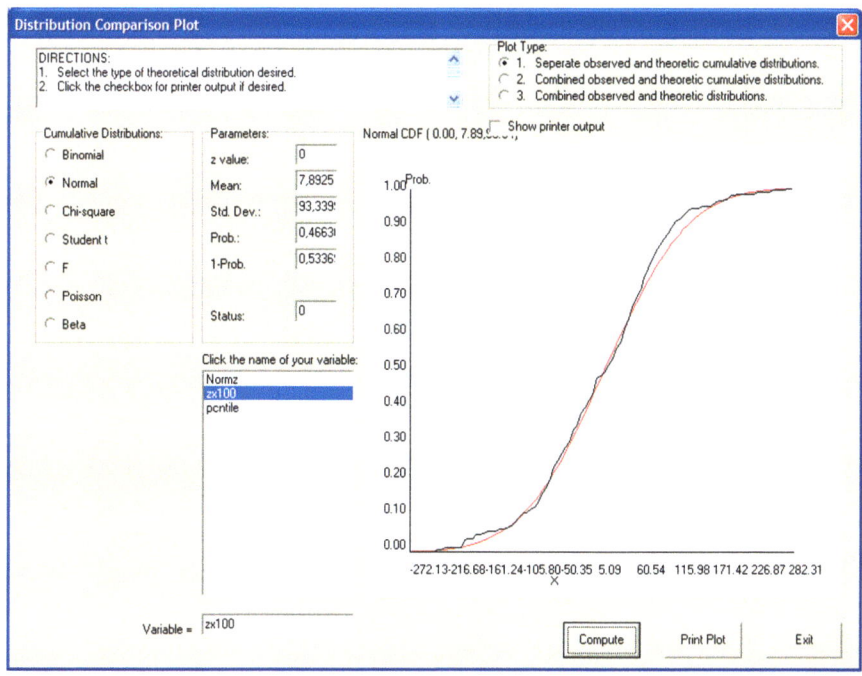

Fig. 6.14 Dialog form for examining theoretical and observed distributions

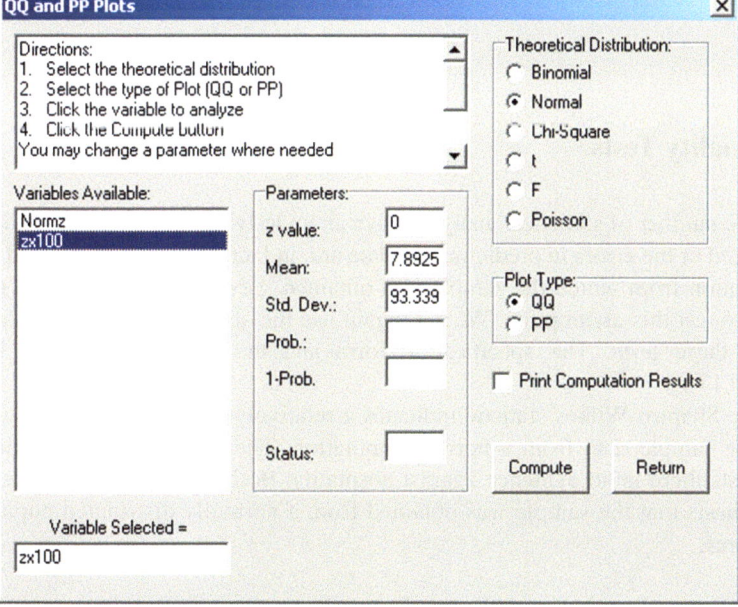

Fig. 6.15 The QQ / PP Plot Specification form

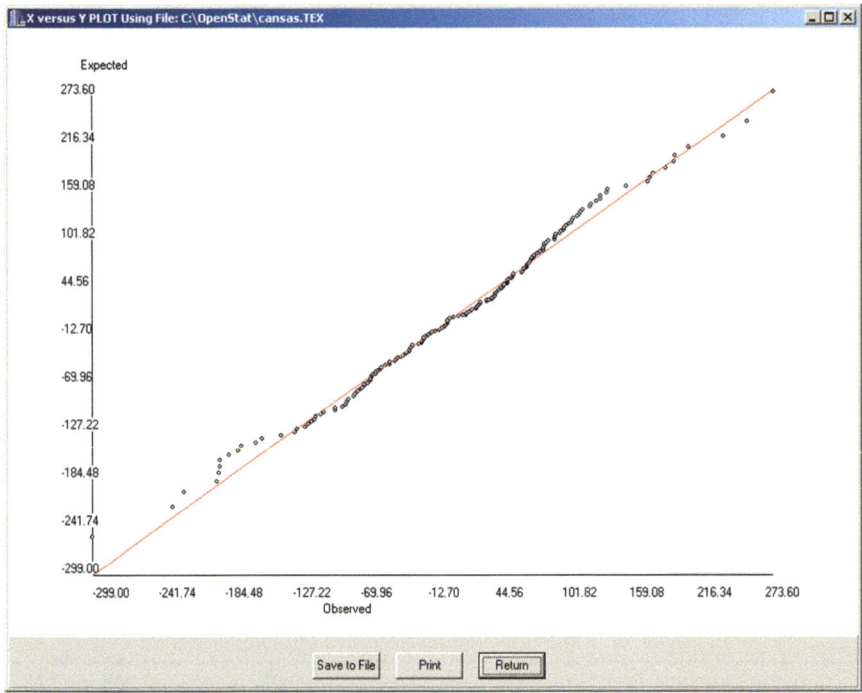

Fig. 6.16 A QQ plot

Normality Tests

A large number of statistical analyses have an underlying assumption that the data analyzed or the errors in predicting the data are, in fact, normally distributed in the population from which the sample was obtained. Several tests have been developed to test this assumption. We will again use the above sample data to demonstrate these tests. The specification form and the results are shown below (Fig. 6.17):

The Shapiro-Wilkes statistic indicates a relatively high probability of obtaining the sample data from a normal population. The Liliefors test statistic also suggests there is no evidence against normality. Both tests lead us to accept the hypothesis that the sample was obtained from a normally distributed population of scores.

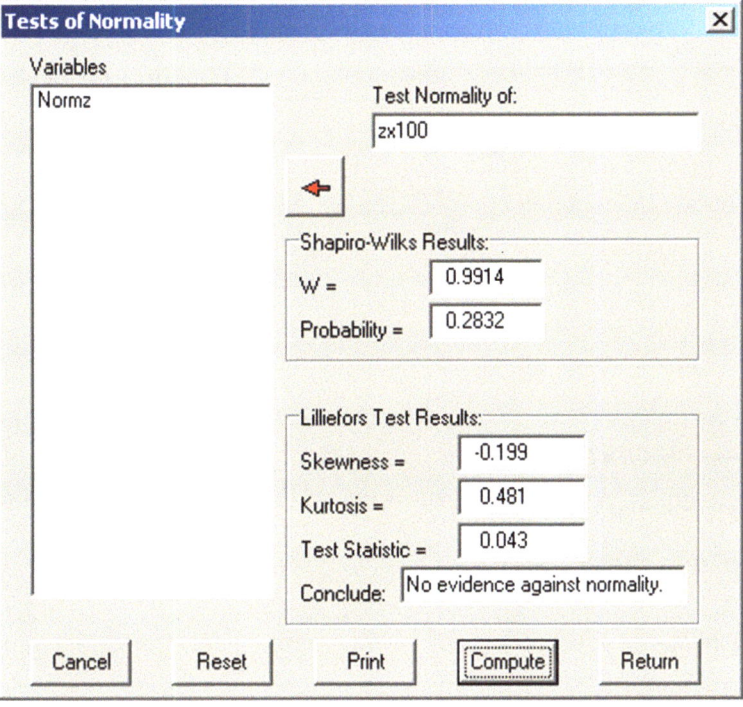

Fig. 6.17 Normality tests

Resistant Line

Tukey (1970, Chap. 10) proposed the three point resistant line as an data analysis tool for quickly fitting a straight line to bivariate data (x and y paired data.) The data are divided into three groups of approximately equal size and sorted on the x variable. The median points of the upper and lower groups are fitted to the middle group to form two slope lines. The resulting slope line is resistant to the effects of extreme scores of either x or y values and provides a quick exploratory tool for investigating the linearity of the data. The ratio of the two slope lines from the upper and lower group medians to the middle group median provides a quick estimate of the linearity which should be approximately 1.0 for linearity. Our example uses the "Cansas.TEX" file. The dialogue for the analysis appears as (Fig. 6.18):

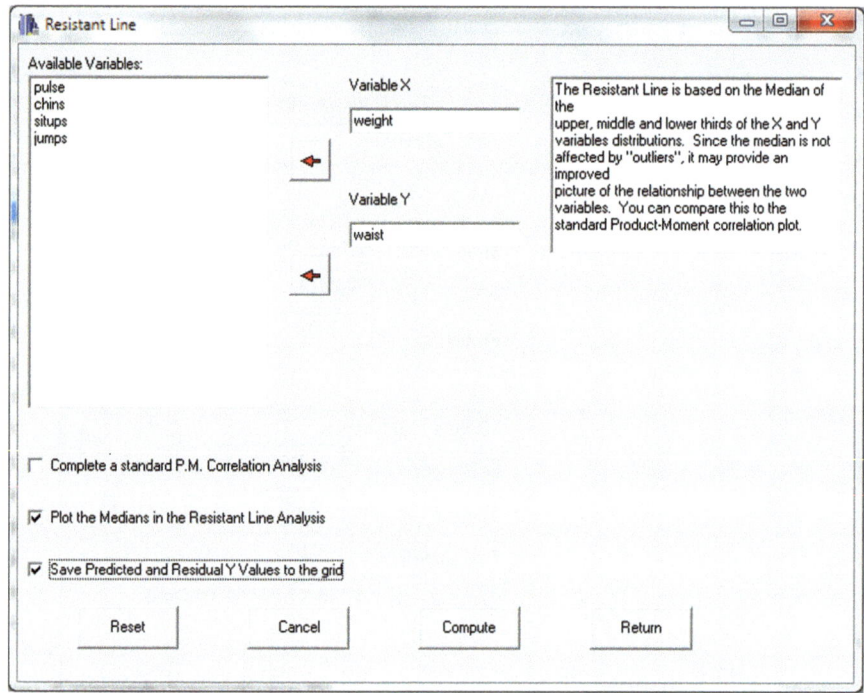

Fig. 6.18 Resistant Line dialog

The results obtained are (Fig. 6.19):

```
Group    X Median      Y Median     Size
  1      155.000       155.000       6
  2      176.000        34.000       8
  3      197.500        36.500       6
```

Half Slopes = -5.762 and 0.116
Slope = -2.788
Ratio of half slopes = -0.020
Equation: y = -2.788 * X + (-566.361)

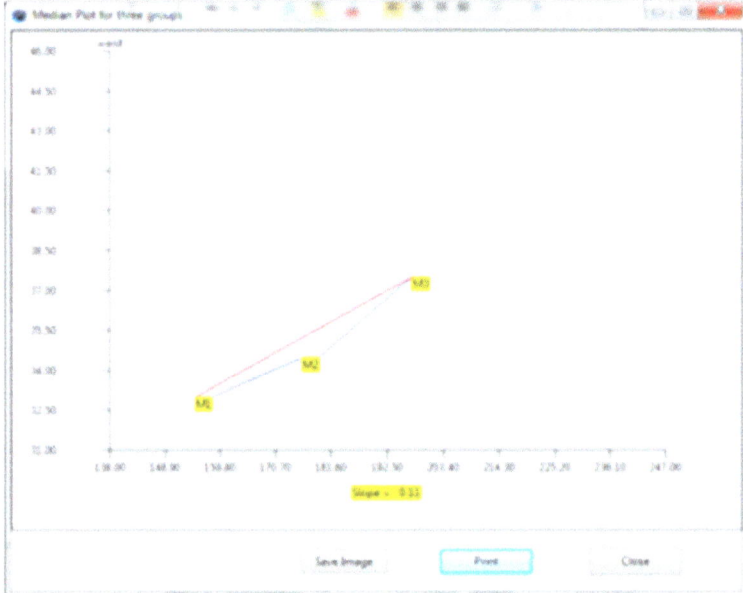

Fig. 6.19 Resistant Line plot

Repeated Measures Bubble Plot

Bubble plots are useful for comparing repeated measures for multiple objects. In our example, we have multiple schools which are being compared across years for student achievement. The size of the bubbles that are plotted represent the ratio of students to teachers. We are using the BubblePlot2.TEX file in the sample data files.

Shown below is the dialog for the bubble plot procedure followed by the plot and the descriptive data of the analysis (Figs. 6.20, 6.21):

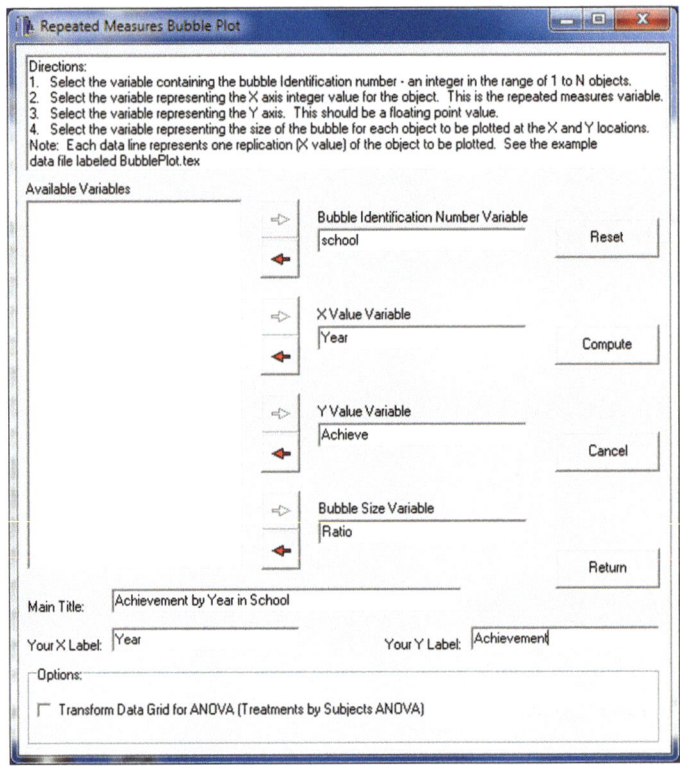

Fig. 6.20 Dialog for the repeated measures bubble plot

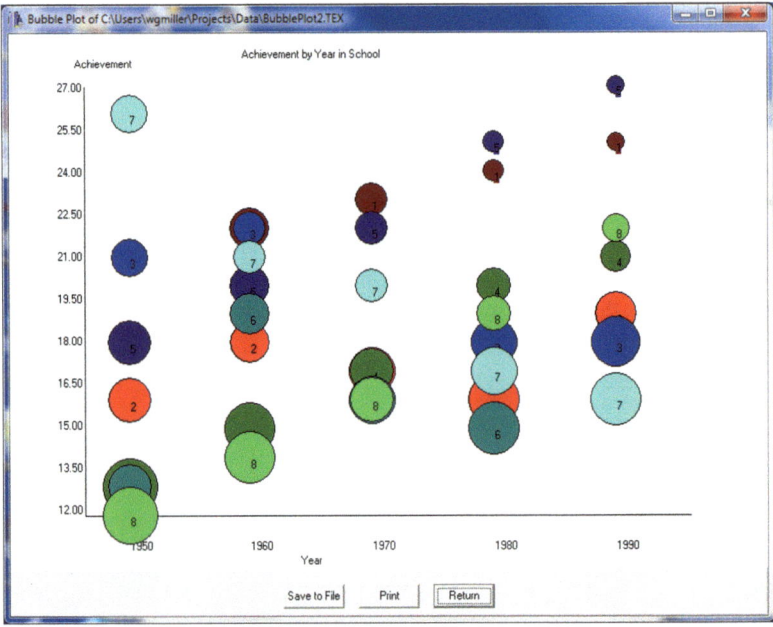

Fig. 6.21 Bubble plot

```
MEANS FOR Y AND SIZE VARIABLES

Grand Mean for Y = 18.925
Grand Mean for Size = 23.125

REPLICATION MEAN Y VALUES (ACROSS OBJECTS)
Replication   1 Mean =    17.125
Replication   2 Mean =    18.875
Replication   3 Mean =    18.875
Replication   4 Mean =    19.250
Replication   5 Mean =    20.500

REPLICATION MEAN SIZE VALUES (ACROSS OBJECTS}
Replication   1 Mean =    25.500
Replication   2 Mean =    23.500
Replication   3 Mean =    22.750
Replication   4 Mean =    22.500
Replication   5 Mean =    21.375

MEAN Y VALUES FOR EACH BUBBLE (OBJECT)
Object    1 Mean =    22.400
Object    2 Mean =    17.200
Object    3 Mean =    19.800
Object    4 Mean =    17.200
Object    5 Mean =    22.400
Object    6 Mean =    15.800
Object    7 Mean =    20.000
Object    8 Mean =    16.600

MEAN SIZE VALUES FOR EACH BUBBLE (OBJECT)
Object    1 Mean =    19.400
Object    2 Mean =    25.200
Object    3 Mean =    23.000
Object    4 Mean =    24.600
Object    5 Mean =    19.400
Object    6 Mean =    25.800
Object    7 Mean =    23.200
Object    8 Mean =    24.400
```

Smooth Data by Averaging

Measurements made on multiple objects often contain "noise" or error variations that mask the trend of data. One method for reducing this "noise" is to smooth the data by averaging the data points. In this method, three contiguous data points are averaged to obtain a new value for the first of the three points. The next point is

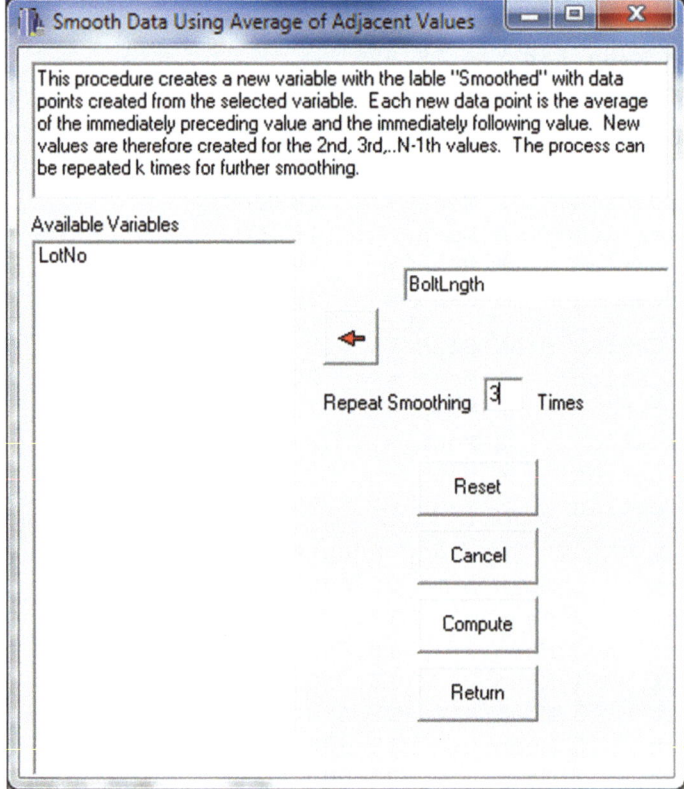

Fig. 6.22 Dialog for smoothing data by averaging

the average of three points, etc. across all points. Only the first and last data points are left unchanged. To illustrate this procedure, we will use the file labeled "boltsize.TEX". The dialog is shown followed by a comparison of the original data with the smoothed data using the procedure to compare two distributions (Figs. 6.22, 6.23, 6.24):

X Versus Multiple Y Plot

You may have collected multiple measurements for a group of objects and wish to compare these measurements in a plot. This procedure lets you select a variable for the X axis and multiple Y variables to plot as points or lines. To illustrate we have selected a file labeled "multiplemeas.TEX" and have plotted a group of repeated measures against the first one. The dialog is shown below followed by the plot (Figs. 6.25, 6.26):

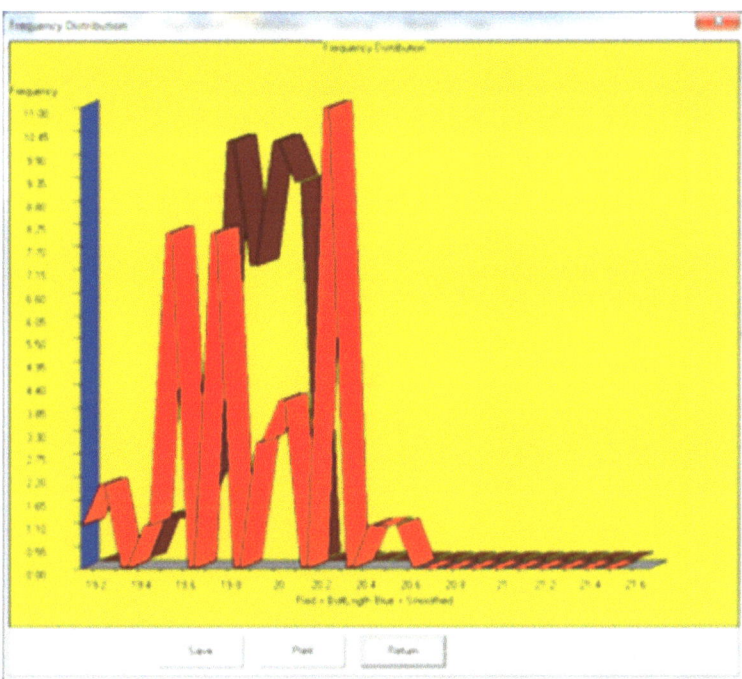

Fig. 6.23 Smoothed data frequency distribution plot

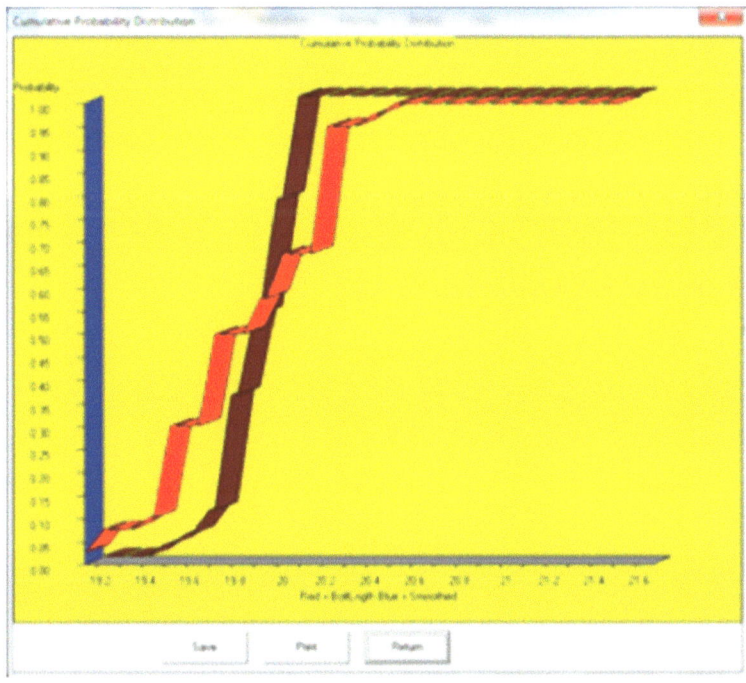

Fig. 6.24 Cumulative frequency of smoothed data

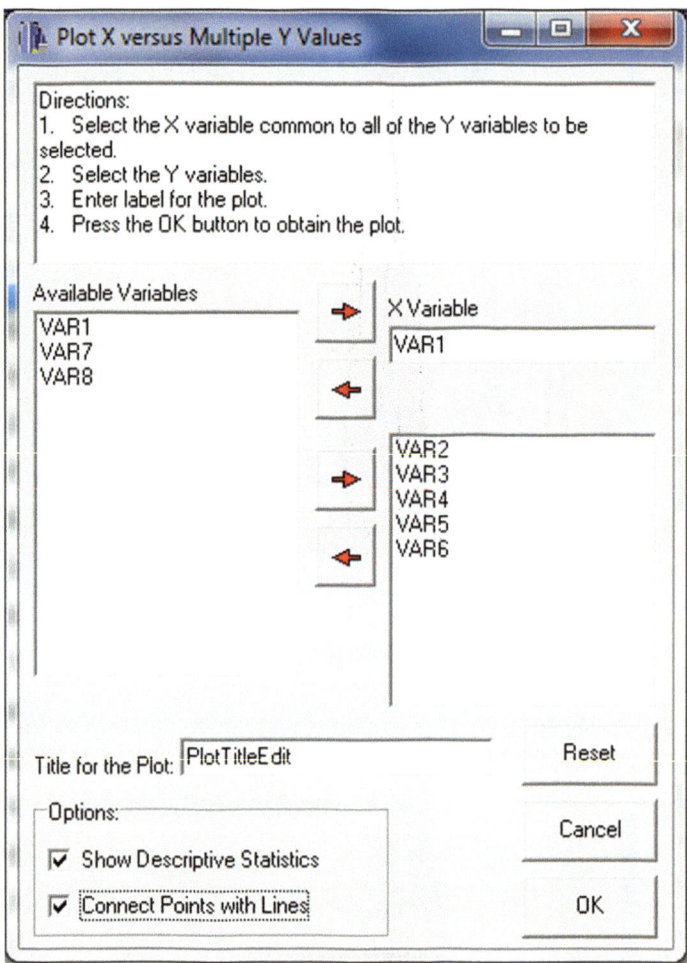

Fig. 6.25 Dialog for an X versus multiple Y plot

```
X VERSUS MULTIPLE Y VALUES PLOT
CORRELATION MATRIX

Correlations
        VAR2    VAR3    VAR4    VAR5    VAR6    VAR1
VAR2    1.000   0.255   0.542   0.302   0.577   0.325
VAR3    0.255   1.000  -0.048   0.454   0.650   0.763
VAR4    0.542  -0.048   1.000   0.125  -0.087   0.005
VAR5    0.302   0.454   0.125   1.000   0.527   0.304
VAR6    0.577   0.650  -0.087   0.527   1.000   0.690
VAR1    0.325   0.763   0.005   0.304   0.690   1.000

Means

Variables    VAR2   VAR3   VAR4   VAR5   VAR6   VAR1
             8.894  9.682  5.021  9.721  9.451  6.639

Standard Deviations

Variables    VAR2    VAR3    VAR4    VAR5    VAR6    VAR1
             12.592  16.385  17.310  13.333  16.157  11.834

No. of valid cases = 30
```

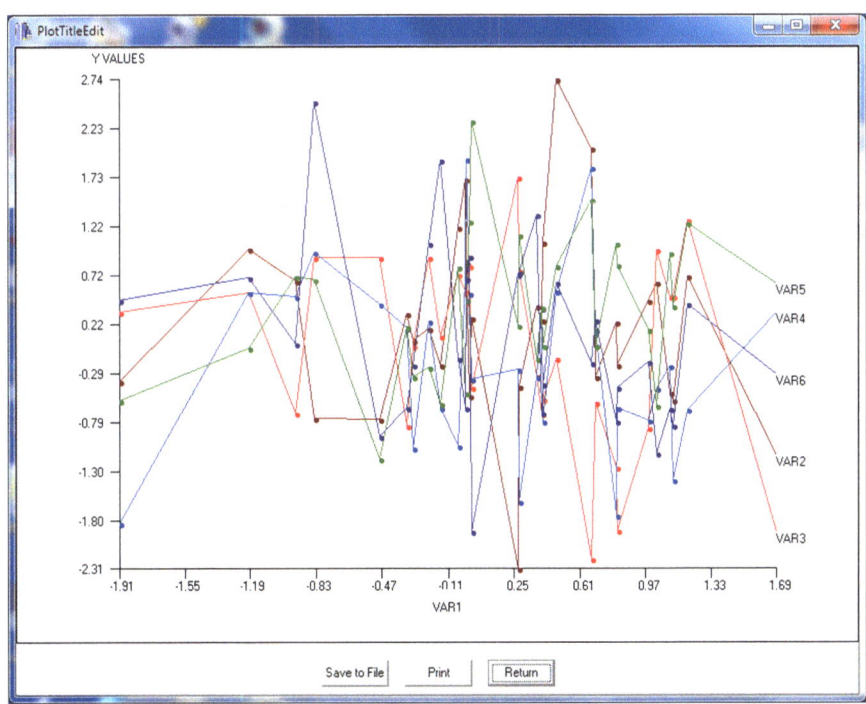

Fig. 6.26 X versus multiple Y plot

Compare Observed to a Theoretical Distribution

Observed data may be distributed in a manner similar to a variety of theoretical distributions. This procedure lets you plot the observed scores against various theoretical distributions to see if the data tends to be more similar to one than another. We will demonstrate using a set of simulated data that we created to follow an approximately normal distribution. We smoothed the data using the smoothing procedure and then compared the smoothed data to the normal distribution by means of this procedure. Shown below is the dialog utilized and the resulting plot of the data (Figs. 6.27, 6.28):

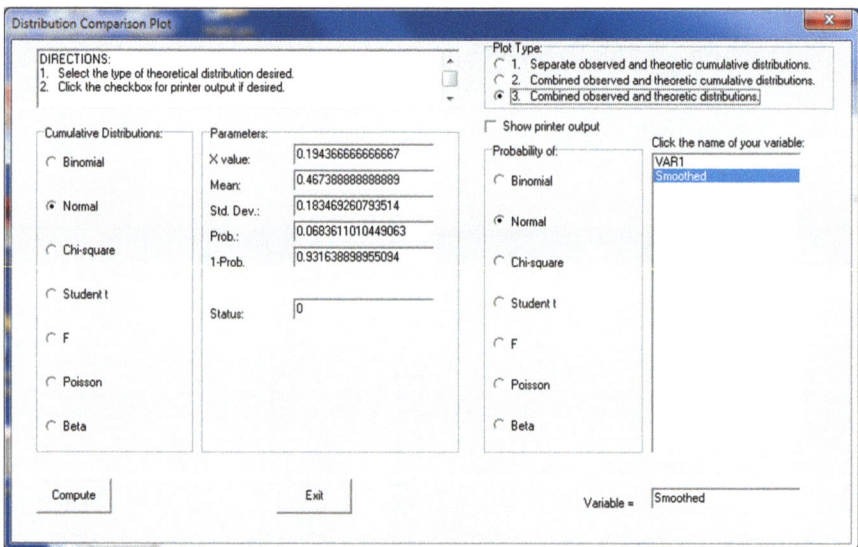

Fig. 6.27 Dialog for comparing observed and theoretical distributions

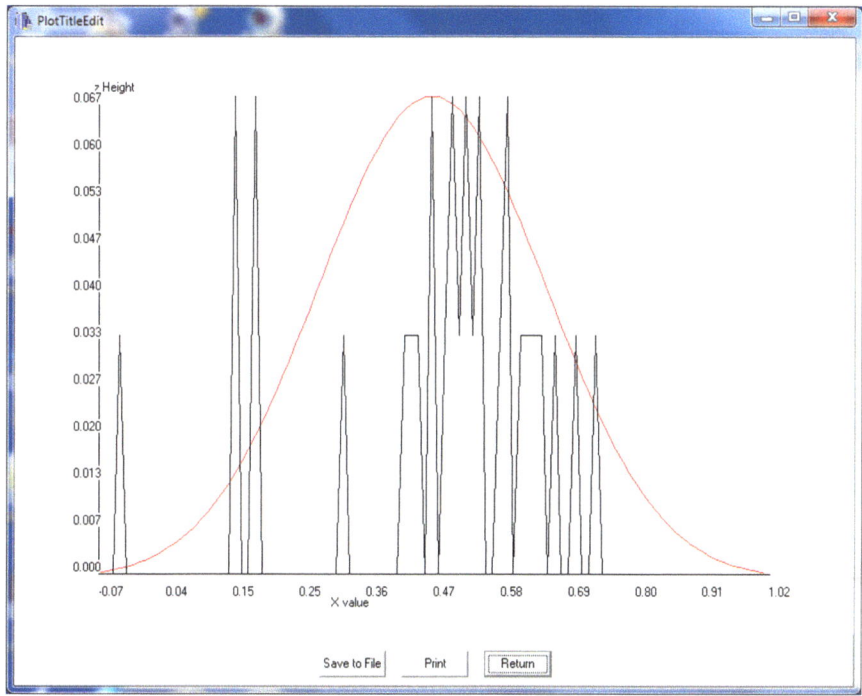

Fig. 6.28 Comparison of an observed and theoretical distribution

Multiple Groups X versus Y Plot

You may have observed objects within groups such as male and female (coded 0 and 1 for example) and wish to plot the relationship between two other measures for those groups. To demonstrate this procedure we will use the sample data file labeled "anova2.TEX" and plot the lines for the relationship of the dependent variable x and the covariate2 in the file. The dialog is shown below followed by the plot (Figs. 6.29, 6.30):

Fig. 6.29 Dialog for multiple groups X versus Y plot

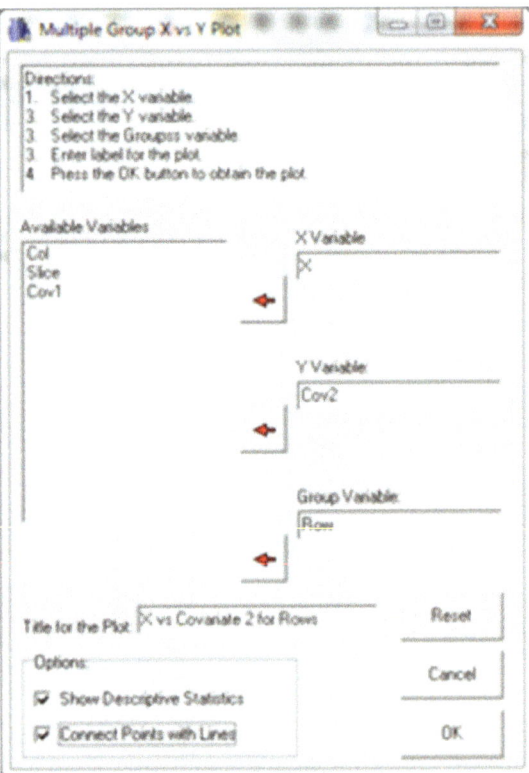

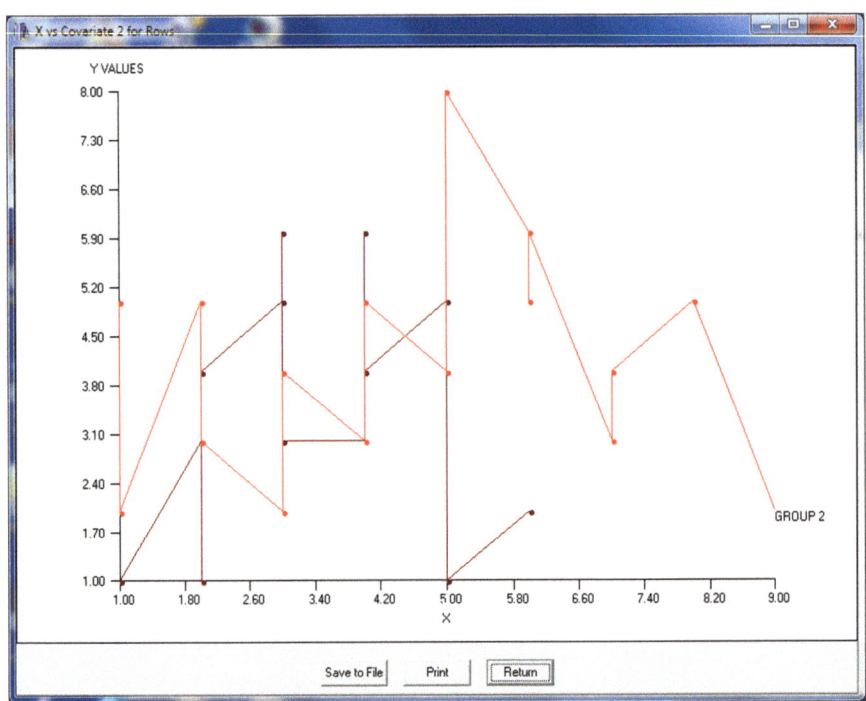

Fig. 6.30 X versus Y plot for multiple groups

```
X VERSUS Y FOR GROUPS PLOT
VARIABLE    MEAN    STANDARD DEVIATION
   X        4.083   1.962
   Y        3.917   1.628
```

Chapter 7
Correlation

The Product Moment Correlation

It seems most living creatures observe relationships, perhaps as a survival instinct. We observe signs that the weather is changing and prepare ourselves for the winter season. We observe that when seat belts are worn in cars that the number of fatalities in car accidents decrease. We observe that students that do well in one subject tend to perform will in other subjects. This chapter explores the linear relationship between observed phenomena.

If we make systematic observations of several phenomena using some scales of measurement to record our observations, we can sometimes see the relationship between them by "plotting" the measurements for each pair of measures of the observations. As a hypothetical example, assume you are a commercial artist and produce sketches for advertisement campaigns. The time given to produce each sketch varies widely depending on deadlines established by your employer. Each sketch you produce is ranked by five marketing executives and an average ranking produced (rank 1 = best, rank 5 = poorest.) You suspect there is a relationship between time given (in minutes) and the average quality ranking obtained. You decide to collect some data and observe the following:

Average rank (Y)	Minutes (X)
3.8	10
2.6	35
4.0	5
1.8	42
3.0	30
2.6	32
2.8	31
3.2	26
3.6	11
2.8	33

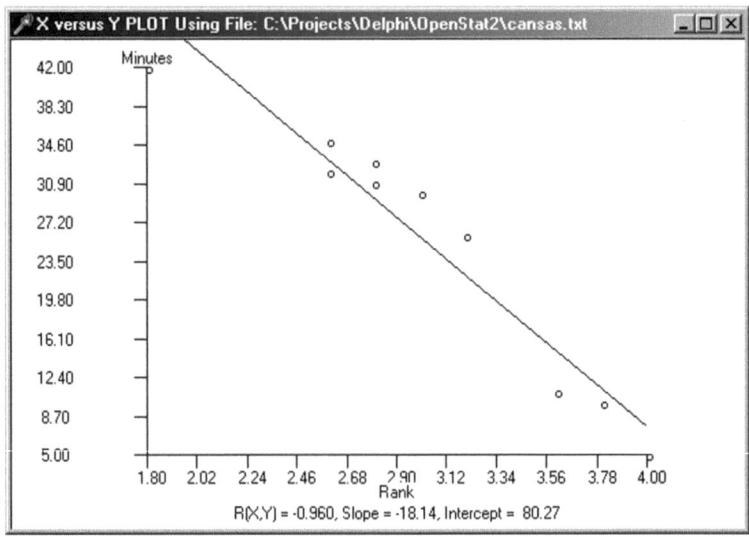

Fig. 7.1 Correlation regression line

Using OpenStat Descriptive menu's Plot X vs. Y procedure to plot these values yields the scatter-plot shown above following page. Is there a relationship between the time and ranks? (Fig. 7.1).

Testing Hypotheses for Relationships Among Variables: Correlation

To further understand and learn to interpret the product–moment correlation, OpenStat provides a means of simulating pairs of data, plotting those pairs, drawing the "best-fitting line" to the data points and showing the marginal distributions of the X and Y variables. Go to the Simulation menu and click on the Bivariate Scatter Plot. The figure below shows a simulation for a population correlation of −.95 with population means and variances as shown. A sample of 100 cases are generated. Actual sample means and standard deviations will vary (as sample statistics do!) from the population values specified (Fig. 7.2).

```
POPULATION PARAMETERS FOR THE SIMULATION
Mean X :=    100.000, Std. Dev. X :=     15.000
Mean Y :=    100.000, Std. Dev. Y :=     15.000
Product-Moment Correlation :=    -0.900
Regression line slope :=     -0.900, constant :=    190.000
SAMPLE STATISTICS FOR 100 OBSERVATIONS FROM THE POPULATION
Mean X :=     99.988, Std. Dev. X :=     14.309
Mean Y :=    100.357, Std. Dev. Y :=     14.581
Product-Moment Correlation :=    -0.915
Regression line slope :=     -0.932, constant :=    193.577
```

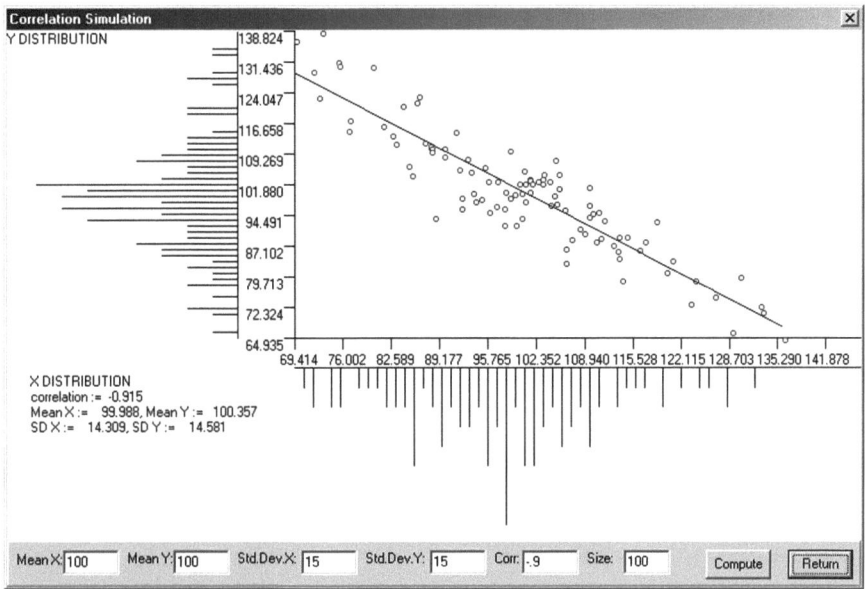

Fig. 7.2 Simulated bivariate scatterplot

Simple Linear Regression

The product–moment correlation discussed in the previous section is an index of the linear relationship between two continuous variables. But what is the nature of that linear relationship? That is, what is the slope of the line and where does the line intercept the vertical (Y variable) axis? This unit will examine the straight line "fit" to data points representing observations with two variables. We will also examine how this straight line may be used for prediction purposes as well as describing the relationship to the product–moment correlation coefficient.

OpenStat contains a procedure for completing a z test for data like that presented above.

Under the Statistics menu, move your mouse down to the Comparisons submenu, and then to the option entitled "One Sample Tests". When the form below displays, click on the Correlation button and enter the sample value .5, the population value .6, and the sample size 50. Change the confidence level to 90.0 %.

Shown below is the z-test for the above data (Figs. 7.3, 7.4):

```
ANALYSIS OF A SAMPLE CORRELATION
Sample Correlation = 0.600
Population Correlation = 0.500
Sample Size = 50
z Transform of sample correlation = 0.693
z Transform of population correlation = 0.549
Standard error of transform = 0.146
z test statistic = 0.986 with probability 0.838
z value required for rejection = 1.645
Confidence Interval for sample correlation = ( 0.425, 0.732)
```

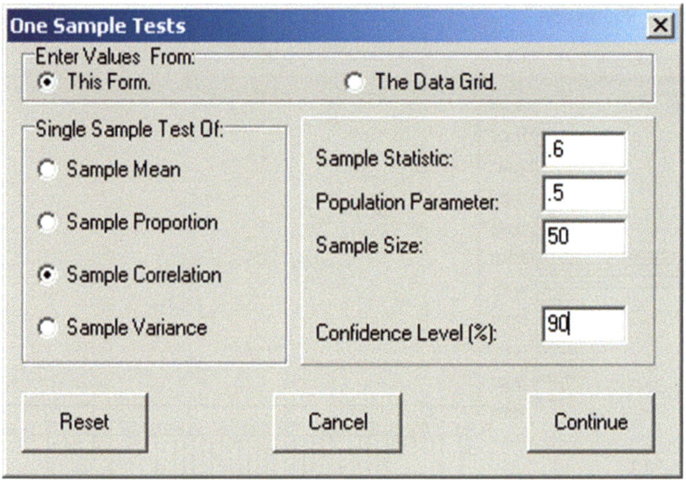

Fig. 7.3 Single sample tests form for correlations

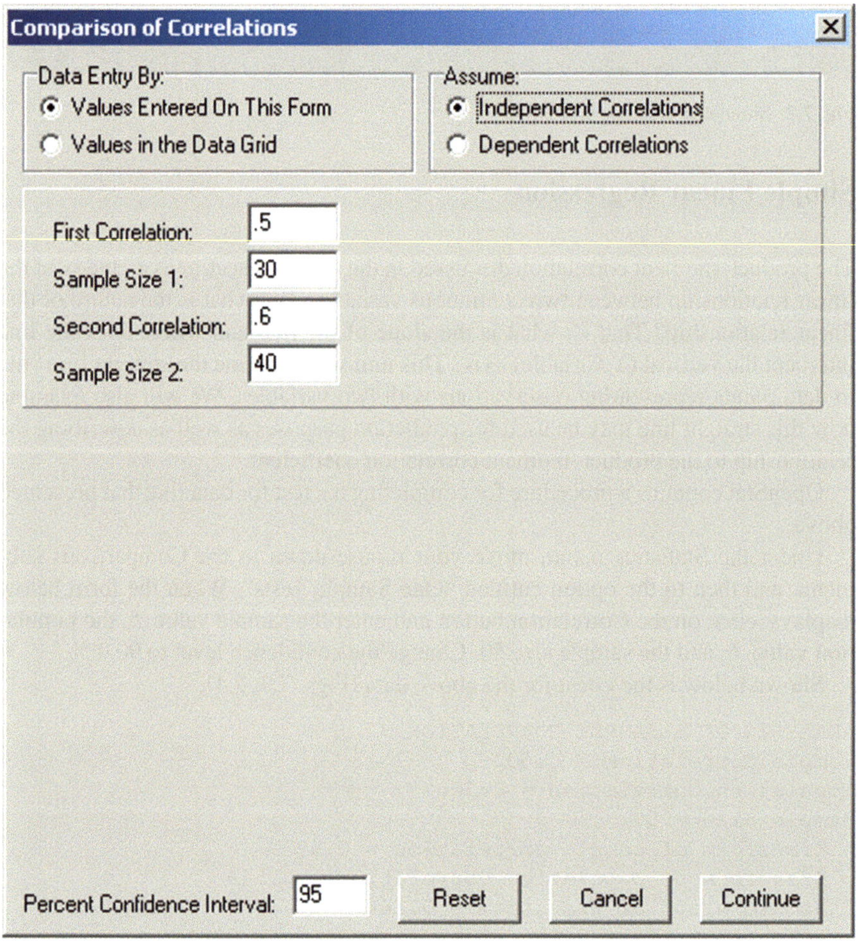

Fig. 7.4 Comparison of two independent correlations

Simple Linear Regression

Testing Equality of Correlations in Two Populations

```
COMPARISON OF TWO CORRELATIONS
Correlation one = 0.500
Sample size one = 30
Correlation two = 0.600
Sample size two = 40
Difference between correlations = -0.100
Confidence level selected = 95
z for Correlation One = 0.549
z for Correlation Two = 0.693
z difference = -0.144
Standard error of difference = 0.253
z test statistic = -0.568
Probability > |z| = 0.715
z Required for significance = 1.960
Note: above is a two-tailed test.
Confidence Limits = (-0.565, 0.338)
```

Differences Between Correlations in Dependent Samples

Again, OpenStat provides the computations for the difference between dependent correlations as shown in the figure below (Fig. 7.5):

```
COMPARISON OF TWO CORRELATIONS
Correlation x with y = 0.400
Correlation x with z = 0.600
Correlation y with z = 0.700
Sample size = 50
Confidence Level Selected = 95.0
Difference r(x,y) - r(x,z) = -0.200
t test statistic = -2.214
Probability > |t| = 0.032
t value for significance = 2.012
```

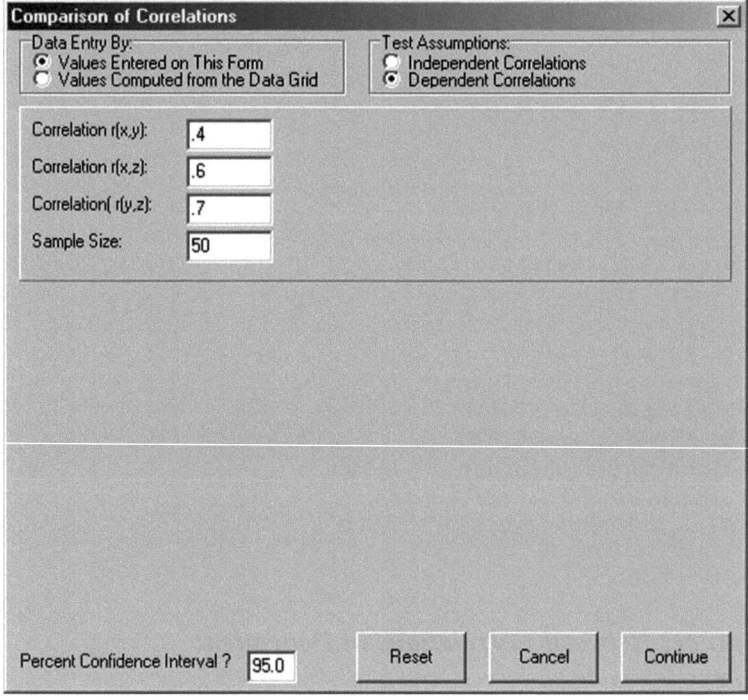

Fig. 7.5 Comparison of correlations for dependent samples

Binary Receiver Operating Characteristics

Two or more groups, for example a control group and treatment groups, may be compared by a variety of means such as with analysis of variance, a t-test or a nonparametric test. It is often of interest to know that point in comparing the groups which minimizes false positive results and maximizes true effects. This procedure produces a graph which plots false positives against true positives for the two groups. In our example, five groups are examined for possible presence of an abnormal medical condition. A count of negative or positive observation of this condition is recorded and analyzed. The file we have selected to demonstrate this procedure is labeled "binaryroc.TEX" and contains five groups (cases) with counts of the normal and positive results. The dialog for the analysis is shown below followed by the results and plot (Figs. 7.6, 7.7):

Binary Receiver Operating Characteristics

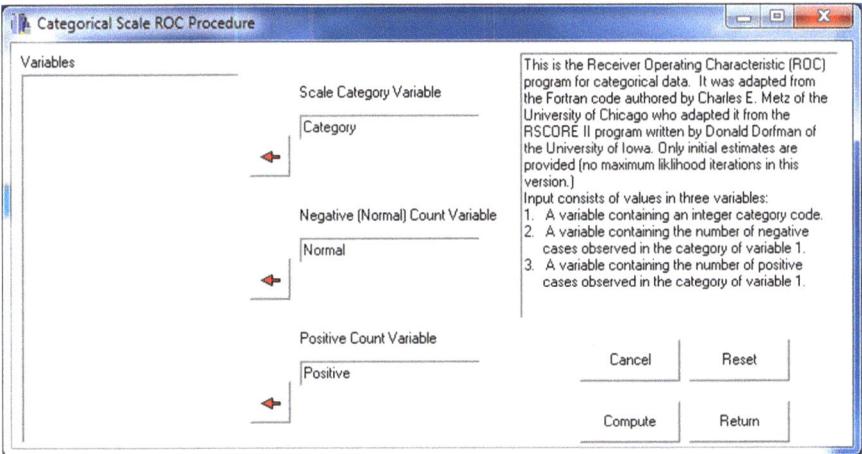

Fig. 7.6 Dialog for the ROC analysis

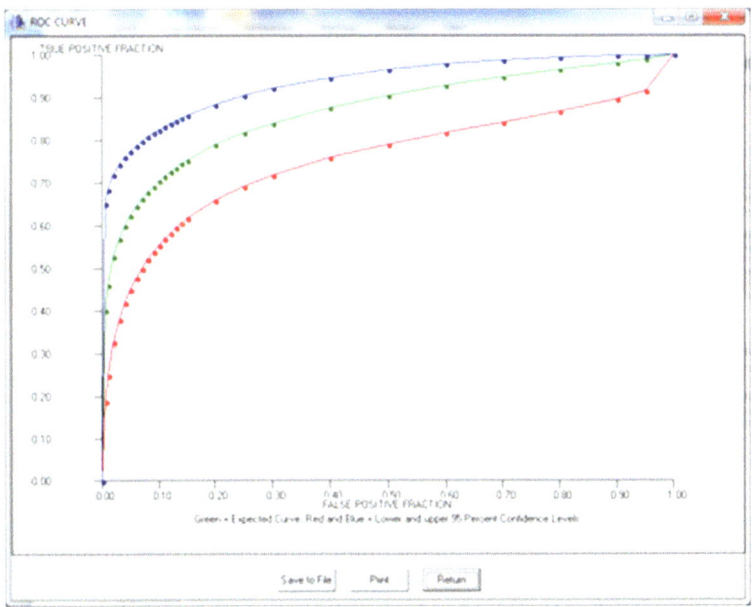

Fig. 7.7 ROC plot

CASES FOR FILE C:\Users\wgmiller\Projects\Data\BinaryROC.TEX

UNITS	Category	Normal	Positive
CASE 1	1	30	5
CASE 2	2	19	6
CASE 3	3	8	5
CASE 4	4	2	12
CASE 5	5	1	22

Categorical ROC Analysis Results
No. of Cases = 5

No. of Categories = 5
Low category = 5, Highest category = 1
Total negative count = 60
Total positive count = 50

TOTAL
CATEGORY COUNT

1	35
2	25
3	13
4	14
5	23

Observed Operating Points

NORMAL	POSITIVE
0.0000	0.0000
0.0167	0.4400
0.0500	0.6800
0.1833	0.7800
0.5000	0.9000
1.0000	1.0000

INITIAL VALUES OF PARAMETERS: A = 1.3281, B = 0.6292
i = 1   Z(i) = -0.0000
i = 2   Z(i) = 0.9027
i = 3   Z(i) = 1.6449
i = 4   Z(i) = 2.1280

LOGL = -143.8050
GOODNES OF FIT CHI-SQUARE = 110.0000 WITH 2 D.F. p = 0.0000
Final values of parameters: A = 1.3155 B = 0.6071
z(1) = -0.2013
Z(2) = 1.0547
Z(3) = 1.7149
Z(4) = 2.1485
LOGL = -146.8468

```
GOODNES OF FIT CHI-SQUARE = 110.0000 WITH 2 D.F. p = 0.0000
Correlation Matrix:
A       1.0000   0.6397   0.3730   0.2853   0.0742  -0.0706
B       0.6397   1.0000   0.2097  -0.0848  -0.4566  -0.6404
Z(1)    0.3730   0.2097   1.0000   0.5289   0.2423   0.1130
Z(2)    0.2853  -0.0848   0.5289   1.0000   0.6195   0.4638
Z(3)    0.0742  -0.4566   0.2423   0.6195   1.0000   0.8299
Z(4)   -0.0706  -0.6404   0.1130   0.4638   0.8299   1.0000
AREA =  0.8696   Std.Dev. (AREA) =  0.0381
```

Estimated Binormal ROC Curve with Lower and Upper
Bounds on Asymetric 95 onfidence Interval for
True-Positive Fraction at each specified
False-Positive fraction:

```
  FPF       TPF         (Lower bound, Upper bound)
  0.005    0.4020         0.1878,    0.6516
  0.010    0.4615         0.2504,    0.6842
  0.020    0.5274         0.3277,    0.7203
  0.030    0.5689         0.3795,    0.7435
  0.040    0.5997         0.4190,    0.7611
  0.050    0.6243         0.4509,    0.7755
  0.060    0.6449         0.4777,    0.7879
  0.070    0.6626         0.5008,    0.7988
  0.080    0.6781         0.5210,    0.8085
  0.090    0.6920         0.5389,    0.8174
  0.100    0.7045         0.5550,    0.8256
  0.110    0.7160         0.5695,    0.8331
  0.120    0.7265         0.5828,    0.8402
  0.130    0.7362         0.5950,    0.8468
  0.140    0.7453         0.6063,    0.8531
  0.150    0.7537         0.6167,    0.8590
  0.200    0.7895         0.6597,    0.8844
  0.250    0.8175         0.6923,    0.9048
  0.300    0.8406         0.7184,    0.9216
  0.400    0.8773         0.7590,    0.9474
  0.500    0.9058         0.7907,    0.9658
  0.600    0.9291         0.8178,    0.9789
  0.700    0.9488         0.8427,    0.9881
  0.800    0.9661         0.8676,    0.9944
  0.900    0.9818         0.8962,    0.9983
  0.950    0.9897         0.9156,    0.9994
```

ESTIMATES OF EXPECTED OPERATING POINTS ON FITTED ROC
CURVE, WITH LOWER AND UPPER BOUNDS OF ASYMMETRIC 95%
CONFIDENCE INTERVALS ALONG THE CURVE FOR THOSE POINTS:

```
EXPECTED OPERATING POINT      LOWER BOUND             UPPER BOUND
(FPF , TPF)                  ( FPF , TPF)            ( FPF , TPF )
{0.0158, 0.5045)              (0.0024, 0.3468)        (0.0693, 0.6614)
{0.0432, 0.6081)              (0.0136, 0.4900)        (0.1109, 0.7170)
{0.1458, 0.7502)              (0.0801, 0.6783)        (0.2403, 0.8125)
{0.5798, 0.9247)              (0.4543, 0.8936)        (0.6976, 0.9484)
```

Partial and Semi_Partial Correlations

Partial Correlation

OpenStat provides a procedure for obtaining partial and semi-partial correlations. You can select the Analyses/Correlation/Partial procedure. We have used the cansas.tab file to demonstrate how to obtain partial and semi-partial correlations as shown below (Fig. 7.8):

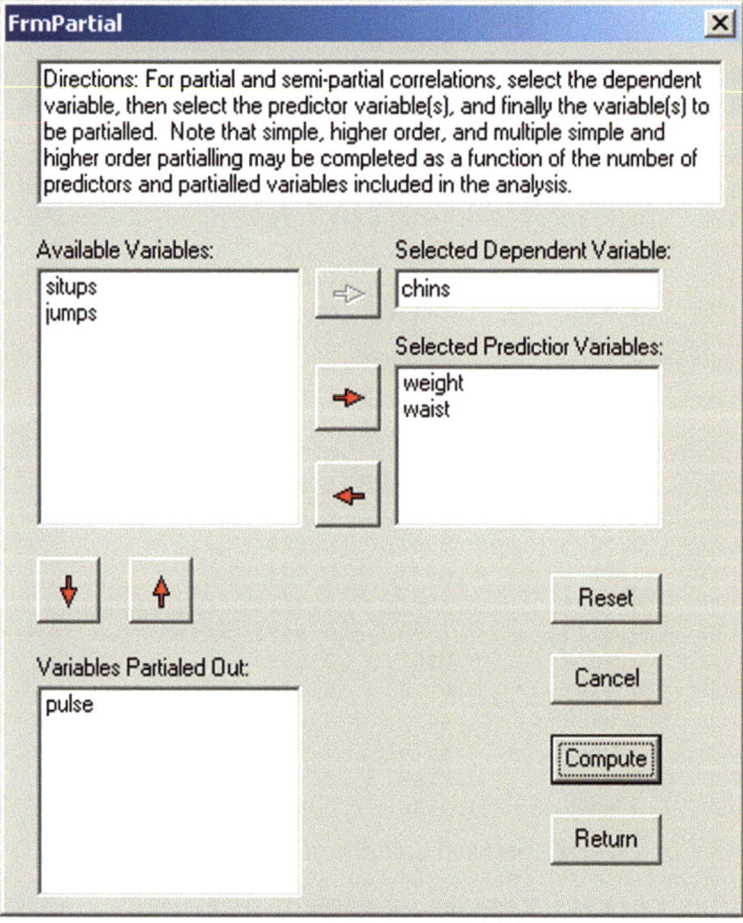

Fig. 7.8 Form for calculating partial and semi-partial correlations

Partial and Semi-Partial Correlation Analysis
Dependent variable = chins

Predictor VarList:
Variable 1 = weight
Variable 2 = waist

Control Variables:
Variable 1 = pulse

Higher order partialling at level = 2

CORRELATION MATRIX

Correlations

	chins	weight	waist	pulse
chins	1.000	-0.390	-0.552	0.151
weight	-0.390	1.000	0.870	-0.366
waist	-0.552	0.870	1.000	-0.353
pulse	0.151	-0.366	-0.353	1.000

Means

Variables	chins	weight	waist	pulse
	9.450	178.600	35.400	56.100

Standard Deviations

Variables	chins	weight	waist	pulse
	5.286	24.691	3.202	7.210

No. of valid cases = 20

Squared Multiple Correlation with all Variables = 0.340

Standardized Regression Coefficients:
   weight = 0.368
   waist = -0.882
   pulse = -0.026

Squared Multiple Correlation with control Variables = 0.023

Standardized Regression Coefficients:
   pulse = 0.151

Partial Correlation = 0.569

Semi-Partial Correlation = 0.563

F =   3.838 with probability = 0.0435, D.F.1 = 2 and D.F.2 = 16

Autocorrelation

Now let us look at an example of auto-correlation. We will use a file named strikes.tab. The file contains a column of values representing the number of strikes which occurred each month over a 30 month period. Select the auto-correlation procedure from the Correlations sub-menu of the Analyses main menu. Below is a representation of the form as completed to obtain auto-correlations, partial auto-correlations, and data smoothing using both moving average smoothing and polynomial regression smoothing (Fig. 7.9):

When we click the Compute button, we first obtain a dialog form for setting the parameters of our moving average. In that form we first enter the number of values to include in the average from both sides of the current average value. We selected 2. Be sure and press the Enter key after entering the order value. When you do, two theta values will appear in a list box. When you click on each of those thetas, you will see a default value appear in a text box. This is the weight to assign the leading

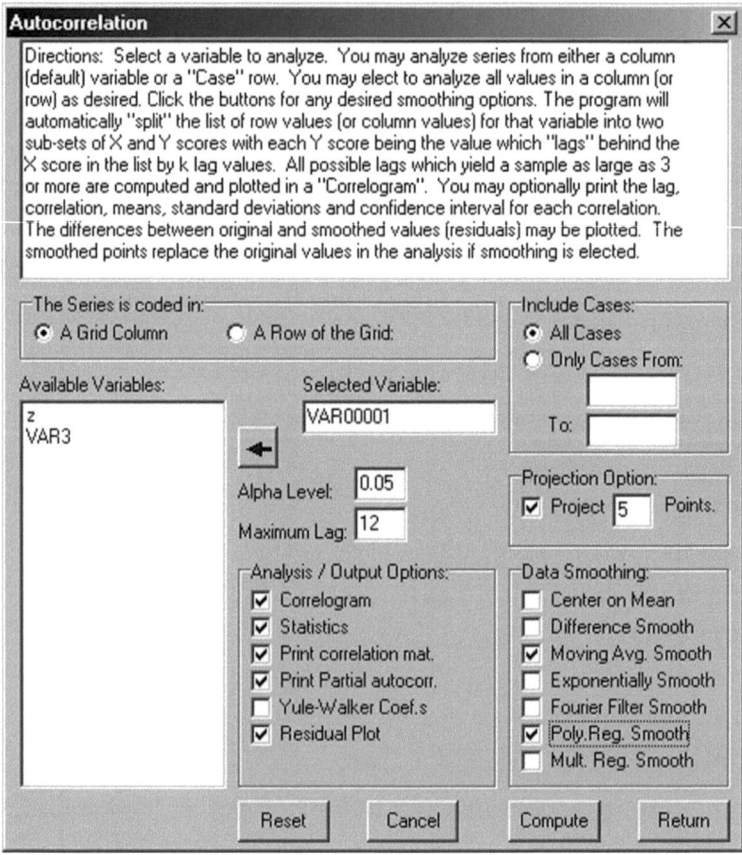

Fig. 7.9 The Autocorrelation form

Fig. 7.10 Moving Average form

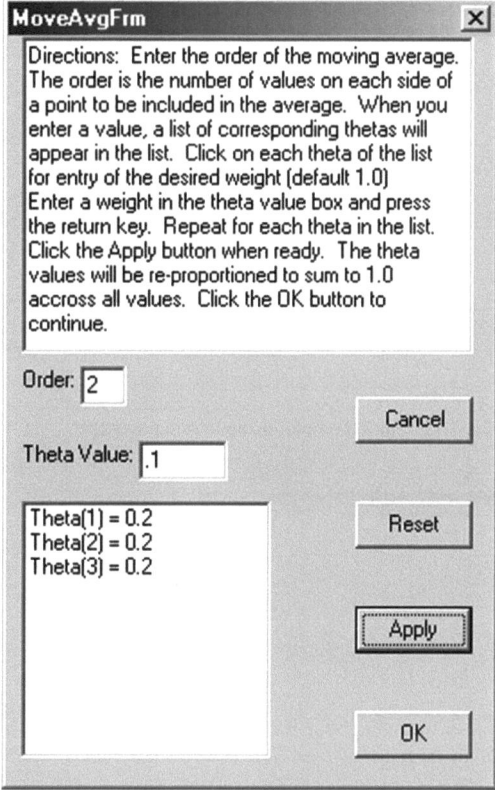

and trailing averages (first or second in our example.) In our example we have accepted the default value for both thetas (simply press the Return key to accept the default or enter a value and press the Return key.) Now press the Apply button. When you do this, the weights for all of the values (the current mean and the 1, 2, ... order means) are recalculated. You can then press the OK button to proceed with the process (Fig. 7.10).

The procedure then plots the original (30) data points and their moving average smoothed values. Since we also asked for a projection of 5 points, they too are plotted. The plot should look like that shown below (Fig. 7.11):

We notice that there seems to be a "wave" type of trend with a half-cycle of about 15 months. When we press the Return button on the plot of points we next get the following (Fig. 7.12):

This plot shows the original points and the difference (residual) of the smoothed values from the original. At this point, the procedure replaces the original points with the smoothed values. Press the Return button and you next obtain the following (Fig. 7.13):

This is the form for specifying our next smoothing choice, the polynomial regression smoothing. We have elected to use a polynomial value of 2 which will result in a model for a data point $Y_{t-1} = B * t^2 + C$ for each data point. Click the OK button to proceed. You then obtain the following result (Fig. 7.14):

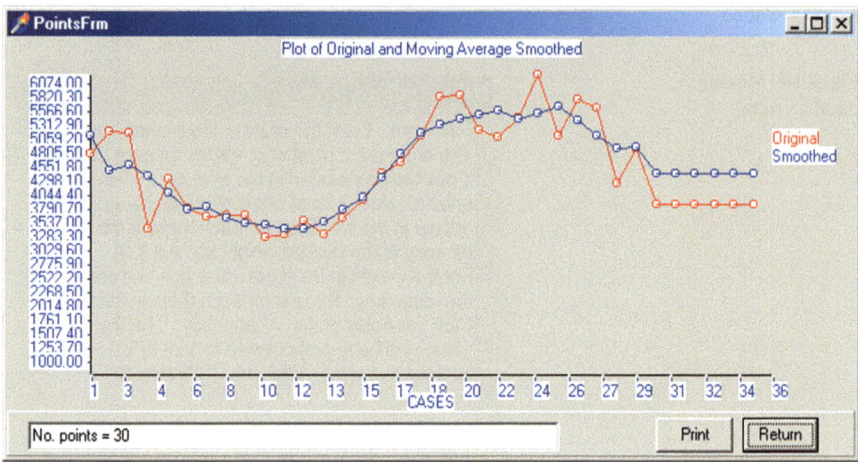

Fig. 7.11 Smoothed plot using moving average

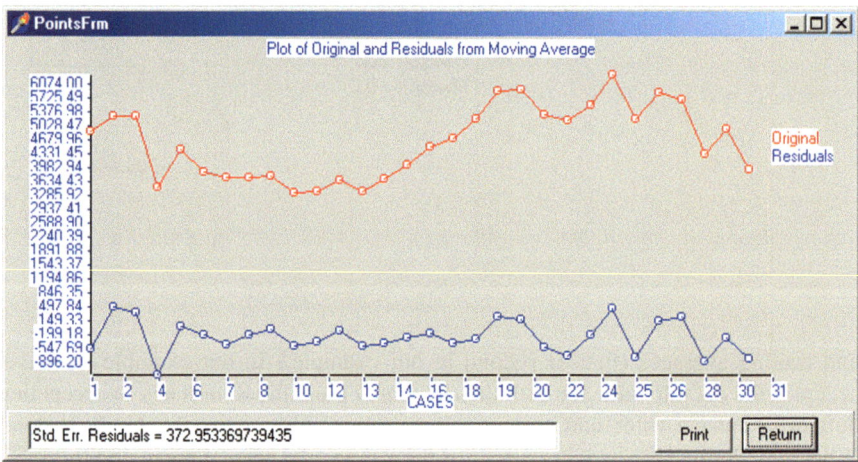

Fig. 7.12 Plot of residuals obtained using moving averages

Fig. 7.13 Polynomial regression smoothing form

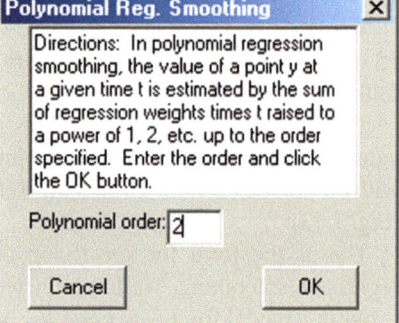

It appears that the use of the second order polynomial has "removed" the cyclic trend we saw in the previously smoothed data points. Click the return key to obtain the next output as shown below (Fig. 7.15):

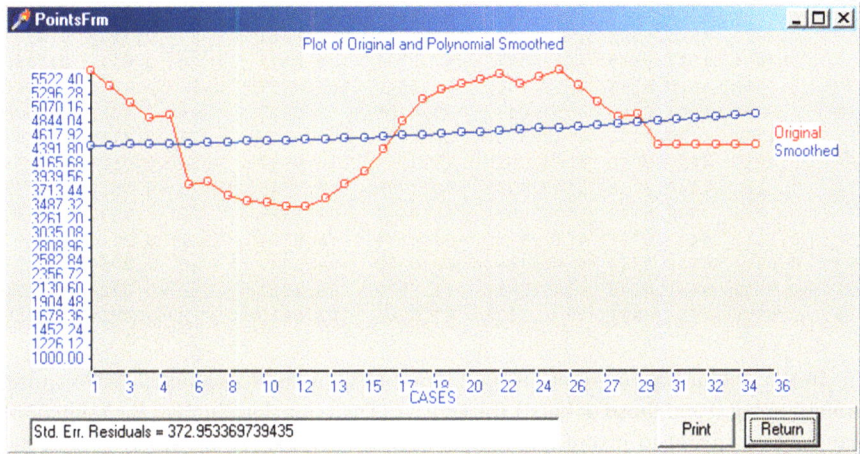

Fig. 7.14 Plot of polynomial smoothed points

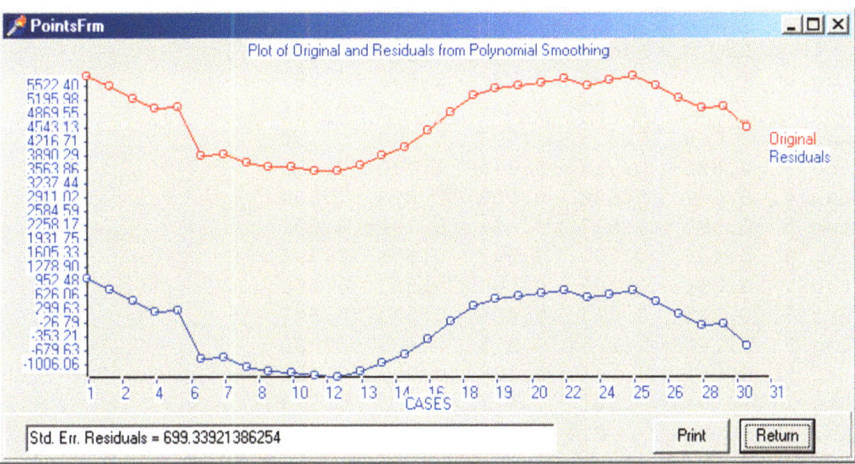

Fig. 7.15 Plot of residuals from polynomial smoothing

This result shows the previously smoothed data points and the residuals obtained by subtracting the polynomial smoothed points from those previous points. Click the Return key again to see the next output shown below:

```
Overall mean = 4532.604, variance = 11487.241
Lag     Rxy      MeanX      MeanY      Std.Dev.X  Std.Dev.Y  Cases    LCL      UCL
0     1.0000   4532.6037  4532.6037   109.0108   109.0108     30    1.0000   1.0000
1     0.8979   4525.1922  4537.3814   102.9611   107.6964     29    0.7948   0.9507
2     0.7964   4517.9688  4542.3472    97.0795   106.2379     28    0.6116   0.8988
3     0.6958   4510.9335  4547.5011    91.3660   104.6337     27    0.4478   0.8444
4     0.5967   4504.0864  4552.8432    85.8206   102.8825     26    0.3012   0.7877
5     0.4996   4497.4274  4558.3734    80.4432   100.9829     25    0.1700   0.7287
6     0.4050   4490.9565  4564.0917    75.2340    98.9337     24    0.0524   0.6679
7     0.3134   4484.6738  4569.9982    70.1928    96.7340     23   -0.0528   0.6053
8     0.2252   4478.5792  4576.0928    65.3196    94.3825     22   -0.1470   0.5416
9     0.1410   4472.6727  4582.3755    60.6144    91.8784     21   -0.2310   0.4770
10    0.0611   4466.9544  4588.8464    56.0772    89.2207     20   -0.3059   0.4123
11   -0.0139   4461.4242  4595.5054    51.7079    86.4087     19   -0.3723   0.3481
12   -0.0836   4456.0821  4602.3525    47.5065    83.4415     18   -0.4309   0.2852
```

In the output above we are shown the auto-correlations obtained between the values at lag 0 and those at lags 1 through 12. The procedure limited the number of lags automatically to insure a sufficient number of cases upon which to base the correlations. You can see that the upper and lower 95 % confidence limits increases as the number of cases decreases. Click the Return button on the output form to continue the process.

```
Matrix of Lagged Variable: VAR00001 with 30 valid cases.
Variables
          Lag 0    Lag 1    Lag 2    Lag 3    Lag 4
Lag 0     1.000    0.898    0.796    0.696    0.597
Lag 1     0.898    1.000    0.898    0.796    0.696
Lag 2     0.796    0.898    1.000    0.898    0.796
Lag 3     0.696    0.796    0.898    1.000    0.898
Lag 4     0.597    0.696    0.796    0.898    1.000
Lag 5     0.500    0.597    0.696    0.796    0.898
Lag 6     0.405    0.500    0.597    0.696    0.796
Lag 7     0.313    0.405    0.500    0.597    0.696
Lag 8     0.225    0.313    0.405    0.500    0.597
Lag 9     0.141    0.225    0.313    0.405    0.500
Lag 10    0.061    0.141    0.225    0.313    0.405
Lag 11   -0.014    0.061    0.141    0.225    0.313
Lag 12   -0.084   -0.014    0.061    0.141    0.225
```

Variables

	Lag 5	Lag 6	Lag 7	Lag 8	Lag 9
Lag 0	0.500	0.405	0.313	0.225	0.141
Lag 1	0.597	0.500	0.405	0.313	0.225
Lag 2	0.696	0.597	0.500	0.405	0.313
Lag 3	0.796	0.696	0.597	0.500	0.405
Lag 4	0.898	0.796	0.696	0.597	0.500
Lag 5	1.000	0.898	0.796	0.696	0.597
Lag 6	0.898	1.000	0.898	0.796	0.696
Lag 7	0.796	0.898	1.000	0.898	0.796
Lag 8	0.696	0.796	0.898	1.000	0.898
Lag 9	0.597	0.696	0.796	0.898	1.000
Lag 10	0.500	0.597	0.696	0.796	0.898
Lag 11	0.405	0.500	0.597	0.696	0.796
Lag 12	0.313	0.405	0.500	0.597	0.696

Variables

	Lag 10	Lag 11	Lag 12
Lag 0	0.061	-0.014	-0.084
Lag 1	0.141	0.061	-0.014
Lag 2	0.225	0.141	0.061
Lag 3	0.313	0.225	0.141
Lag 4	0.405	0.313	0.225
Lag 5	0.500	0.405	0.313
Lag 6	0.597	0.500	0.405
Lag 7	0.696	0.597	0.500
Lag 8	0.796	0.696	0.597
Lag 9	0.898	0.796	0.696
Lag 10	1.000	0.898	0.796
Lag 11	0.898	1.000	0.898
Lag 12	0.796	0.898	1.000

The above data presents the inter-correlations among the 12 lag variables. Click the output form's Return button to obtain the next output:

Partial Correlation Coefficients with 30 valid cases.

Variables	Lag 0	Lag 1	Lag 2	Lag 3	Lag 4
	1.000	0.898	-0.051	-0.051	-0.052

Variables	Lag 5	Lag 6	Lag 7	Lag 8	Lag 9
	-0.052	-0.052	-0.052	-0.052	-0.051

Variables	Lag 10	Lag 11
	-0.051	-0.051

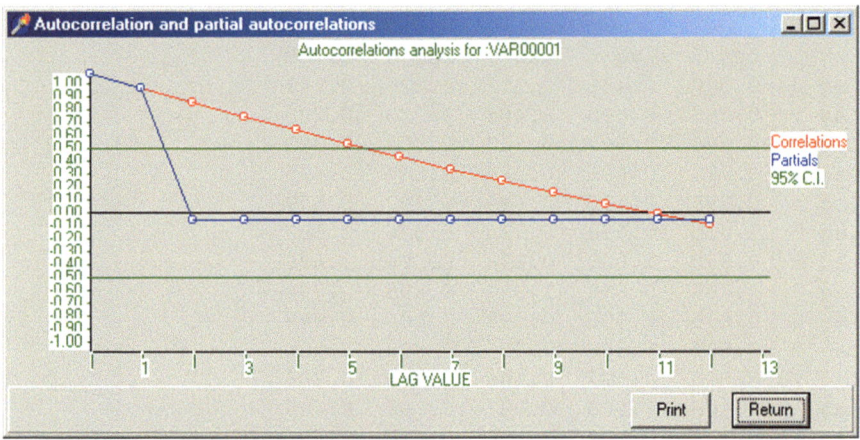

Fig. 7.16 Auto and partial autocorrelation plot

The partial auto-correlation coefficients represent the correlation between lag 0 and each remaining lag with previous lag values partialled out. For example, for lag 2 the correlation of −0.051 represents the correlation between lag 0 and lag 2 with lag 1 effects removed. Since the original correlation was 0.796, removing the effect of lag 1 made a considerable impact. Again click the Return button on the output form. Next you should see the following results (Fig. 7.16):

This plot or "correlogram" shows the auto-correlations and partial auto-correlations obtained in the analysis. If only "noise" were present, the correlations would vary around zero. The presence of large values is indicative of trends in the data.

Chapter 8
Comparisons

One Sample Tests

OpenStat provides the ability to perform tests of hypotheses based on a single sample. Typically the user is interested in testing the hypothesis that

1. A sample mean does not differ from a specified hypothesized mean,
2. A sample proportion does not differ from a specified population proportion,
3. A sample correlation does not differ from a specified population correlation, or
4. A sample variance does not differ from a specified population variance.

The One Sample Test for means, proportions, correlations and variances is started by selecting the Comparisons option under the Statistics menu and moving the mouse to the One Sample Tests option which you then click with the left mouse button. If you do this you will then see the specification form for your comparison as seen below. In this form there is a button corresponding to each of the above type of comparison. You click the one of your choice. There are also text boxes in which you enter the sample statistics for your test and select the confidence level desired for the test. We will illustrate each test. In the first one we will test the hypothesis that a sample mean of 105 does not differ from a hypothesized population mean of 100. The standard deviation is estimated to be 15 and our sample size is 20 (Fig. 8.1).

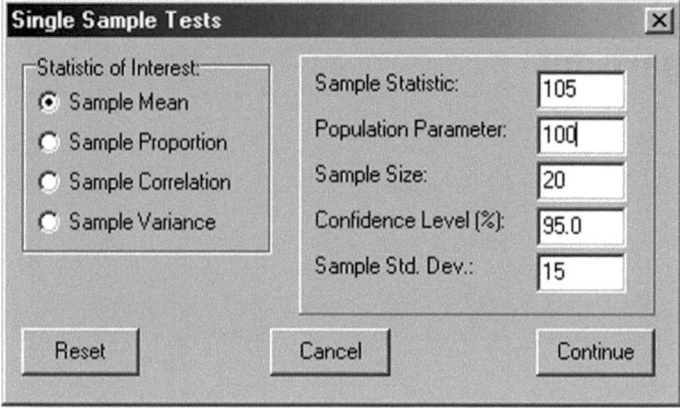

Fig. 8.1 Single Sample Tests Dialog form

When we click the Continue button on the form we then obtain our results in an output form as shown below:

```
ANALYSIS OF A SAMPLE MEAN

Sample Mean = 105.000
Population Mean = 100.000
Sample Size = 20
Standard error of Mean = 3.354
t test statistic = 1.491 with probability 0.152
t value required for rejection = 2.093
Confidence Interval = (97.979,112.021)
```

We notice that our sample mean is "captured" in the 95% confidence interval and this would lead us to accept the null hypothesis that the sample is not different from that expected by chance alone from a population with mean 100.

Now let us perform a test of a sample proportion. Assume we have an elective high school course in Spanish I. We notice that the proportion of 30 students in the class that are female is only 0.4 (12 students) yet the population of high school students in composed of 50% male and 50% female. Is the proportion of females enrolled in the class representative of a random sample from the population? To test the hypothesis that the proportion of .4 does not differ from the population proportion of .5 we click the proportion button of the form and enter our sample data as shown below (Fig. 8.2):

One Sample Tests

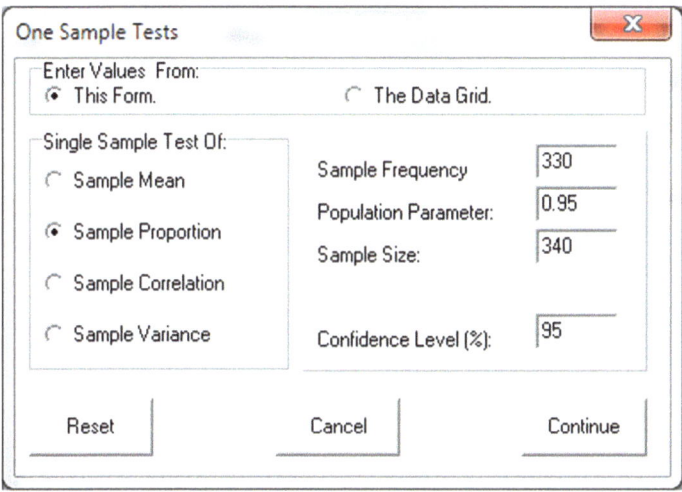

Fig. 8.2 Single Sample Proportion test

When we click the Continue button we see the results as shown below:

```
ANALYSIS OF A SAMPLE PROPORTION

Two tailed test at the 0.950 confidence level
Sample Proportion = 0.9705882
Population Proportion = 0.9500000
Sample Size = 340
Standard error of sample proportion = 0.0091630
z test statistic = 2.2469 with probability > z = 0.0123
z test statistic = 2.2469 with probability < z = 0.9877
z value required for rejection = 2.4673
Confidence Interval = (0.9526290,0.9885474)
```

We note that the z statistic obtained for our sample has a fairly low probability of occurring by chance when drawn from a population with a proportion of .5 so we are led to reject the null hypothesis.

We examined the test for a hypothesis about a sample correlation being obtained from a population with a given correlation. See the Correlation chapter (Chap. 7) to review that test.

It occurs to a teacher that perhaps her Spanish students are from a more homogeneous population than that of the validation study reported in a standardized Spanish aptitude test. If that were the case, the correlation she observed might well be attenuated due to the differences in variances. In her class of 30 students she observed a sample variance of 25 while the validation study for the instrument reported a variance of 36. Let's examine the test for the hypothesis that her sample variance does not differ significantly from the "population" value. Again we invoke the One Sample Test from the Univariate option of the Analyses menu and complete the form as shown below (Fig. 8.3):

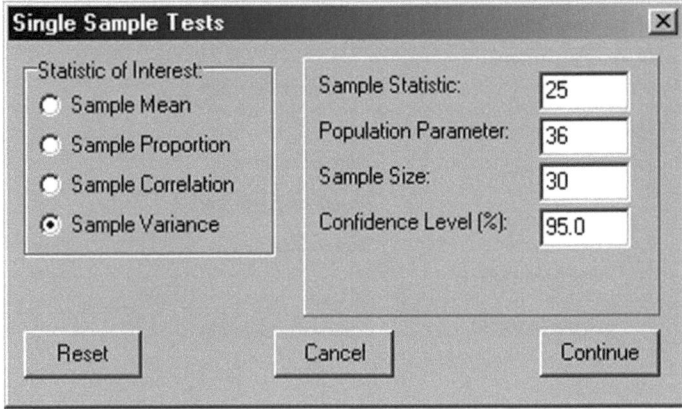

Fig. 8.3 Single Sample Variance test

Upon clicking the Continue button our teacher obtains the following results in the output form:

```
ANALYSIS OF A SAMPLE VARIANCE

Sample Variance = 25.000
Population Variance = 36.000
Sample Size = 30
Chi-square statistic = 20.139 with probability > chisquare =
0.889 and D.F. = 29
Chi-square value required for rejection = 16.035
Chi-square Confidence Interval = (45.725,16.035)
Variance Confidence Interval = (15.856,45.215)
```

The chi-square statistic obtained leads our teacher to accept the hypothesis of no difference between her sample variance and the population variance. Note that the population variance is clearly within the 95% confidence interval for the sample variance.

Proportion Differences

A most common research question arises when an investigator has obtained two sample proportions. One asks whether or not the two sample proportions are really different considering that they are based on observations drawn randomly from a population. For example, a school nurse observes during the flu season that 13 eighth grade students are absent due to flu symptoms while only 8 of the ninth grade students are absent. The class sizes of the two grades are 110 and 121 respectively. The nurse decides to test the hypothesis that the two proportions (.118 and .066) do not differ significantly using the OpenStat program. The first step is to start the

Proportion Differences

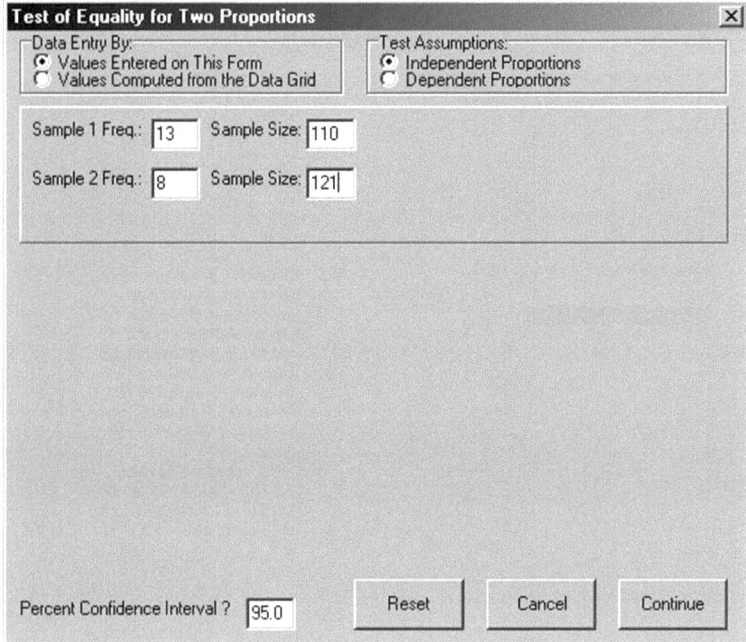

Fig. 8.4 Test of equality of two proportions

Proportion Differences procedure by clicking on the Analyses menu, moving the mouse to the Univariate option and the clicking on the Proportion Differences option. The specification form for the test then appears. We will enter the required values directly on the form and assume the samples are independent random samples from a population of eighth and ninth grade students (Fig. 8.4).

When the nurse clicks the Continue button the following results are shown in the Output form:

```
COMPARISON OF TWO PROPORTIONS

Test for Difference Between Two Independent Proportions

Entered Values

Sample 1: Frequency = 13 for 110 cases.
Sample 2: Frequency = 8 for 121 cases.
Proportion 1 = 0.118, Proportion 2 = 0.066, Difference = 0.052
Standard Error of Difference = 0.038
Confidence Level selected = 95.0
z test statistic = 1.375 with probability = 0.0846
z value for confidence interval = 1.960
Confidence Interval: ( -0.022, 0.126)
```

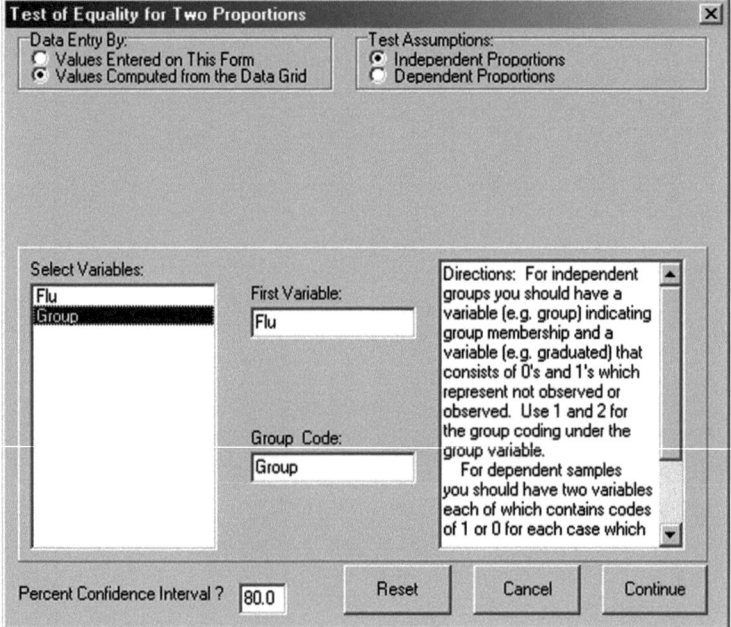

Fig. 8.5 Test of Equality of Two Proportions form

The nurse notices that the value of zero is within the 95% confidence interval as therefore accepts the null hypothesis that the two proportions are not different than that expected due to random sampling variability. What would the nurse conclude had the 80.0% confidence level been chosen?

If the nurse had created a data file with the above data entered into the grid such as:

CASE/VAR	FLU	GROUP
CASE 1	0	1
CASE 2	1	1
I.		--
CASE 110	0	1
CASE 111	0	2
–		--
CASE 231	1	2

then the option would have been to analyze data in a file.

In this case, the absence or presence of flu symptoms for the student are entered as zero (0) or one (1) and the grade is coded as 1 or 2. If the same students, say the eighth grade students, are observed at weeks 10 and 15 during the semester, than the test assumptions would be changed to Dependent Proportions. In that case the form changes again to accommodate two variables coded zero and one to reflect the observations for each student at weeks 10 and 15 (Fig. 8.5).

t-Tests

Among the comparison techniques the "Student" t-test is one of the most commonly employed. One may test hypotheses regarding the difference between population means for independent or dependent samples which meet or do not meet the assumptions of homogeneity of variance. To complete a t-test, select the t-test option from the Comparisons sub-menu of the Statistics menu. You will see the form below (Fig. 8.6):

Notice that you can enter values directly on the form or from a file read into the data grid. If you elect to read data from the data grid by clicking the button corresponding to "Values Computed from the Data Grid" you will see that the form is modified as shown below (Fig. 8.7).

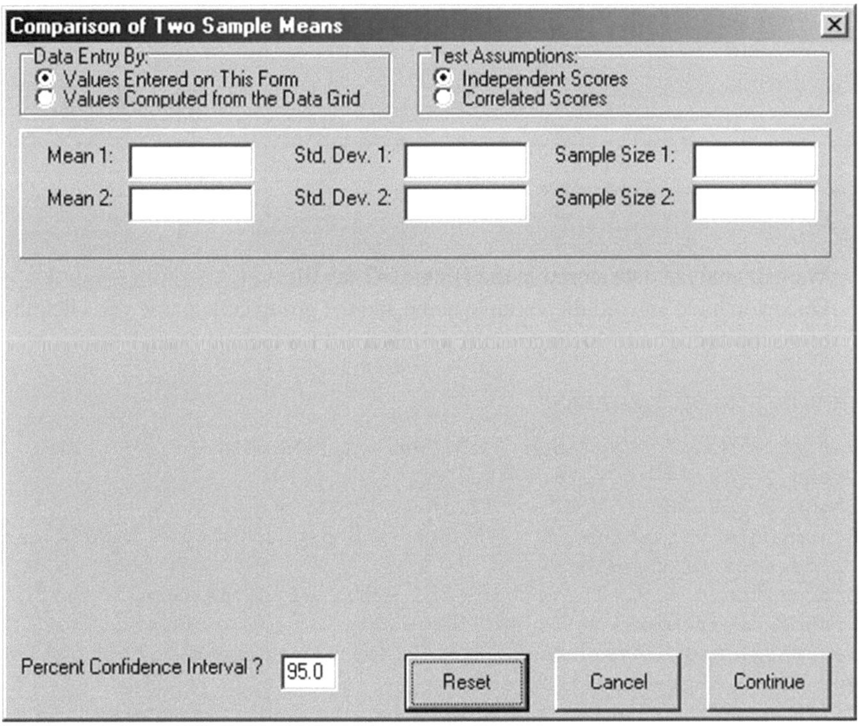

Fig. 8.6 Comparison of Two Sample Means form

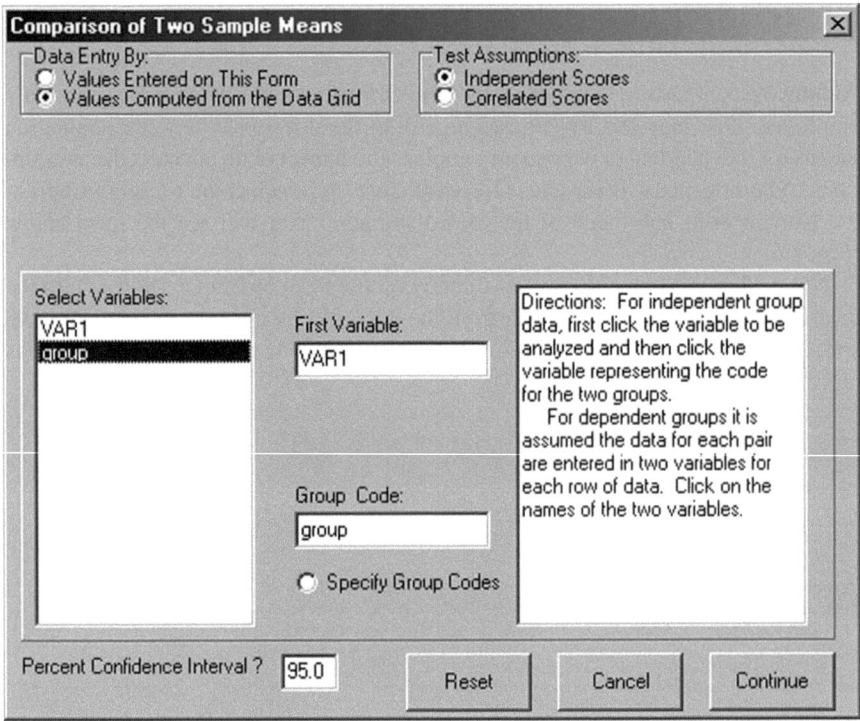

Fig. 8.7 Comparison of two sample means

We will analyze data stored in the Hinkle247.tab file.

Once you have entered the variable name and the group code name you click the Continue button. The following results are obtained for the above analysis:

```
COMPARISON OF TWO MEANS

Variable Mean    Variance   Std.Dev.   S.E.Mean  N
Group 1   49.44   107.78     10.38      3.46     9
Group 2   68.88   151.27     12.30      4.35     8
Assuming = variances, t = -3.533 with probability = 0.0030 and
15 degrees of freedom
Difference = -19.43 and Standard Error of difference = 5.50
Confidence interval = ( -31.15, -7.71)
Assuming unequal variances, t = -3.496 with probability = 0.0034
and 13.82 degrees of freedom
Difference = -19.43 and Standard Error of difference = 5.56
Confidence interval = ( -31.37, -7.49)
F test for equal variances = 1.404, Probability = 0.3209
```

The F test for equal variances indicates it is reasonable to assume the sampled populations have equal variances hence we would report the results of the first test.

Since the probability of the obtained statistic is rather small (0.003), we would likely infer that the samples were drawn from two different populations. Note that the confidence interval for the observed difference is reported.

One, Two or Three Way Analysis of Variance

An experiment often involves the observation of some continuous variable under one or more controlled conditions or factors. For example, one might observe two randomly assigned groups of subjects performance under two or more levels of some treatment. The question posed is whether or not the means of the populations under the various levels of treatment are equal. Of course, if there is only two levels of treatment for one factor then we could analyze the data with the t-test described above. In fact, we will analyze the same "Hinkle.txt" file data with the anova program. Select the "One, Two or Three Way ANOVA" option from the Comparisons sub-menu of the Statistics menu. You will see the form below (Fig. 8.8):

Since our first example involves one factor only we will click the VAR1 variable name and click the right arrow button to place it in the Dependent Variable box. We then click the "group" variable label and the right arrow to place it in the Factor 1 Variable box. We will assume the levels represent fixed treatment levels. We will also elect to plot the sample means for each level using three dimension bars. When we click the Continue button we will obtain the results shown below:

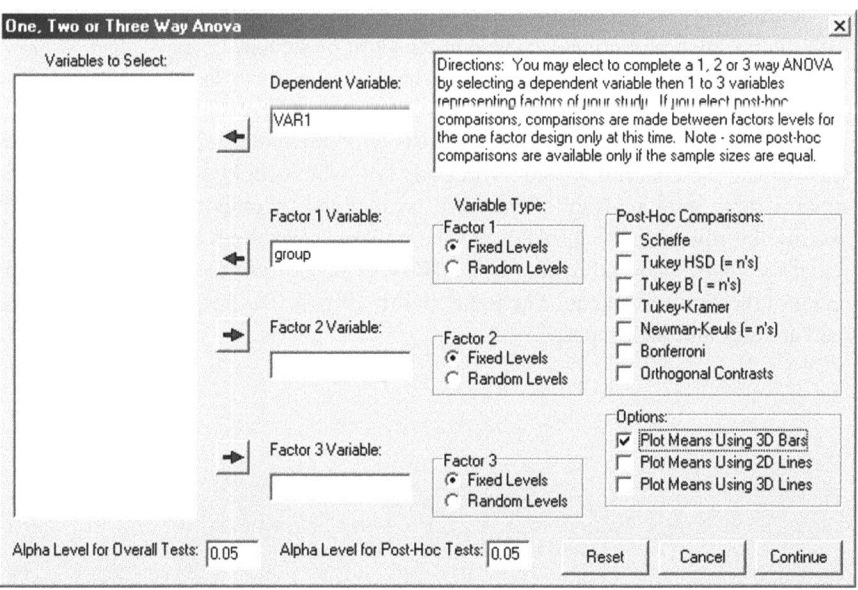

Fig. 8.8 One, two or three way ANOVA dialog

```
ONE WAY ANALYSIS OF VARIANCE RESULTS

Dependent variable is: VAR1, Independent variable is: group
------------------------------------------------------------------
SOURCE    D.F.   SS         MS        F       PROB.>F    OMEGA SQR.
------------------------------------------------------------------
BETWEEN   1      1599.02    1599.02   12.49   0.00       0.40
WITHIN    15     1921.10    128.07
TOTAL     16     3520.12
------------------------------------------------------------------


MEANS AND VARIABILITY OF THE DEPENDENT VARIABLE FOR LEVELS OF
THE INDEPENDENT VARIABLE
------------------------------------------------------------------
GROUP     MEAN      VARIANCE     STD.DEV.     N
------------------------------------------------------------------
  1       49.44     107.78       10.38        9
  2       68.88     151.27       12.30        8
------------------------------------------------------------------
TOTAL     58.59     220.01       14.83        17
------------------------------------------------------------------


TESTS FOR HOMOGENEITY OF VARIANCE
------------------------------------------------------------------
Hartley Fmax test statistic =    1.40 with deg.s freedom: 2 and 8.
Cochran C statistic         =    0.58 with deg.s freedom: 2 and 8.
Bartlett Chi-square         =    0.20 with 1 D.F. Prob. = 0.347
------------------------------------------------------------------
```

In this example, we note that the F statistic (12.49) is simply the square of the previously observed t statistic (within rounding error.) The Bartlett Chi-square test for homogeneity of variance and the Hartley Fmax test also agree approximately with the F statistic for equal variance in the t-test procedure.

The plot of the sample means obtained in our analysis are shown below (Fig. 8.9):

Now let us run an example of an analysis with one fixed and one random factor. We will use the data file named "Threeway.txt" which could also serve to demonstrate a three way analysis of variance (with fixed or random effects.) We will assume the row variable is fixed and the column variable is a random level. We select the One, Two and Three Way ANOVA option from the Comparisons submenu of the Statistics menu. The figure below (Fig. 8.10) shows how we specified the variables and their types:

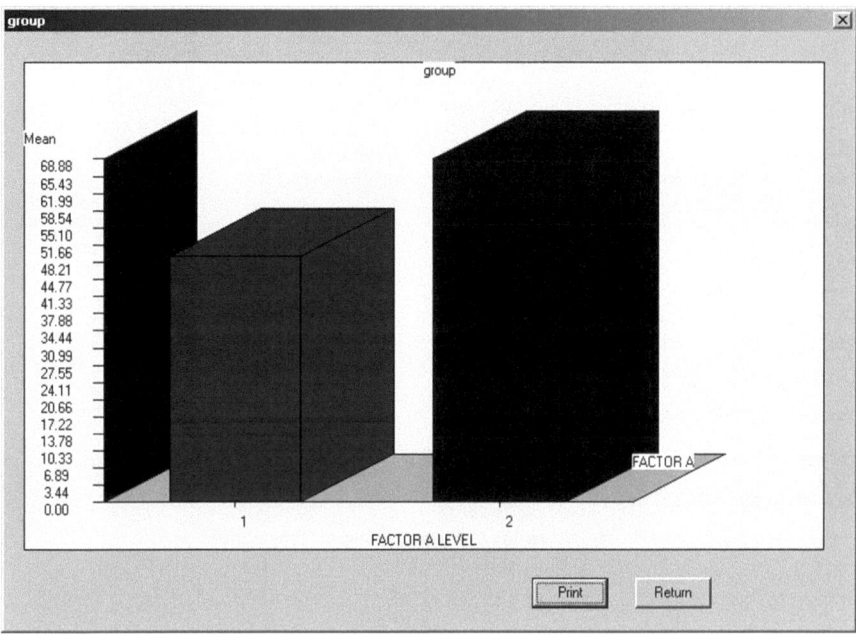

Fig. 8.9 Plot of sample means from a one-way ANOVA

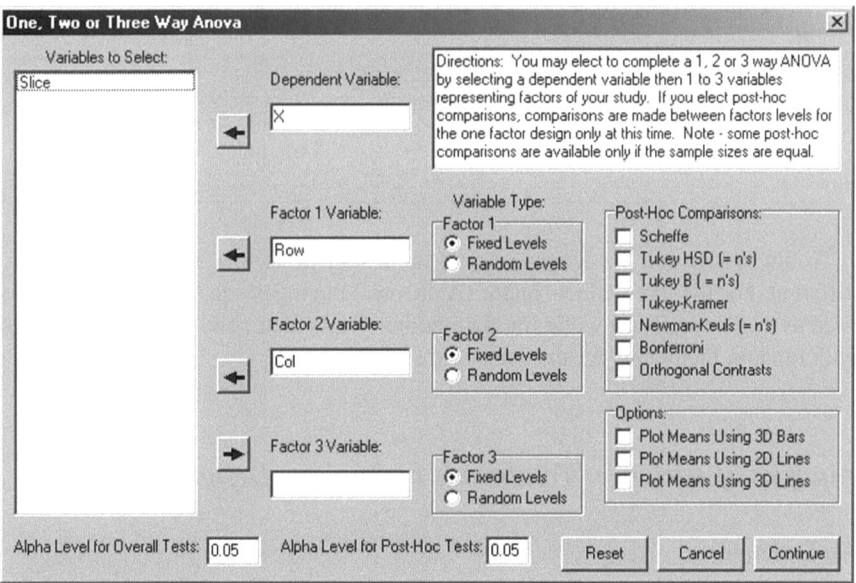

Fig. 8.10 Specifications for a two-way ANOVA

Now when we click the Continue button we obtain:

```
Two Way Analysis of Variance

Variable analyzed: X

Factor A (rows) variable: Row (Fixed Levels)
Factor B (columns) variable: Col (Fixed Levels)

SOURCE            D.F.    SS        MS       F        PROB.> F   Omega Squared

Among Rows         1    12.250    12.250    5.765    0.022        0.074
Among Columns      1    42.250    42.250   19.882    0.000        0.293
Interaction        1    12.250    12.250    5.765    0.022        0.074
Within Groups     32    68.000     2.125
Total             35   134.750     3.850

Omega squared for combined effects = 0.441

Note: Denominator of F ratio is MSErr

Descriptive Statistics

GROUP    Row    Col.    N     MEAN    VARIANCE   STD.DEV.
Cell      1     1       9    3.000    1.500      1.225
Cell      1     2       9    4.000    1.500      1.225
Cell      2     1       9    3.000    3.000      1.732
Cell      2     2       9    6.333    2.500      1.581
Row       1            18    3.500    1.676      1.295
Row       2            18    4.667    5.529      2.351
Col       1            18    3.000    2.118      1.455
Col       2            18    5.167    3.324      1.823
TOTAL                  36    4.083    3.850      1.962

TESTS FOR HOMOGENEITY OF VARIANCE
-----------------------------------------------------------------------
Hartley Fmax test statistic = 2.00 with deg.s freedom: 4 and 8.
Cochran C statistic = 0.35 with deg.s freedom: 4 and 8.
Bartlett Chi-square statistic = 3.34 with 3 D.F. Prob. = 0.658
-----------------------------------------------------------------------
```

You will note that the denominator of the F statistic for the two main effects are different. For the fixed effects factor (A or rows) the mean square for interaction is used as the denominator while for the random effects factor and interaction of fixed with random factors the mean square within cells is used.

Analysis of Variance: Treatments by Subjects Design

An Example

To perform a Treatments by Subjects analysis of variance, we will use a sample data file labeled "ABRData.txt" which you can find as a ".tab" type of file in your sample of data files. We open the file and select the option "Within Subjects Anova" in the

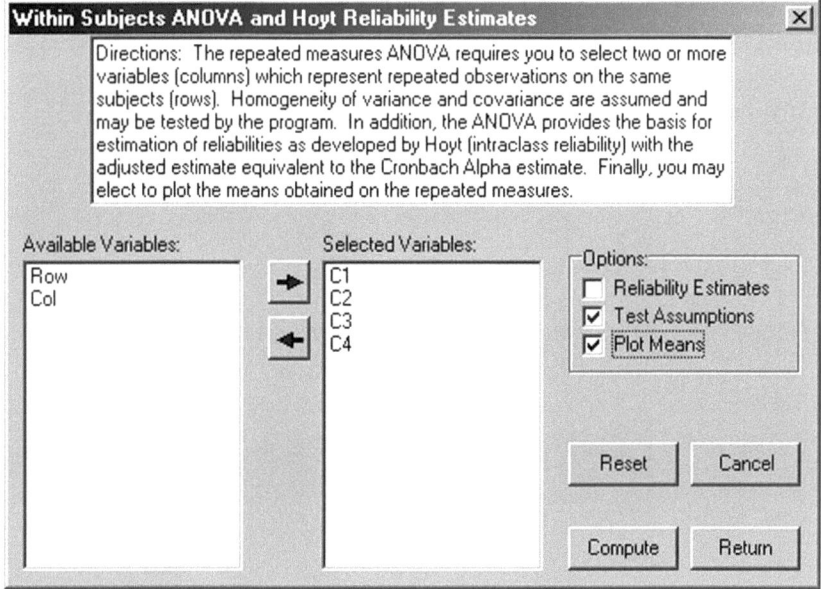

Fig. 8.11 Within subjects ANOVA dialog

Comparisons sub-menu under the Statistics menu. The figure above (Fig. 8.11) is then completed as shown:

Notice that the repeated measures are the columns labeled C1 through C4. You will also note that this same procedure will report intraclass reliability estimates if elected. If you now click the Compute button, you obtain the results shown below:

```
Treatments by Subjects (AxS) ANOVA Results.

Data File = C:\Projects\Delphi\OpenStat\ABRData.txt

-------------------------------------------------------------------
SOURCE              DF    SS          MS           F         Prob. > F
-------------------------------------------------------------------
SUBJECTS            11    181.000     330.500
WITHIN SUBJECTS     36    1077.000    29.917
   TREATMENTS        3    991.500     330.500      127.561   0.000
   RESIDUAL         33    85.500      2.591
-------------------------------------------------------------------
TOTAL               47    1258.000    26.766
-------------------------------------------------------------------

TREATMENT (COLUMN) MEANS AND STANDARD DEVIATIONS
VARIABLE    MEAN      STD.DEV.
C1          16.500    2.067
C2          11.500    2.431
C3           7.750    2.417
C4           4.250    2.864

Mean of all scores = 10.000 with standard deviation = 5.174
```

```
BOX TEST FOR HOMOGENEITY OF VARIANCE-COVARIANCE MATRIX
SAMPLE COVARIANCE MATRIX with 12 valid cases.

Variables
            C1       C2       C3       C4
    C1    4.273    2.455    1.227    1.318
    C2    2.455    5.909    4.773    5.591
    C3    1.227    4.773    5.841    5.432
    C4    1.318    5.591    5.432    8.205

ASSUMED POP. COVARIANCE MATRIX with 12 valid cases.

Variables
            C1       C2       C3       C4
    C1    6.057    0.693    0.693    0.693
    C2    0.114    5.977    0.614    0.614
    C3    0.114    0.103    5.914    0.551
    C4    0.114    0.103    0.093    5.863

Determinant of variance-covariance matrix = 81.7
Determinant of homogeneity matrix = 1.26E3
ChiSquare = 108.149 with 8 degrees of freedom
Probability of larger chisquare = 9.66E-7
```

One Between, One Repeated Design

An Example Mixed Design

We select the AxS ANOVA option in the Comparisons sub-menu of the Statistics menu and complete the specifications on the form as show below (Fig. 8.12):

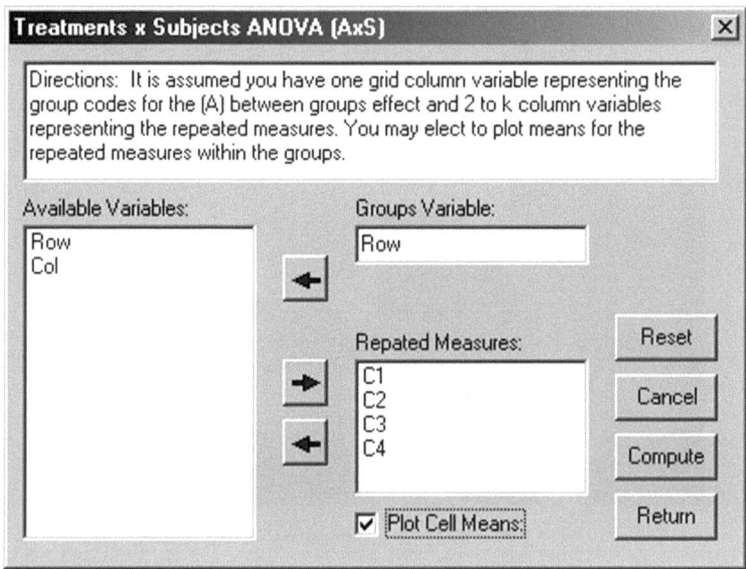

Fig. 8.12 Treatment by subjects ANOVA dialog

When the Compute button is clicked you should see these results:

```
ANOVA With One Between Subjects and One Within Subjects Treatments
-----------------------------------------------------------------
Source                  df    SS          MS         F         Prob.
-----------------------------------------------------------------
Between                 11    181.000
   Groups (A)            1     10.083     10.083     0.590     0.4602
   Subjects w.g.        10    170.917     17.092
Within Subjects         36   1077.000
   B Treatments          3    991.500    330.500   128.627     0.0000
   A X B inter.          3      8.417      2.806     1.092     0.3677
   B X S w.g.           30     77.083      2.569

TOTAL                   47   1258.000
-----------------------------------------------------------------
Means
TRT.   B 1     B 2     B 3     B 4     TOTAL
  A
  1   16.167  11.000   7.833   3.167    9.542
  2   16.833  12.000   7.667   5.333   10.458
TOTAL 16.500  11.500   7.750   4.250   10.000
```

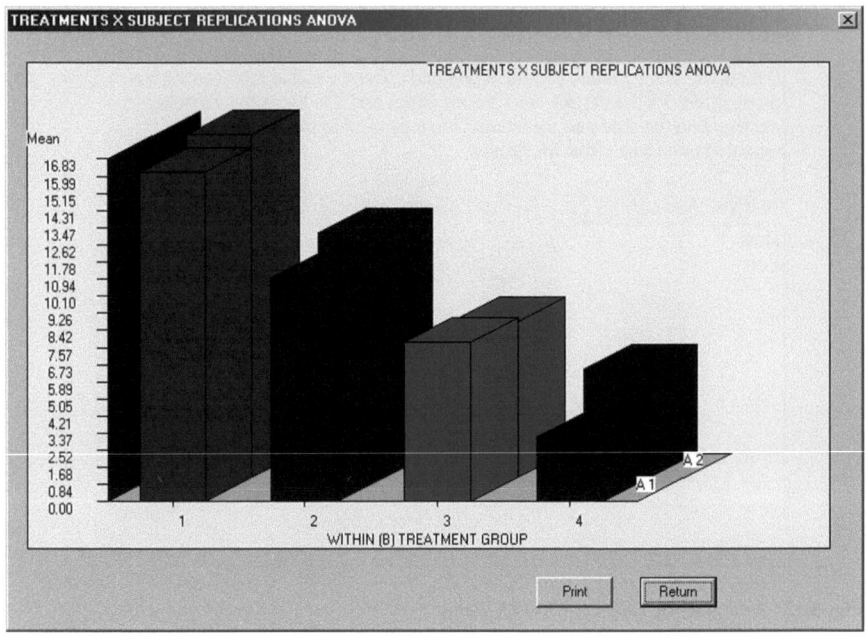

Fig. 8.13 Plot of treatment by subjects ANOVA means

```
Standard Deviations
TRT.    B 1     B 2     B 3     B 4    TOTAL
 A
 1      2.714   2.098   2.714   1.835   5.316
 2      1.329   2.828   2.338   3.445   5.099
TOTAL   2.067   2.431   2.417   2.864   5.174
```

Notice there appears to be no significant difference between the two groups of subjects but that within the groups, the first two treatment means appear to be significantly larger than the last two.

Since we elected to plot the means, we would also obtain the figure shown above (Fig. 8.13):

The graphics again demonstrate the greatest differences appear to be among the repeated measures and not the groups (A1 and A2).

You may also have a design with two between-groups factors and repeated measures within each cell composed of subjects randomly assigned to the factor A and factor B level combinations. If you have such a design, you can employ the AxBxR Anova procedure in the OpenStat package.

Two Factor Repeated Measures Analysis

Repeated measures designs have the advantage that the error terms are typically smaller that designs using independent groups of observations. This was true for the Student t-test using matched or correlated scores. On the down-side, repeated measures on the same objects pose a special problem, particularly when the objects are human subjects. The main problem is "practice" or "learning" effects that may be greater for one treatment level than another. These effects are completely confounded with the actual treatment effects. While random or counter-balanced assignment of the treatments may reduce the cumulative effects to some degree, it does not remove the effects specific to a given treatment. It is also assumed that the covariance matrices are equal among the treatment levels. Users of these designs with human subjects should be careful to minimize the practice effects. This can sometimes be done by having subjects do tasks that are similar to those in the actual experiment before beginning trials of the experiment.

In this analysis, subjects (or objects) are observed (measured) under two different treatment levels (Factors A and B levels) . For example, there might be two levels of a Factor A and three levels of a Factor B for a total of $2 \times 3 = 6$ treatment level combinations. Each subject would be observed 6 times in all. There must be the same subjects in each of the combinations.

The data file analyzed must consist of 4 columns of information for each observation: a variable containing an integer identification code for the subject (1..N), an integer from 1 to A for the treatment level of A, an integer from 1 to B for the treatment level of the Factor B, and a floating point variable for the observation (measurement).

A sample file (tworepeated.tex or tworepeated.TAB) was created from the example given by Quinn McNemar in his text book "Psychological Statistics", fourth edition, John Wiley and Sons, Inc., 1969, page 367. The data represent an experiment in which four subjects are observed under two levels of illumination and three levels of Albedo (Factors A and B.) The data file therefore contains 24 observations ($4 \times 2 \times 3$.) The analysis is initiated by loading the file and clicking on the "Two Within Subjects" option in the Analyses of Variance menu. The form which appears is shown below. Notice that the options have been selected to plot means of the two main effects and the interaction effects. An option has also been clicked to obtain post-hoc comparisons among the 6 means for the treatment combinations. When the "Compute" button is clicked the following output is obtained (Figs. 8.14, 8.15, 8.16, 8.17):

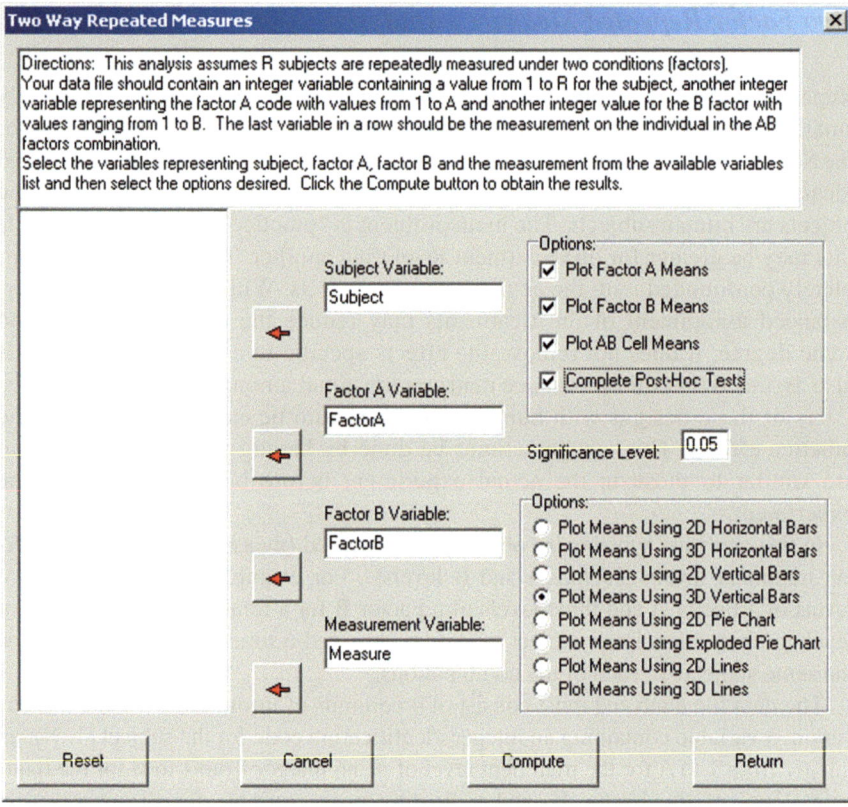

Fig. 8.14 Dialog for the two-way repeated measures ANOVA

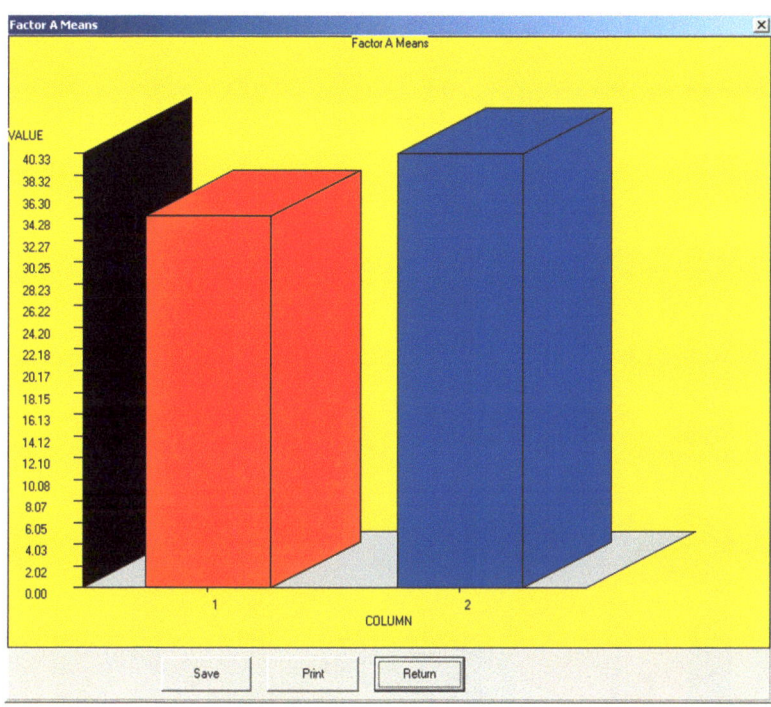

Fig. 8.15 Plot of factor A means in the two-way repeated measures analysis

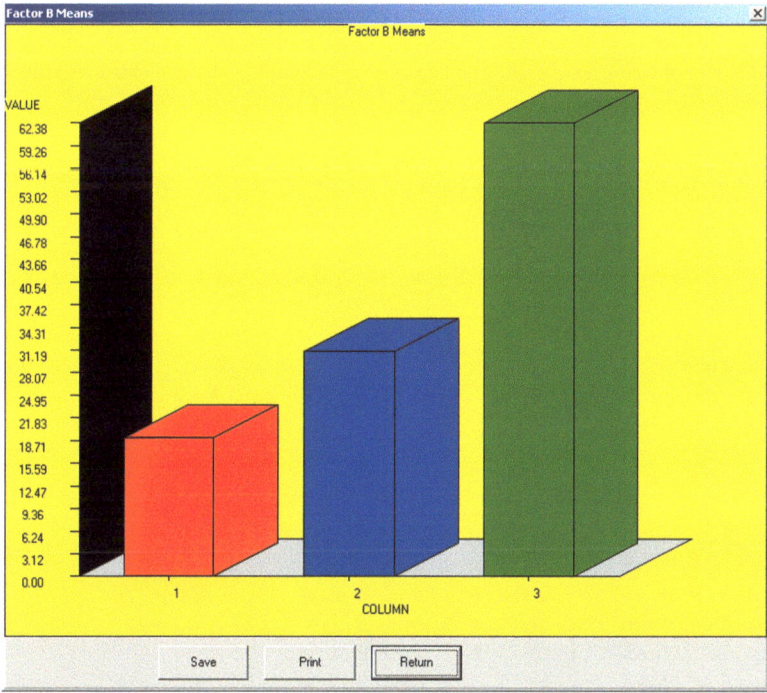

Fig. 8.16 Plot of factor B in the two-way repeated measures analysis

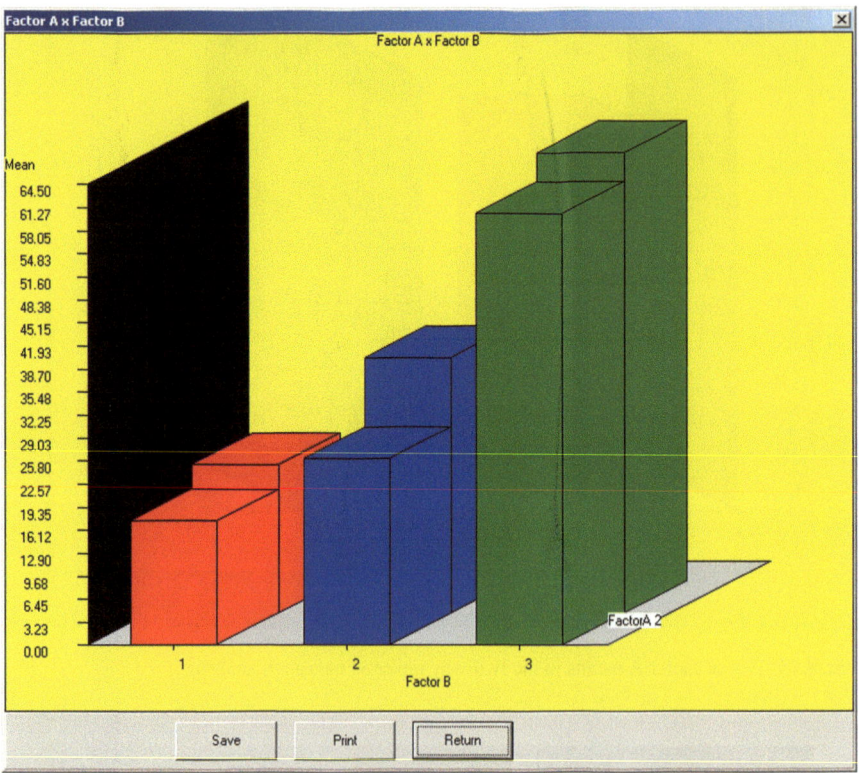

Fig. 8.17 Plot of factor A and factor B interaction in the two-way repeated measures analysis

```
------------------------------------------------------------------
SOURCE                DF     SS          MS          F       Prob.>F
------------------------------------------------------------------
Factor A              1       204.167    204.167    9.853    0.052
Factor B              2      8039.083   4019.542   24.994    0.001
Subjects              3      1302.833    434.278
A x B Interaction     2        46.583     23.292    0.803    0.491
A x S Interaction     3        62.167     20.722
B x S Interaction     6       964.917    160.819
A x B x S Inter.      6       174.083     29.01
------------------------------------------------------------------
Total                23     10793.833
------------------------------------------------------------------

Group 1 : Mean for cell A 1 and B 1 =    17.250
Group 2 : Mean for cell A 1 and B 2 =    26.000
Group 3 : Mean for cell A 1 and B 3 =    60.250
Group 4 : Mean for cell A 2 and B 1 =    20.750
Group 5 : Mean for cell A 2 and B 2 =    35.750
Group 6 : Mean for cell A 2 and B 3 =    64.500
```

```
Means for Factor A
Group 1 Mean =     34.500
Group 2 Mean =     40.333

Means for Factor B
Group 1 Mean =     19.000
Group 2 Mean =     30.875
Group 3 Mean =     62.375
```

The above results reflect possible significance for the main effects of Factors A and B but not for the interaction. The F ratio of the Factor A is obtained by dividing the mean square for Factor A by the mean square for interaction of subjects with Factor A. In a similar manner, the F ratio for Factor B is the ratio of the mean square for Factor B to the mean square of the interaction of Factor B with subjects. Finally, the F ratio for the interaction of Factor A with Factor B uses the triple interaction of A with B with Subjects as the denominator.

Between 5 and 6 of the post-hoc comparisons were not significant among the 15 possible comparisons among means using the 0.05 level for rejection of the hypothesis of no difference.

Nested Factors Analysis of Variance Design

Shown below is an example of a nested analysis using the file ABNested.tab.. When you select this analysis, you see the dialog below (Fig. 8.18):

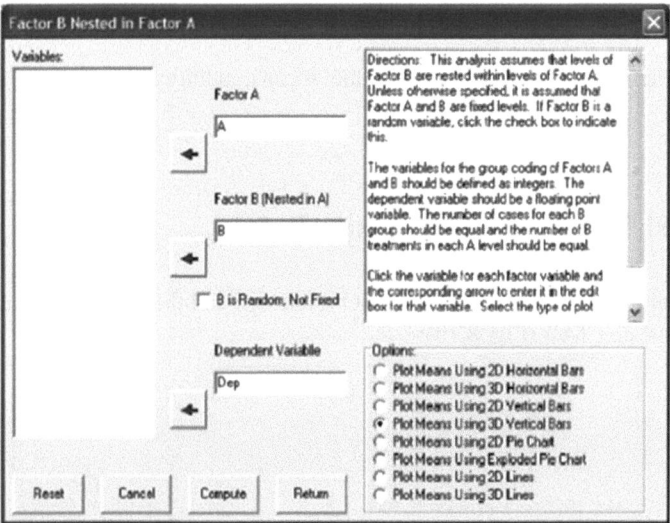

Fig. 8.18 The nested ANOVA dialog

The results are shown below:

```
NESTED ANOVA by Bill Miller

File Analyzed: C:\Documents and Settings\Owner\My Documents\
Projects\Clanguage\OpenStat\ABNested.tab
CELL MEANS
A LEVEL         B LEVEL         MEAN            STD.DEV.
   1               1            2.667           1.528
   1               2            3.333           1.528
   1               3            4.000           1.732
   2               4            3.667           1.528
   2               5            4.000           1.000
   2               6            5.000           1.000
   3               7            3.667           1.155
   3               8            5.000           1.000
   3               9            6.333           0.577

A MARGIN MEANS
A LEVEL     MEAN       STD.DEV.
   1        3.333      1.500
   2        4.222      1.202
   3        5.000      1.414

GRAND MEAN = 4.185

ANOVA TABLE
SOURCE    D.F.  SS       MS      F       PROB.
A          2    12.519   6.259   3.841   0.041
B(A)       6    16.222   2.704   1.659   0.189
w.cells   18    29.333   1.630
Total     26    58.074
```

Of course, if you elect to plot the means, additional graphical output is included.

A, B and C Factors with B Nested in A

Shown below is the dialog for this ANOVA design and the results of analyzing the file ABCNested.TAB (Fig. 8.19):

A, B and C Factors with B Nested in A

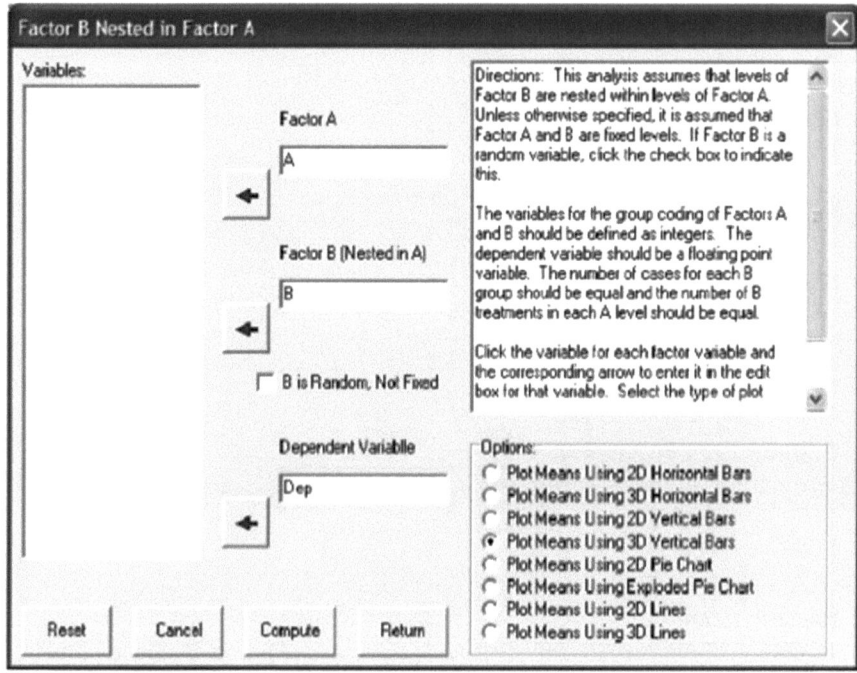

Fig. 8.19 Three factor nested ANOVA dialog

The results are:

```
NESTED ANOVA by Bill Miller
File Analyzed: C:\Documents and Settings\Owner\My Documents\
Projects\Clanguage\OpenStat\ABCNested.TAB

CELL MEANS
A LEVEL   B LEVEL   C LEVEL    MEAN      STD.DEV.
   1         1         1       2.667     1.528
   1         1         2       3.333     1.155
   1         2         1       3.333     1.528
   1         2         2       3.667     2.082
   1         3         1       4.000     1.732
   1         3         2       5.000     1.732
   2         4         1       3.667     1.528
   2         4         2       4.667     1.528
   2         5         1       4.000     1.000
   2         5         2       4.667     0.577
   2         6         1       5.000     1.000
   2         6         2       3.000     1.000
   3         7         1       3.667     1.155
   3         7         2       2.667     1.155
   3         8         1       5.000     1.000
```

```
              3         8         2      6.000   1.000
              3         9         1      6.667   1.155
              3         9         2      6.333   0.577
A MARGIN MEANS
A LEVEL    MEAN      STD.DEV.
    1      3.667     1.572
    2      4.167     1.200
    3      5.056     1.731

B MARGIN MEANS
B LEVEL    MEAN      STD.DEV.
    1      3.000     1.265
    2      3.500     1.643
    3      4.500     1.643
    4      4.167     1.472
    5      4.333     0.816
    6      4.000     1.414
    7      3.167     1.169
    8      5.500     1.049
    9      6.500     0.837

C MARGIN MEANS
C LEVEL    MEAN      STD.DEV.
    1      4.222     1.577
    2      4.370     1.644

AB MEANS
A LEVEL    B LEVEL    MEAN     STD.DEV.
    1         1       3.000    1.265
    1         2       3.500    1.643
    1         3       4.500    1.643
    2         4       4.167    1.472
    2         5       4.333    0.816
    2         6       4.000    1.414
    3         7       3.167    1.169
    3         8       5.500    1.049
    3         9       6.500    0.837

AC MEANS
A LEVEL    C LEVEL    MEAN     STD.DEV.
    1         1       3.333    1.500
    1         2       4.000    1.658
    2         1       4.222    1.202
    2         2       4.111    1.269
    3         1       5.111    1.616
    3         2       5.000    1.936

GRAND MEAN = 4.296
```

```
ANOVA TABLE
SOURCE       D.F.   SS        MS       F        PROB.
A            2      17.815    8.907    5.203    0.010
B(A)         6      42.444    7.074    4.132    0.003
C            1       0.296    0.296    0.173    0.680
AxC          2       1.815    0.907    0.530    0.593
B(A) x C     6      11.556    1.926    1.125    0.368
w.cells     36      61.630    1.712
Total       53     135.259
```

Latin and Greco-Latin Square Designs

We have prepared an example file for you to analyze with OpenStat. Open the file labeled LatinSqr.TAB in your set of sample data files. We have entered four cases for each unit in our design for instructional mode, college and home residence. Once you have loaded the file, select the Latin squares designs option under the sub-menu for comparisons under the Analyses menu. You should see the form below for selecting the Plan 1 analysis (Fig. 8.20).

When you have selected Plan 1 for the analysis, click the OK button to continue. You will then see the form below for entering the specifications for your analysis. We have entered the variables for factors A, B and C and entered the number of cases for each unit (Fig. 8.21):

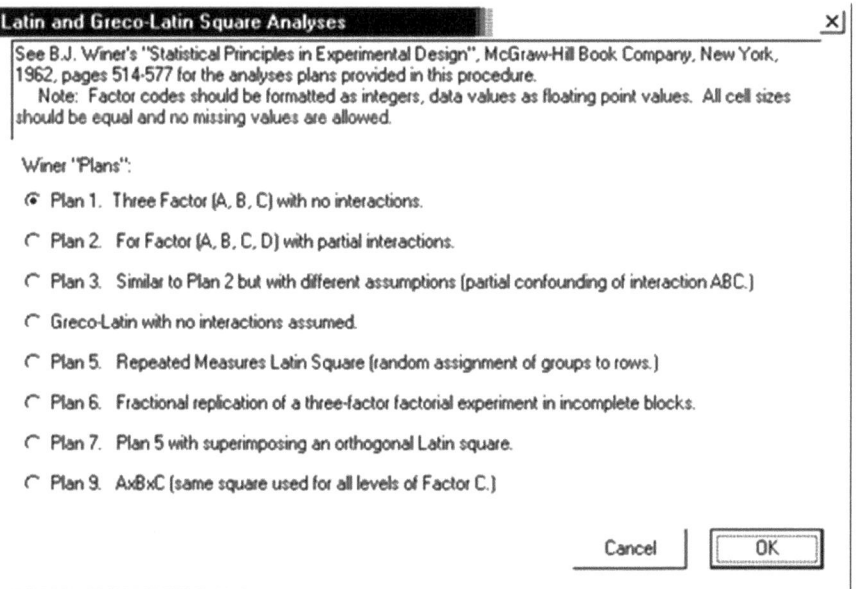

Fig. 8.20 Latin and Greco-Latin squares dialog

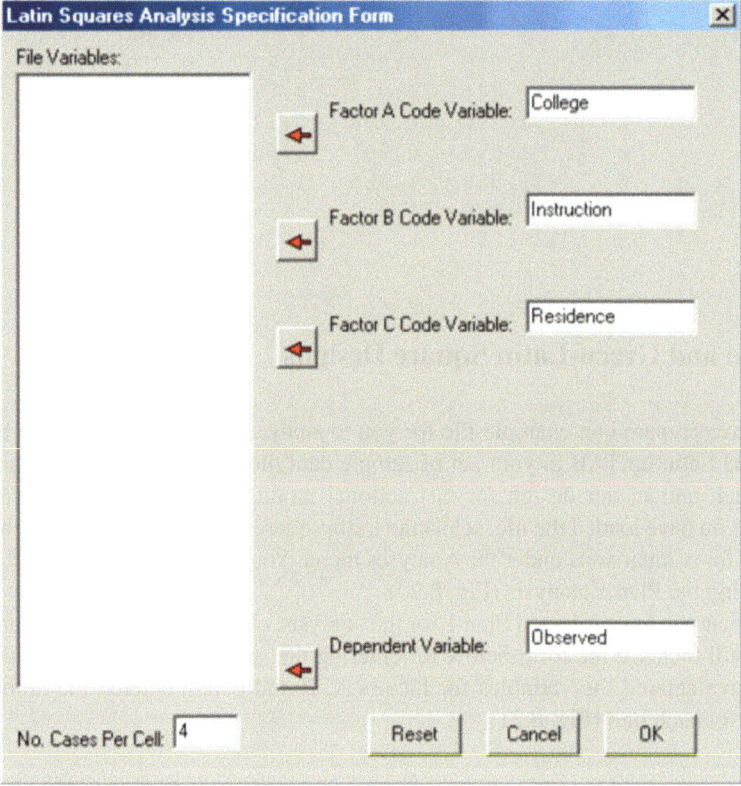

Fig. 8.21 Latin squares analysis dialog

We have completed the entry of our variables and the number of cases and are ready to continue.

When you press the OK button, the following results are presented on the output page:

```
Latin Square Analysis Plan 1 Results
-----------------------------------------------------------------
Source          SS         DF       MS         F          Prob.>F
-----------------------------------------------------------------
Factor A        92.389     2        46.194     12.535     0.000
Factor B        40.222     2        20.111     5.457      0.010
Factor C        198.722    2        99.361     26.962     0.000
Residua         133.389    2        16.694     4.530      0.020
Within          99.500     27       3.685
Total           464.222    35
-----------------------------------------------------------------
```

Latin and Greco-Latin Square Designs

```
Experimental Design
-------------------------------
Instruction    1      2      3
-------------------------------
 College
   1          C2     C3     C1
   2          C3     C1     C2
   3          C1     C2     C3
-------------------------------

Cell means and totals
----------------------------------------------------
Instruction    1         2         3        Total
----------------------------------------------------
 College
   1         2.750    10.750     3.500     5.667
   2         8.250     2.250     1.250     3.917
   3         1.500     1.500     2.250     1.750
Total        4.167     4.833     2.333     3.778
----------------------------------------------------


----------------------------------------------------
Residence      1         2         3        Total
----------------------------------------------------
             2.417     1.833     7.083     3.778
----------------------------------------------------
```

A partial test of the interaction effects can be made by the ratio of the MS for residual to the MS within cells. In our example, it appears that our assumptions of no interaction effects may be in error. In this case, the main effects may be confounded by interactions among the factors. The results may never the less suggest differences do exist and we should complete another balanced experiment to determine the interaction effects.

Plan 2

We have included the file "LatinSqr2.TAB" as an example for analysis. Load the file in the grid and select the Latin Square Analyses, Plan 2 design. The form below shows the entry of the variables and the sample size for the analysis (Fig. 8.22):

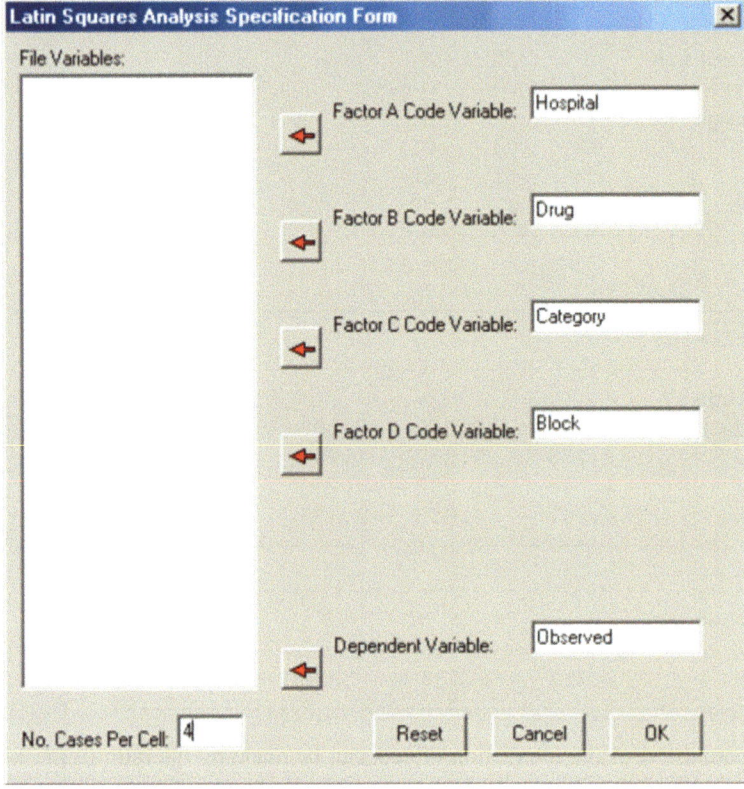

Fig. 8.22 Four factor Latin square design dialog

When you click the OK button, you will see the following results:

```
Latin Square Analysis Plan 2 Results
---------------------------------------------------------------
Source           SS        DF       MS           F        Prob.>F
---------------------------------------------------------------
Factor A     148.028       2      74.014      20.084      0.000
Factor B       5.444       2       2.722       0.739      0.483
Factor C      66.694       2      33.347       9.049      0.000
Factor D      18.000       1      18.000       4.884      0.031
A x D         36.750       2      18.375       4.986      0.010
B x D         75.000       2      37.500      10.176      0.000
C x D        330.750       2     165.375      44.876      0.000
Residual      66.778       4      16.694       4.530      0.003
Within       199.000      54       3.685
Total        946.444      71
---------------------------------------------------------------
```

Latin and Greco-Latin Square Designs

```
Experimental Design for block 1
-----------------------------
    Drug        1      2      3
-----------------------------
  Hospital
    1          C2     C3     C1
    2          C3     C1     C2
    3          C1     C2     C3
-----------------------------
Experimental Design for block 2
-----------------------------
    Drug        1      2      3
-----------------------------
  Hospital
    1          C2     C3     C1
    2          C3     C1     C2
    3          C1     C2     C3
-----------------------------

BLOCK 1

Cell means and totals
-------------------------------------------------
    Drug         1         2         3       Total
-------------------------------------------------
  Hospital
    1          2.750    10.750    3.500     5.667
    2          8.250     2.250    1.250     3.917
    3          1.500     1.500    2.250     1.750
Total          4.167     4.833    2.333     4.278
-------------------------------------------------

BLOCK 2

Cell means and totals
-------------------------------------------------
    Drug         1         2         3       Total
-------------------------------------------------
  Hospital
    1          9.250     2.250    3.250     4.917
    2          3.750     4.500   11.750     6.667
    3          2.500     3.250    2.500     2.750
Total          5.167     3.333    5.833     4.278
-------------------------------------------------


-------------------------------------------------
    Category     1         2         3       Total
-------------------------------------------------
               2.917     4.958    4.958     4.278
-------------------------------------------------
```

Notice that the interactions with Factor D are obtained. The residual however indicates that some of the other interactions confounded with the main factors may be significant and, again, we do not know the portion of the differences among the main effects that are potentially due to interactions among A, B, and C.

Plan 3 Latin Squares Design

The file "LatinSqr3.tab" contains an example of data for the Plan 3 analysis. Following the previous plans, we show below the specifications for the analysis and results from analyzing this data (Fig. 8.23):

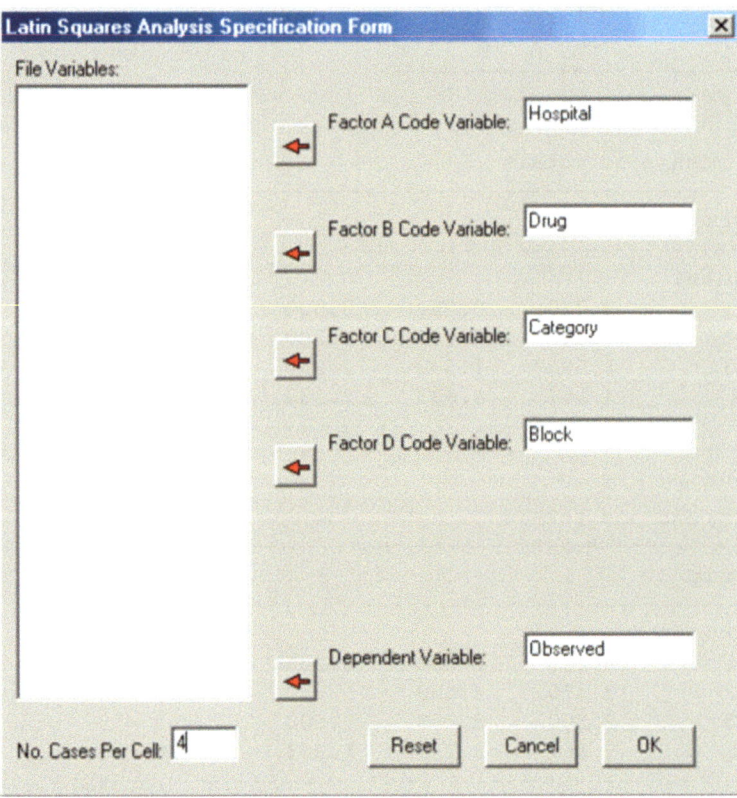

Fig. 8.23 Another Latin Square Specification form

Latin and Greco-Latin Square Designs

```
Latin Square Analysis Plan 3 Results
----------------------------------------------------------------
Source            SS          DF       MS          F       Prob.>F
----------------------------------------------------------------
Factor A        26.963         2     13.481      3.785     0.027
Factor B       220.130         2    110.065     30.902     0.000
Factor C       213.574         2    106.787     29.982     0.000
Factor D        19.185         2      9.593      2.693     0.074
A x B           49.148         4     12.287      3.450     0.012
A x C          375.037         4     93.759     26.324     0.000
B x C           78.370         4     19.593      5.501     0.001
A x B x C      118.500         6     19.750      5.545     0.000
Within         288.500        81      3.562
Total         1389.407       107
----------------------------------------------------------------

Experimental Design for block 1
-------------------------------
   Drug       1      2      3
-------------------------------
 Hospital
    1        C1     C2     C3
    2        C2     C3     C1
    3        C3     C1     C2
-------------------------------

Experimental Design for block 2
-------------------------------
   Drug       1      2      3
-------------------------------
 Hospital
    1        C2     C3     C1
    2        C3     C1     C2
    3        C1     C2     C3
-------------------------------

Experimental Design for block 3
-------------------------------
   Drug       1      2      3
-------------------------------
 Hospital
    1        C3     C1     C2
    2        C1     C2     C3
    3        C2     C3     C1
-------------------------------
```

BLOCK 1

Cell means and totals

Drug	1	2	3	Total
Hospital				
1	2.750	1.250	1.500	1.833
2	3.250	4.500	2.500	3.417
3	10.250	8.250	2.250	6.917
Total	5.417	4.667	2.083	4.074

BLOCK 2

Cell means and totals

Drug	1	2	3	Total
Hospital				
1	10.750	8.250	2.250	7.083
2	9.250	11.750	3.250	8.083
3	3.500	1.750	1.500	2.250
Total	7.833	7.250	2.333	4.074

BLOCK 3

Cell means and totals

Drug	1	2	3	Total
Hospital				
1	3.500	2.250	1.500	2.417
2	2.250	3.750	2.500	2.833
3	2.750	1.250	1.500	1.833
Total	2.833	2.417	1.833	4.074

Means for each variable

Hospital	1	2	3	Total
	3.778	4.778	3.667	4.074

Drug	1	2	3	Total
	5.361	4.778	2.083	4.074

Latin and Greco-Latin Square Designs

```
-----------------------------------------------
Category          1         2         3       Total
-----------------------------------------------
                4.056     5.806     2.361     4.074
-----------------------------------------------


-----------------------------------------------
Block             1         2         3       Total
-----------------------------------------------
                4.500     4.222     3.500     4.074
-----------------------------------------------
```

Here, the main effect of factor D is partially confounded with the ABC interaction.

Analysis of Greco-Latin Squares

The specifications for the analysis are entered as (Fig. 8.24):

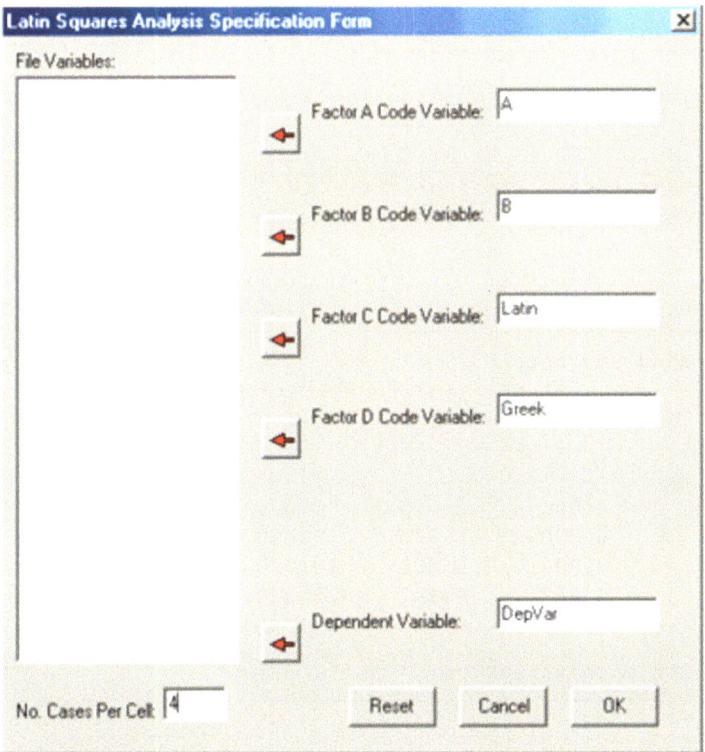

Fig. 8.24 Latin Square Design form

The results are obtained as:

```
Greco-Latin Square Analysis (No Interactions)
-------------------------------------------------------------
Source          SS        DF      MS          F       Prob.>F
-------------------------------------------------------------
Factor A       64.889      2     32.444      9.733     0.001
Factor B       64.889      2     32.444      9.733     0.001
Latin Sqr.     24.889      2     12.444      3.733     0.037
Greek Sqr.     22.222      2     11.111      3.333     0.051
Residual         -         -        -          -         -
Within         90.000     27      3.333
Total         266.889     35
-------------------------------------------------------------

Experimental Design for Latin Square
------------------------------
      B     1      2      3
------------------------------
      A
1           C1     C2     C3
2           C2     C3     C1
3           C3     C1     C2
------------------------------

Experimental Design for Greek Square
------------------------------
      B     1      2      3
------------------------------
      A
1           C1     C2     C3
2           C3     C1     C2
3           C2     C3     C1
------------------------------

Cell means and totals
-----------------------------------------------------
      B      1         2         3        Total
-----------------------------------------------------
      A
1          4.000     6.000     7.000      5.667
2          6.000    12.000     8.000      8.667
3          7.000     8.000    10.000      8.333
Total      5.667     8.667     8.333      7.556
-----------------------------------------------------

Means for each variable
-----------------------------------------------------
      A      1         2         3        Total
-----------------------------------------------------
           5.667     8.667     8.333      7.556
-----------------------------------------------------
```

Latin and Greco-Latin Square Designs

```
------------------------------------------------------
   B          1         2         3        Total
------------------------------------------------------
            5.667     8.667     8.333      7.556
------------------------------------------------------


------------------------------------------------------
  Latin       1         2         3        Total
------------------------------------------------------
            6.667     7.333     8.667      7.556
------------------------------------------------------


------------------------------------------------------
  Greek       1         2         3        Total
------------------------------------------------------
            8.667     7.000     7.000      7.556
------------------------------------------------------
```

Notice that in the case of 3 levels that the residual degrees of freedom are 0 hence no term is shown for the residual in this example. For more than 3 levels the test of the residuals provides a partial check on the assumptions of negligible interactions. The residual is sometimes combined with the within cell variance to provide an over-all estimate of variation due to experimental error.

Plan 5 Latin Square Design

The specifications for the analysis of the sample file "LatinPlan5.TAB" is shown below (Fig. 8.25):

If you examine the sample file, you will notice that the subject Identification numbers (1,2,3,4) for the subjects in each group are the same even though the subjects in each group are different from group to group. The same ID is used in each group because they become "subscripts" for several arrays in the program. The results for our sample data are shown below:

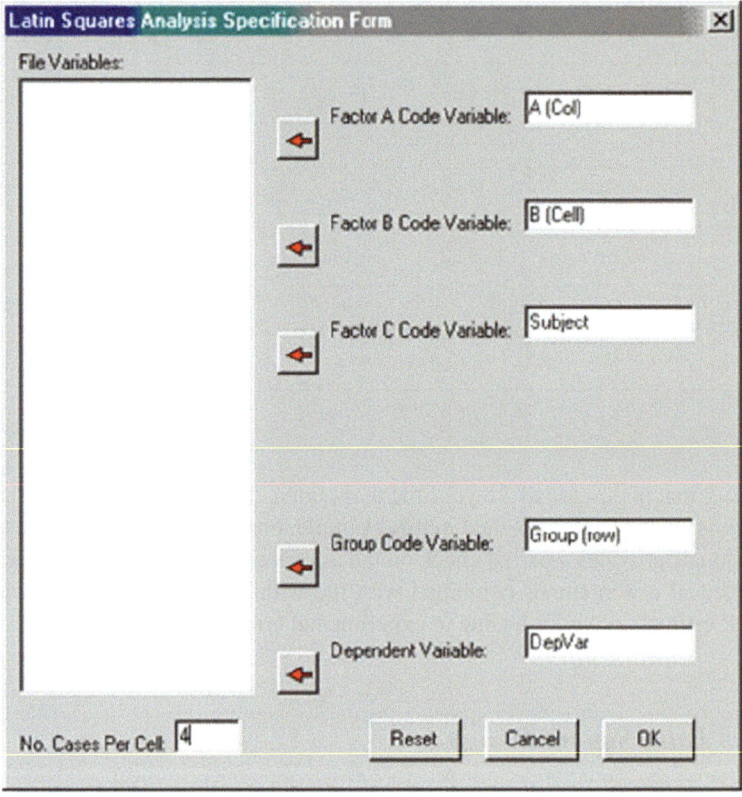

Fig. 8.25 Latin Square Plan 5 Specifications form

```
Sums for ANOVA Analysis
Group (rows) times A Factor (columns) sums with 36 cases.
Variables
               1         2         3        Total
      1     14.000    19.000    18.000     51.000
      2     15.000    18.000    16.000     49.000
      3     14.000    21.000    18.000     53.000
Total       43.000    58.000    52.000    153.000
Group (rows) times B (cells Factor) sums with 36 cases.
Variables
               1         2         3        Total
      1     19.000    18.000    14.000     51.000
      2     15.000    18.000    16.000     49.000
      3     18.000    14.000    21.000     53.000
Total       52.000    50.000    51.000    153.000
```

Latin and Greco-Latin Square Designs

```
Groups (rows) times Subjects (columns) matrix with 36 cases.
Variables
                1          2          3          4         Total
     1       13.000     11.000     13.000     14.000      51.000
     2       10.000     14.000     10.000     15.000      49.000
     3       13.000      9.000     17.000     14.000      53.000
Total        36.000     34.000     40.000     43.000     153.000
```

Latin Squares Repeated Analysis Plan 5 (Partial Interactions)

```
-------------------------------------------------------------
Source              SS          DF       MS        F      Prob.>F
-------------------------------------------------------------
Betw.Subj.        20.083        11
  Groups           0.667         2      0.333    0.155    0.859
  Subj.w.g.       19.417         9      2.157

Within Sub        36.667        24
  Factor A         9.500         2      4.750    3.310    0.060
  Factor B         0.167         2      0.083    0.058    0.944
  Factor AB        1.167         2      0.583    0.406    0.672
  Error w.        25.833        18      1.435
Total             56.750        35
-------------------------------------------------------------
```

Experimental Design for Latin Square

```
-----------------------------
A (Col)      1     2     3
-----------------------------
Group (row)
    1        B3    B1    B2
    2        B1    B2    B3
    3        B2    B3    B1
-----------------------------
```

Cell means and totals

```
----------------------------------------------------
A (Col)        1          2          3        Total
----------------------------------------------------
Group (row)
    1        3.500      4.750      4.500      4.250
    2        3.750      4.500      4.000      4.083
    3        3.500      5.250      4.500      4.417
Total        3.583      4.833      4.333      4.250
----------------------------------------------------
```

Means for each variable

```
----------------------------------------------------
A (Col)        1          2          3        Total
----------------------------------------------------
             4.333      4.167      4.250      4.250
----------------------------------------------------
```

```
B (Cell)        1           2           3           Total

              4.250       4.083       4.417       4.250


Group (row)     1           2           3           Total

              4.250       4.083       4.417       4.250
```

Plan 6 Latin Squares Design

LatinPlan6.TAB is the name of a sample file which you can analyze with the Plan 6 option of the Latin squares analysis procedure. Shown below is the specification form for the analysis of the data in that file (Fig. 8.26):

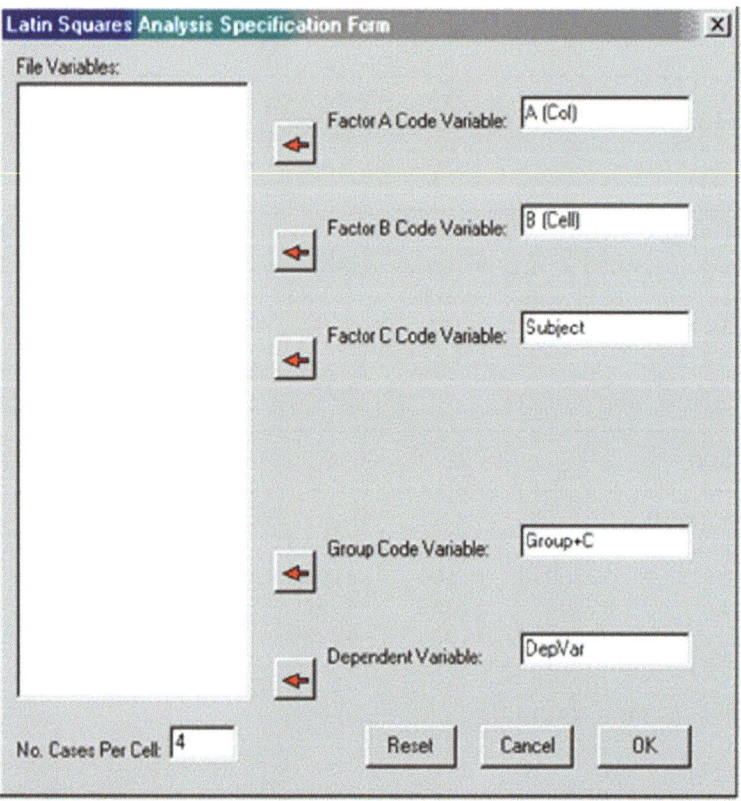

Fig. 8.26 Latin square plan 6 specification

Latin and Greco-Latin Square Designs

The results obtained when you click the OK button are shown below:

```
Latin Squares Repeated Analysis Plan 6
Sums for ANOVA Analysis
Group - C (rows) times A Factor (columns) sums with 36 cases.
Variables
                1          2          3        Total
    1        23.000     16.000     22.000     61.000
    2        22.000     14.000     18.000     54.000
    3        24.000     21.000     21.000     66.000
Total        69.000     51.000     61.000    181.000
Group - C (rows) times B (cells Factor) sums with 36 cases.
Variables
                1          2          3        Total
    1        16.000     22.000     23.000     61.000
    2        22.000     14.000     18.000     54.000
    3        21.000     24.000     21.000     66.000
Total        59.000     60.000     62.000    181.000
Group - C (rows) times Subjects (columns) matrix with 36
cases.
Variables
                1          2          3          4        Total
    1        16.000     14.000     13.000     18.000     61.000
    2        12.000     13.000     14.000     15.000     54.000
    3        18.000     19.000     11.000     18.000     66.000
Total        46.000     46.000     38.000     51.000    181.000
Latin Squares Repeated Analysis Plan 6
-----------------------------------------------------------
Source            SS         DF        MS       F       Prob.>F
-----------------------------------------------------------
Betw.Subj.      26.306       11
  Factor C       6.056        2      3.028    1.346     0.308
  Subj.w.g.     20.250        9      2.250

Within Sub      70.667       24
  Factor A      13.556        2      6.778    2.259     0.133
  Factor B       0.389        2      0.194    0.065     0.937
  Residual       2.722        2      1.361    0.454     0.642
  Error w.      54.000       18      3.000
Total           96.972       35
-----------------------------------------------------------
```

```
Experimental Design for Latin Square
-------------------------------
A    (Col)    1     2     3
-------------------------------
G    C
1    1        B3    B1    B2
2    2        B1    B2    B3
3    3        B2    B3    B1
-------------------------------

Cell means and totals
-----------------------------------------------------
A (Col)       1         2         3         Total
-----------------------------------------------------
Group+C
   1        5.750     4.000     5.500     5.083
   2        5.500     3.500     4.500     4.500
   3        6.000     5.250     5.250     5.500
Total       5.750     4.250     5.083     5.028
-----------------------------------------------------

Means for each variable
-----------------------------------------------------
A (Col)       1         2         3         Total
-----------------------------------------------------
            4.917     5.000     5.167     5.028
-----------------------------------------------------


-----------------------------------------------------
B (Cell)      1         2         3         Total
-----------------------------------------------------
            5.083     4.500     5.500     5.028
-----------------------------------------------------


-----------------------------------------------------
Group+C       1         2         3         Total
-----------------------------------------------------
            5.083     4.500     5.500     5.028
-----------------------------------------------------
```

Plan 7 for Latin Squares

Shown below is the specification for analysis of the sample data file labeled LatinPlan7.TAB and the results of the analysis (Fig. 8.27):

Latin and Greco-Latin Square Designs

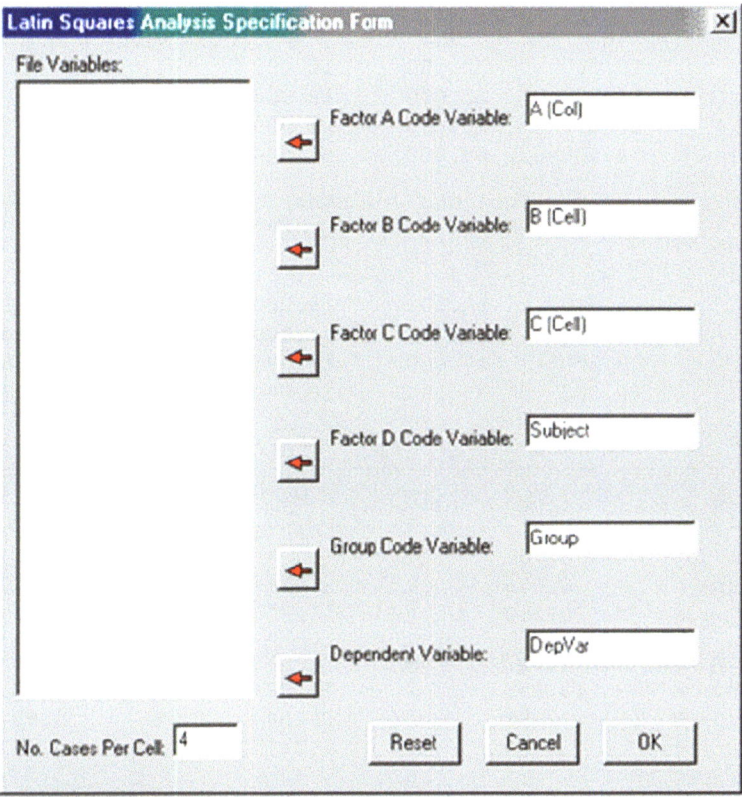

Fig. 8.27 Latin Squares Repeated Analysis Plan 7 form

```
Sums for ANOVA Analysis
Group (rows) times A Factor (columns) sums with 36 cases.
Variables
               1          2          3         Total
    1       23.000     16.000     22.000      61.000
    2       22.000     14.000     18.000      54.000
    3       24.000     21.000     21.000      66.000
Total       69.000     51.000     61.000     181.000
Group (rows) times B (cells Factor) sums with 36 cases.
Variables
               1          2          3         Total
    1       23.000     16.000     22.000      61.000
    2       18.000     22.000     14.000      54.000
    3       21.000     21.000     24.000      66.000
Total       62.000     59.000     60.000     181.000
Group (rows) times C (cells Factor) sums with 36 cases.
```

```
Variables
                1           2           3         Total
      1     23.000      22.000      16.000      61.000
      2     14.000      22.000      18.000      54.000
      3     21.000      21.000      24.000      66.000
  Total     58.000      65.000      58.000     181.000
```

Group (rows) times Subjects (columns) sums with 36 cases.

```
Variables
                1           2           3           4         Total
      1     16.000      14.000      13.000      18.000      61.000
      2     12.000      13.000      14.000      15.000      54.000
      3     18.000      19.000      11.000      18.000      66.000
  Total     46.000      46.000      38.000      51.000     181.000
```

Latin Squares Repeated Analysis Plan 7 (superimposed squares)

```
Source              SS          DF          MS          F       Prob.>F
-------------------------------------------------------------
Betw.Subj.       26.306         11
  Groups          6.056          2       3.028       1.346       0.308
  Subj.w.g.      20.250          9       2.250

Within Sub       70.667         24
  Factor A       13.556          2       6.778       2.259       0.133
  Factor B        0.389          2       0.194       0.065       0.937
  Factor C        2.722          2       1.361       0.454       0.642
  residual          -             0          -
  Error w.       54.000         18       3.000
Total            96.972         35
-------------------------------------------------------------
```

Experimental Design for Latin Square

```
A (Col)         1       2       3
------------------------------------
   Group
   5.         BC11    BC23    BC32
   5.         BC22    BC31    BC13
   5.         BC33    BC12    BC21
------------------------------------
```

Cell means and totals

```
A (Col)         1          2          3         Total
------------------------------------------------------
   Group
      1      5.750      4.000      5.500      5.083
      2      5.500      3.500      4.500      4.500
      3      6.000      5.250      5.250      5.500
  Total      5.750      4.250      5.083      5.028
------------------------------------------------------
```

```
Means for each variable
-------------------------------------------------
A (Col)          1          2          3      Total
-------------------------------------------------
              5.750      4.250      5.083      5.028
-------------------------------------------------


-------------------------------------------------
B (Cell)         1          2          3      Total
-------------------------------------------------
              5.167      4.917      5.000      5.028
-------------------------------------------------


-------------------------------------------------
C (Cell)         1          2          3      Total
-------------------------------------------------
              4.833      5.417      4.833      5.028
-------------------------------------------------


-------------------------------------------------
Group            1          2          3      Total
-------------------------------------------------
              5.083      4.500      5.500      5.028
-------------------------------------------------
```

Plan 9 Latin Squares

The sample data set labeled "LatinPlan9.TAB" is used for the following analysis. The specification form shown below has the variables entered for the analysis. When you click the OK button, the results obtained are as shown following the form (Fig. 8.28).

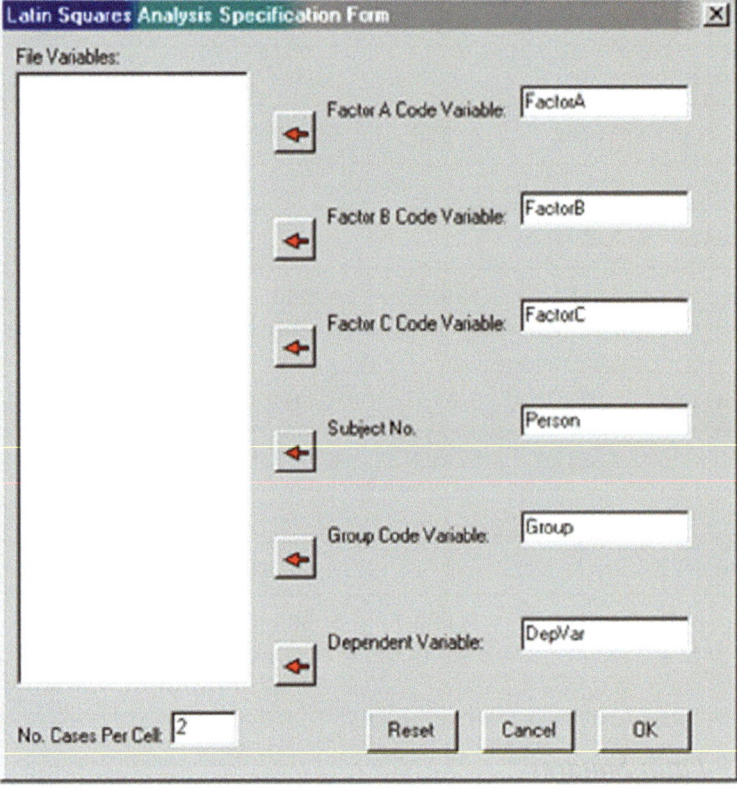

Fig. 8.28 Latin Squares Repeated Analysis Plan 9 form

```
Sums for ANOVA Analysis
ABC matrix
C level 1
                1         2         3
    1       13.000     3.000     9.000
    2        6.000     9.000     3.000
    3       10.000    14.000    15.000
C level 2
                1         2         3
    1       18.000    14.000    18.000
    2       19.000    24.000    20.000
    3        8.000    11.000    10.000
```

```
C level 3
                1           2           3
    1        17.000      12.000      20.000
    2        14.000      13.000       9.000
    3        15.000      12.000      17.000
AB sums with 18 cases.

Variables
                    1               2               3           Total
    1            48.000          29.000          47.000         124.000
    2            39.000          46.000          32.000         117.000
    3            33.000          37.000          42.000         112.000
 Total          120.000         112.000         121.000         353.000
AC sums with 18 cases.

Variables
                    1               2               3           Total
    1            25.000          50.000          49.000         124.000
    2            18.000          63.000          36.000         117.000
    3            39.000          29.000          44.000         112.000
Total            82.000         142.000         129.000         353.000
BC sums with 18 cases.

Variables
                    1               2               3           Total
    1            29.000          45.000          46.000         120.000
    2            26.000          49.000          37.000         112.000
    3            27.000          48.000          46.000         121.000
 Total           82.000         142.000         129.000         353.000
RC sums with 18 cases.

Variables
                    1               2               3           Total
    1            16.000          42.000          36.000          94.000
    2            37.000          52.000          47.000         136.000
    3            29.000          48.000          46.000         123.000
 Total           82.000         142.000         129.000         353.000
Group totals with 18 valid cases.

Variables            1           2           3           4           5
                  16.000      37.000      29.000      42.000      52.000

Variables            6           7           8           9       Total
                  48.000      36.000      47.000      46.000     353.000
Subjects sums with 18 valid cases.
```

Variables	1	2	3	4	5
	7.000	9.000	14.000	28.000	15.000
Variables	6	7	8	9	10
	21.000	16.000	21.000	22.000	30.000
Variables	11	12	13	14	15
	28.000	19.000	10.000	19.000	23.000
Variables	16	17	18	Total	
	25.000	28.000	18.000	0.000	

Latin Squares Repeated Analysis Plan 9

Source	SS	DF	MS	F	Prob.>F
Betw.Subj.	267.426	17			
Factor C	110.704	2	55.352	5.058	0.034
Rows	51.370	2	25.685	2.347	0.151
C x row	6.852	4	1.713	0.157	0.955
Subj.w.g.	98.500	9	10.944		
Within Sub	236.000	36			
Factor A	4.037	2	2.019	0.626	0.546
Factor B	2.704	2	1.352	0.420	0.664
Factor AC	146.519	4	36.630	11.368	0.000
Factor BC	8.519	4	2.130	0.661	0.627
AB prime	7.148	2	3.574	1.109	0.351
ABC prime	9.074	4	2.269	0.704	0.599
Error w.	58.000	18	3.222		
Total	503.426	53			

Experimental Design for Latin Square

FactorA	1	2	3
Group			
1	B2	B3	B1
2	B1	B2	B3
3	B3	B1	B2
4	B2	B3	B1
5	B1	B2	B3
6	B3	B1	B2
7	B2	B3	B1
8	B1	B2	B3
9	B3	B1	B2

Latin and Greco-Latin Square Designs

```
Latin Squares Repeated Analysis Plan 9
Means for ANOVA Analysis
ABC matrix
C level 1
```

	1	2	3
1	6.500	1.500	4.500
2	3.000	4.500	1.500
3	5.000	7.000	7.500

C level 2

	1	2	3
1	9.000	7.000	9.000
2	9.500	12.000	10.000
3	4.000	5.500	5.000

C level 3

	1	2	3
1	8.500	6.000	10.000
2	7.000	6.500	4.500
3	7.500	6.000	8.500

AB Means with 54 cases.

Variables

	1	2	3	4
1	8.000	4.833	7.833	6.889
2	6.500	7.667	5.333	6.500
3	5.500	6.167	7.000	6.222
Total	6.667	6.222	6.722	6.537

AC Means with 54 cases.

Variables

	1	2	3	4
1	4.167	8.333	8.167	6.889
2	3.000	10.500	6.000	6.500
3	6.500	4.833	7.333	6.222
Total	4.556	7.889	7.167	6.537

BC Means with 54 cases.

Variables

	1	2	3	4
1	4.833	7.500	7.667	6.667
2	4.333	8.167	6.167	6.222
3	4.500	8.000	7.667	6.722
Total	4.556	7.889	7.167	6.537

RC Means with 54 cases.

```
Variables
                       1              2              3              4
       1             2.667          7.000          6.000          5.222
       2             6.167          8.667          7.833          7.556
       3             4.833          8.000          7.667          6.833
   Total             4.556          7.889          7.167          6.537
Group Means with 54 valid cases.

Variables              1              2              3              4              5
                    2.667          6.167          4.833          7.000          8.667

Variables              6              7              8              9          Total
                    8.000          6.000          7.833          7.667          6.537
Subjects Means with 54 valid cases.

Variables              1              2              3              4              5
                    3.500          4.500          7.000         14.000          7.500

Variables              6              7              8              9             10
                   10.500          8.000         10.500         11.000         15.000

Variables             11             12             13             14             15
                   14.000          9.500          5.000          9.500         11.500

Variables             16             17             18          Total
                   12.500         14.000          9.000          6.537
```

2 or 3 Way Fixed ANOVA with 1 Case Per Cell

There may be an occasion where you have collected data with a single observation within two or three factor combinations. In this case one cannot obtain an estimate of the variance within a single cell of the two or three factor design and thus an estimate of the mean squared error term typically used in a 2 or 3 way ANOVA. The estimate of error must be made using all cell values. To demonstrate, the following data are analyzed:

```
CASES FOR FILE C:\Users\wgmiller\Projects\Data\OneCase2Way.TEX

              0           Row           Col           Dep
       CASE  1            1             1           1.000
       CASE  2            1             2           2.000
       CASE  3            1             3           3.000
       CASE  4            2             1           3.000
       CASE  5            2             2           5.000
       CASE  6            2             3           9.000
```

The dialog for this procedure and the resulting output are shown below (Figs. 8.29, 8.30, 8.31, 8.32):

2 or 3 Way Fixed ANOVA with 1 Case Per Cell

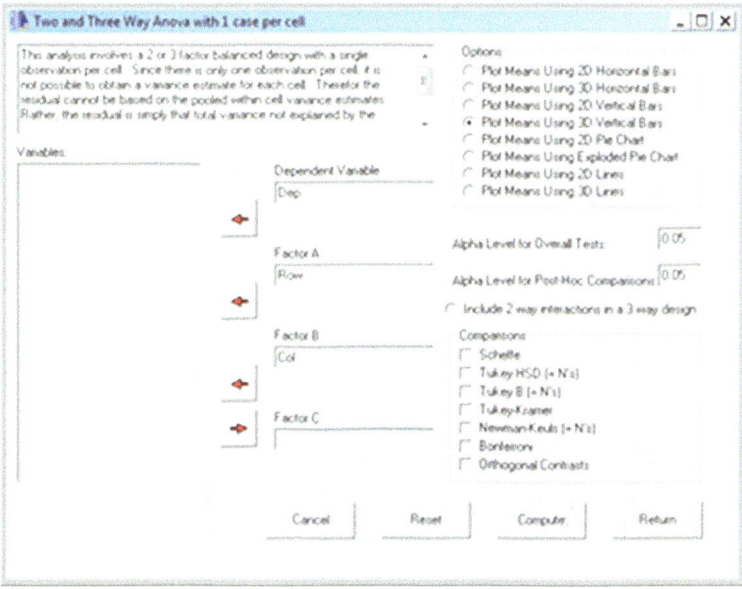

Fig. 8.29 Dialog for 2 or 3 way ANOVA with one case per cell

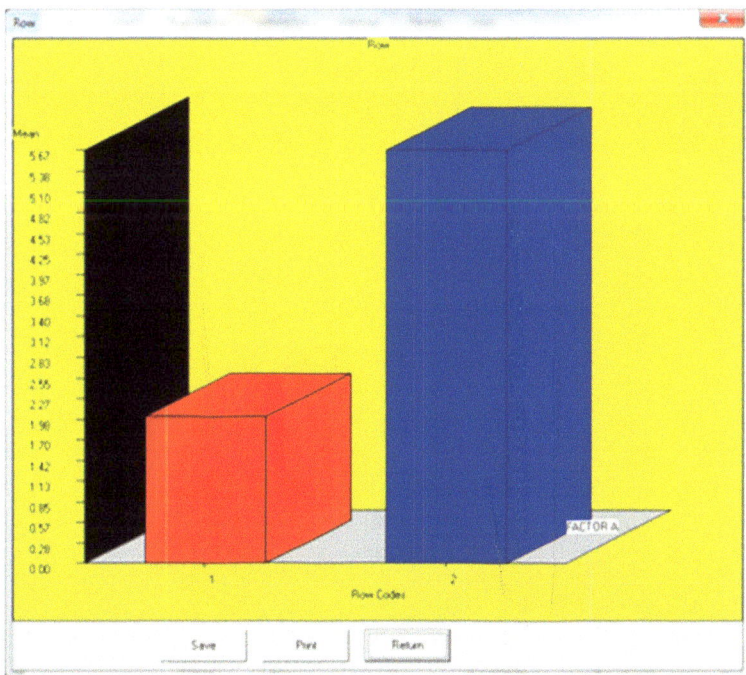

Fig. 8.30 One case ANOVA plot for factor 1

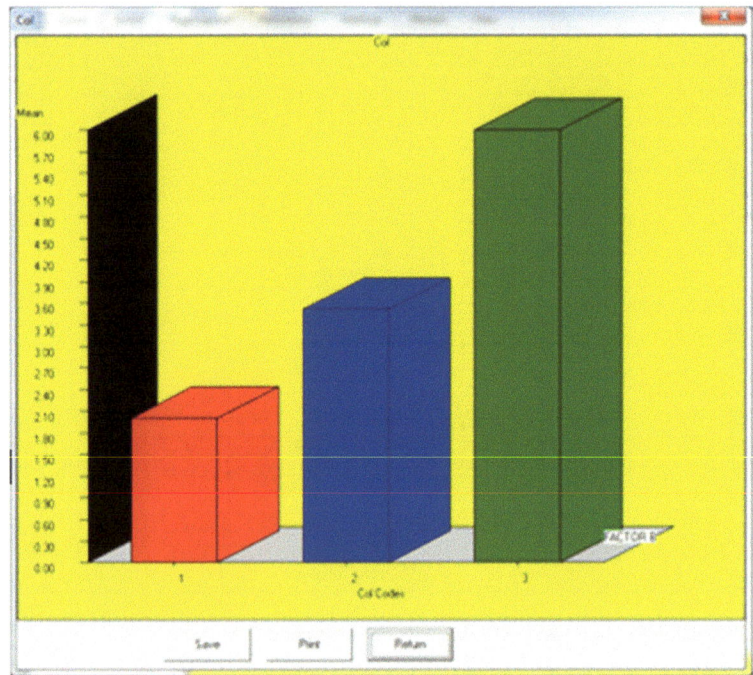

Fig. 8.31 Factor 2 plot for one case ANOVA

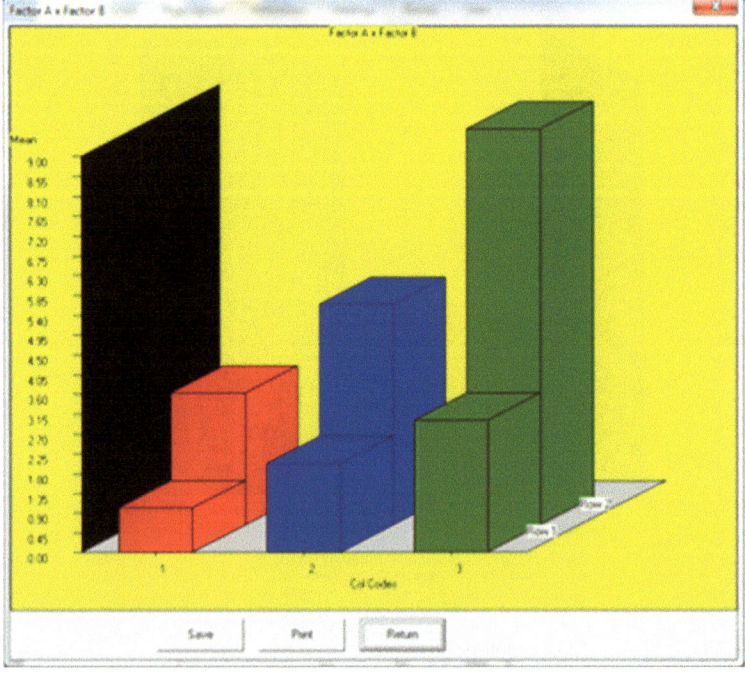

Fig. 8.32 Interaction plot of two factors for one case ANOVA

```
Two Way Analysis of Variance

Variable analyzed: Dep

Factor A (rows) variable: Row
Factor B (columns) variable: Col

SOURCE            D.F.      SS        MS       F     PROB.> F   Omega Squared

Among Rows         1      20.167    20.167   9.308    0.093         0.419
Among Columns      2      16.333     8.167   3.769    0.210         0.279
Residual           2       4.333     2.167
 NonAdditivity     1       4.252     4.252  52.083    0.088
 Balance           1       0.082     0.082
Total              5      40.833     8.167
Omega squared for combined effects = 0.698

Descriptive Statistics

GROUP     Row      Col.      N       MEAN      VARIANCE     STD.DEV.
Cell       1        1        1      1.000       0.000        0.000
Cell       1        2        1      2.000       0.000        0.000
Cell       1        3        1      3.000       0.000        0.000
Cell       2        1        1      3.000       0.000        0.000
Cell       2        2        1      5.000       0.000        0.000
Cell       2        3        1      9.000       0.000        0.000
Row        1                 3      2.000       1.000        1.000
Row        2                 3      5.667       9.333        3.055
Col                 1        2      2.000       2.000        1.414
Col                 2        2      3.500       4.500        2.121
Col                 3        2      6.000      18.000        4.243
TOTAL                        6      3.833       8.167        2.858
```

Two Within Subjects ANOVA

You may have observed the same subjects under two "treatment" factors. As an example, you might have observed subject responses on a visual acuity test before and after consuming an alcoholic beverage. In this case we do not have a "between subjects" analysis but rather a "repeated measures" analysis under two conditions. As an example, we will analyze data from a file labeled "". The data, the dialog and the results are shown below (Figs. 8.33, 8.34, 8.35, 8.36):

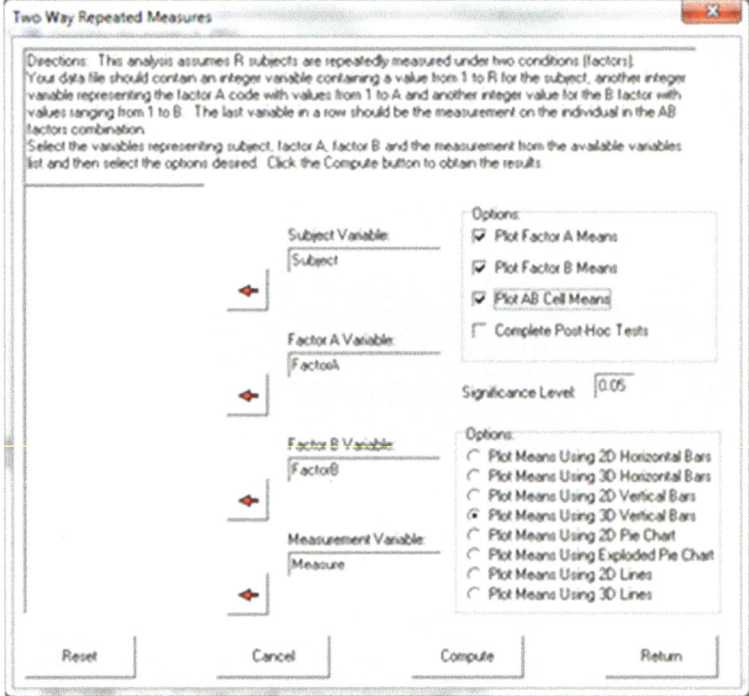

Fig. 8.33 Dialog for two within subjects ANOVA

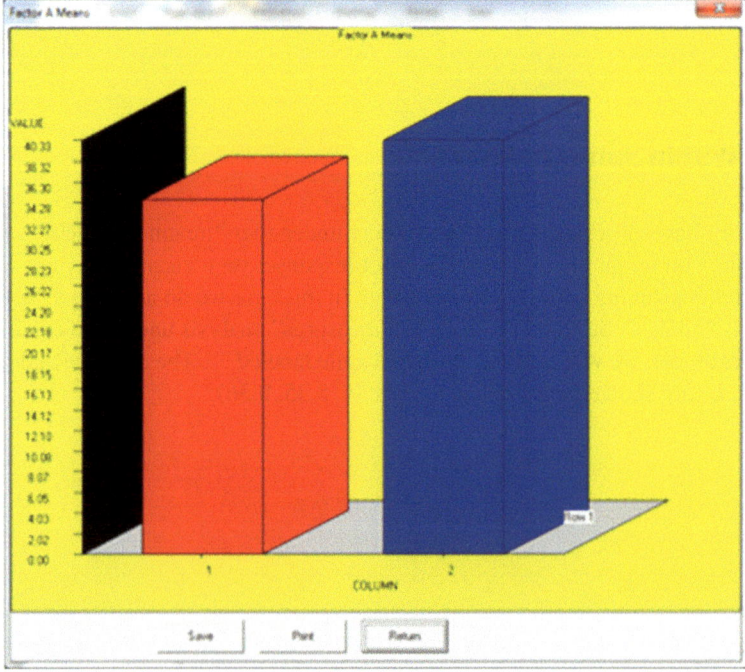

Fig. 8.34 Factor one plot for two within subjects ANOVA

Two Within Subjects ANOVA

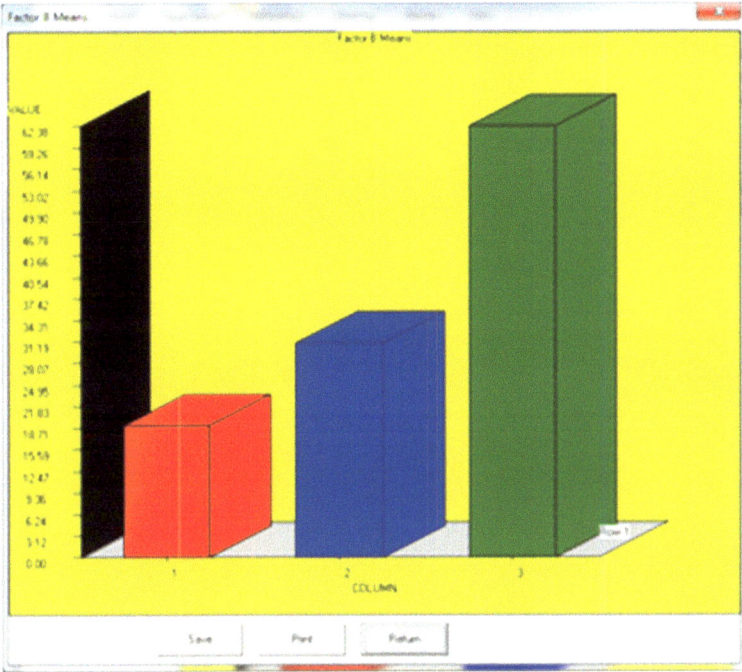

Fig. 8.35 Factor two plot for two within subjects ANOVA

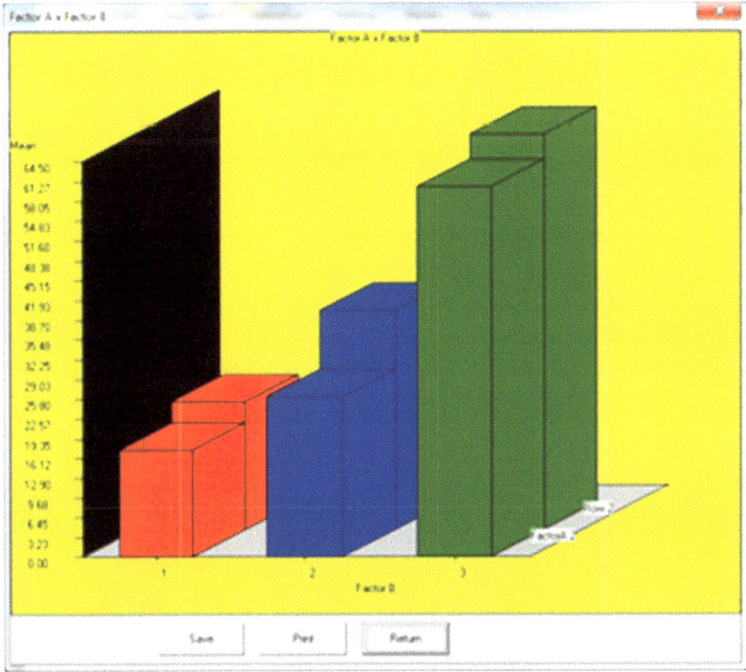

Fig. 8.36 Two way interaction for two within subjects ANOVA

```
SOURCE                  DF          SS          MS         F     Prob.>F
Factor A                 1      204.167     204.167      9.853    0.052
Factor B                 2     8039.083    4019.542     24.994    0.001
Subjects                 3     1302.833     434.278
A x B Interaction        2       46.583      23.292      0.803    0.491
A x S Interaction        3       62.167      20.722
B x S Interaction        6      964.917     160.819
A x B x S Inter.         6      174.083      29.01
Total                   23    10793.833

Group 1 : Mean for cell A 1 and B 1 =       17.250
Group 2 : Mean for cell A 1 and B 2 =       26.000
Group 3 : Mean for cell A 1 and B 3 =       60.250
Group 4 : Mean for cell A 2 and B 1 =       20.750
Group 5 : Mean for cell A 2 and B 2 =       35.750
Group 6 : Mean for cell A 2 and B 3 =       64.500

Means for Factor A
Group 1 Mean  =     34.500
Group 2 Mean  =     40.333

Means for Factor B
Group 1 Mean  =     19.000
Group 2 Mean  =     30.875
Group 3 Mean  =     62.375
```

Analysis of Variance Using Multiple Regression Methods

An Example of an Analysis of Covariance

We will demonstrate the analysis of covariance procedure using multiple regression by loading the file labeled "Ancova2.tab". In this file we have a treatment group code for four groups, a dependent variable (X) and two covariates (Y and Z.) The procedure is started by selection the "Analysis of Covariance by Regression" option in the Comparisons sub-menu under the Statistics menu. Shown below is the completed specification form for our analysis (Fig. 8.37):

Analysis of Variance Using Multiple Regression Methods

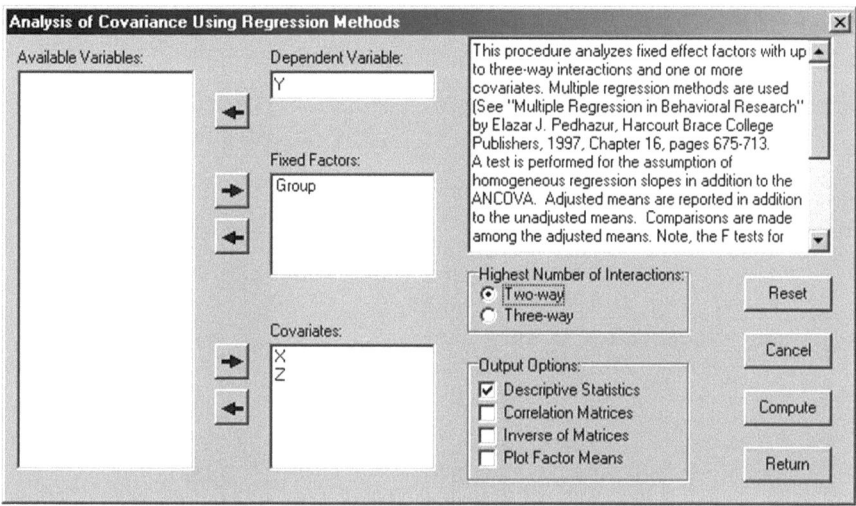

Fig. 8.37 Analysis of covariance dialog

When we click the Compute button, the following results are obtained:

```
ANALYSIS OF COVARIANCE USING MULTIPLE REGRESSION
File Analyzed: C:\Projects\Delphi\OpenStat\Ancova2.txt
Model for Testing Assumption of Zero Interactions with Covariates
MEANS with 40 valid cases.

Variables            X         Z        A1        A2        A3
                 7.125    14.675     0.000     0.000     0.000

Variables         XxA1      XxA2      XxA3      ZxA1      ZxA2
                 0.125     0.025     0.075    -0.400    -0.125

Variables         ZxA3         Y
                -0.200    17.550
VARIANCES with 40 valid cases.

Variables            X         Z        A1        A2        A3
                 4.163    13.866     0.513     0.513     0.513

Variables         XxA1      XxA2      XxA3      ZxA1      ZxA2
                28.010    27.102    27.712   116.759   125.035

Variables         ZxA3         Y
               124.113     8.254
STD. DEV.S with 40 valid cases.
```

Variables	X	Z	A1	A2	A3
	2.040	3.724	0.716	0.716	0.716

Variables	XxA1	XxA2	XxA3	ZxA1	ZxA2
	5.292	5.206	5.264	10.806	11.182

Variables	ZxA3	Y
	11.141	2.873

R	R2	F	Prob.>F	DF1	DF2
0.842	0.709	6.188	0.000	11	28

Adjusted R Squared = 0.594

Std. Error of Estimate = 1.830

Variable	Beta	B	Std.Error	t	Prob.>t
X	0.599	0.843	0.239	3.531	0.001
Z	0.123	0.095	0.138	0.686	0.498
A1	-0.518	-2.077	2.381	-0.872	0.391
A2	0.151	0.606	2.513	0.241	0.811
A3	0.301	1.209	2.190	0.552	0.585
XxA1	-1.159	-0.629	0.523	-1.203	0.239
XxA2	0.714	0.394	0.423	0.932	0.359
XxA3	0.374	0.204	0.334	0.611	0.546
ZxA1	1.278	0.340	0.283	1.200	0.240
ZxA2	-0.803	-0.206	0.284	-0.727	0.473
ZxA3	-0.353	-0.091	0.187	-0.486	0.631

Constant = 10.300

Analysis of Variance for the Model to Test Regression Homogeneity

SOURCE	Deg.F.	SS	MS	F	Prob>F
Explained	11	228.08	20.73	6.188	0.0000
Error	28	93.82	3.35		
Total	39	321.90			

Model for Analysis of Covariance

MEANS with 40 valid cases.

Variables	X	Z	A1	A2	A3
	7.125	14.675	0.000	0.000	0.000

Variables	Y
	17.550

VARIANCES with 40 valid cases.

```
Variables          X          Z         A1        A2        A3
                 4.163     13.866     0.513     0.513     0.513

Variables          Y
                 8.254
```

STD. DEV.S with 40 valid cases.

```
Variables          X          Z         A1        A2        A3
                 2.040      3.724     0.716     0.716     0.716

Variables          Y
                 2.873
       R         R2          F       Prob.>F      DF1       DF2
     0.830     0.689      15.087      0.000        5         34
```

Adjusted R Squared = 0.644

Std. Error of Estimate = 1.715

```
   Variable     Beta         B       Std.Error     t       Prob.>t
      X         0.677      0.954      0.184      5.172      0.000
      Z         0.063      0.048      0.102      0.475      0.638
      A1       -0.491     -1.970      0.487     -4.044      0.000
      A2        0.114      0.458      0.472      0.972      0.338
      A3        0.369      1.482      0.470      3.153      0.003
Constant = 10.046
```

Test for Homogeneity of Group Regression Coefficients
Change in R2 = 0.0192. F =   0.308 Prob.> F = 0.9275 with d.f.
6 and   28

Analysis of Variance for the ANCOVA Model

```
   SOURCE        Deg.F.      SS        MS         F        Prob>F
   Explained       5       221.89    44.38     15.087     0.0000
   Error          34       100.01     2.94
   Total          39       321.90
```

Intercepts for Each Group Regression Equation for Variable: Group

Intercepts with 40 valid cases.

```
Variables     Group 1     Group 2     Group 3     Group 4
               8.076      10.505      11.528      10.076
```

Adjusted Group Means for Group Variables Group

Means with 40 valid cases.
```
Variables     Group 1     Group 2     Group 3     Group 4
              15.580      18.008      19.032      17.579
```

```
Multiple Comparisons Among Group Means
Comparison of Group 1 with Group 2
F = 9.549, probability = 0.004 with degrees of freedom 1 and 34
Comparison of Group 1 with Group 3
F = 19.849, probability = 0.000 with degrees of freedom 1 and 34
Comparison of Group 1 with Group 4
F = 1.546, probability = 0.222 with degrees of freedom 1 and 34
Comparison of Group 2 with Group 3
F = 1.770, probability = 0.192 with degrees of freedom 1 and 34
Comparison of Group 2 with Group 4
F = 3.455, probability = 0.072 with degrees of freedom 1 and 34
Comparison of Group 3 with Group 4
F = 9.973, probability = 0.003 with degrees of freedom 1 and 34
```

Test for Each Source of Variance					
SOURCE	Deg.F.	SS	MS	F	Prob>F
A	3	60.98	20.33	6.911	0.0009
Covariates	2	160.91	80.45	27.352	0.0000
Error	34	100.01	2.94		
Total	39	321.90			

The results reported above begin with a regression model that includes group coding for the four groups (A1, A2 and A3) and again note that the fourth group is automatically identified by members NOT being in one of the first three groups. This model also contains the covariates X and Z as well as the cross-products of group membership and covariates. By comparing this model with the second model created (one which leaves out the cross-products of groups and covariates) we can assess the degree to which the assumptions of homogeneity of covariance among the groups is met. In this particular example, the change in the R2 from the full model to the restricted model was quite small (0.0192) and we conclude that the assumption of homogeneity of covariance is reasonable. The analysis of variance model for the restricted model indicates that the X covariate is probably contributing significantly to the explained variance of the dependent variable Y. The tests for each source of variance at the end of the report confirms that not only are the covariates related to Y but that the group effects are also significant. The comparisons of the group means following adjustment for the covariate effects indicate that group 1 differs from groups 2 and 3 and that group 3 appears to differ from group 4.

Sums of Squares by Regression

The General Linear Model (GLM) procedure is an analysis procedure that encompasses a variety of analyses. It may incorporate multiple linear regression as well as canonical correlation analysis as methods for analyzing the user's data. In some commercial statistics packages the GLM method also incorporates non-linear

Sums of Squares by Regression

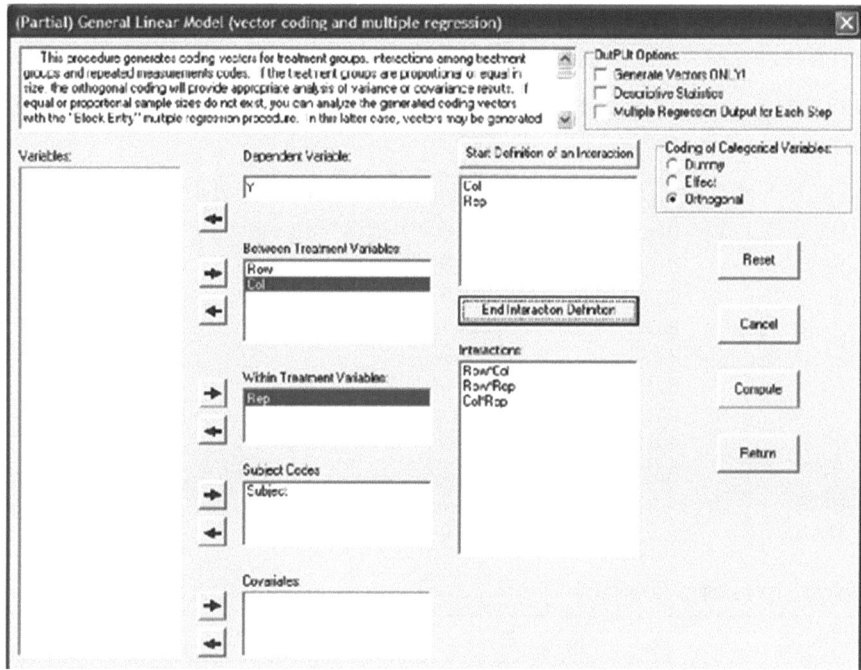

Fig. 8.38 Sum of squares by regression

analyses, maximum-likelihood procedures and a variety of tests not found in the OPENSTAT version of this model. The version in OpenStat is currently limited to a single dependent variable (continuous measure.) You should complete analyses with multiple dependent variables with the Canonical Correlation procedure.

One can complete a variety of analyses of variance with the GLM procedure including multiple factor ANOVA and repeated and mixed model ANOVAs.

The output of the GLM can be somewhat voluminous in that the effects of treatment variables and covariates are analyzed individually by comparing regression models with and without those variables. Several examples are explored below.

When you elect the Sum of Squares by Regression procedure from either the Regression options or the Multivariate options of the Analyses menu, you will see the form shown below. In our first example we will select a data file for completion of a repeated measures analysis of variance that involves two between-groups factors and one within groups factor (the SSRegs2.TAB file.) The data file contains codes for Factor A treatment levels, Factor B treatment levels, the replications factor (Factor C levels), and a code for each subject. In our analysis we will define the two-way and the one three-way interactions that we wish to include in our model. We should then be able to compare our results with the Repeated Measures ANOVA procedure applied to the same data in the file labeled ABRData.TAB (and hopefully see the same results!) (Fig. 8.38).

SUMS OF SQUARES AND MEAN SQUARES BY REGRESSION

TYPE III SS - R2 = Full Model - Restricted Model

VARIABLE	SUM OF SQUARES	D.F.
Row1	10.083	1
Col1	8.333	1
Rep1	150.000	1
Rep2	312.500	1
Rep3	529.000	1
C1R1	80.083	1
R1R1	0.167	1
R2R1	2.000	1
R3R1	6.250	1
R1C1	4.167	1
R2C1	0.889	1
R3C1	7.111	1
ERROR	147.417	35
TOTAL	1258.000	47

TOTAL EFFECTS SUMMARY

SOURCE	SS	D.F.	MS
Row	10.083	1	10.083
Col	8.333	1	8.333
Rep	991.500	3	330.500
Row*Col	80.083	1	80.083
Row*Rep	8.417	3	2.806
Col*Rep	12.167	3	4.056

SOURCE	SS	D.F.	MS
BETWEEN SUBJECTS	181.000	11	
Row	10.083	1	10.083
Col	8.333	1	8.333
Row*Col	80.083	1	80.083
ERROR BETWEEN	82.500	8	10.312
WITHIN SUBJECTS	1077.000	36	
Rep	991.500	3	330.500
Row*Rep	8.417	3	2.806
Col*Rep	12.167	3	4.056
ERROR WITHIN	64.917	27	2.404
TOTAL	1258.000		

Sums of Squares by Regression

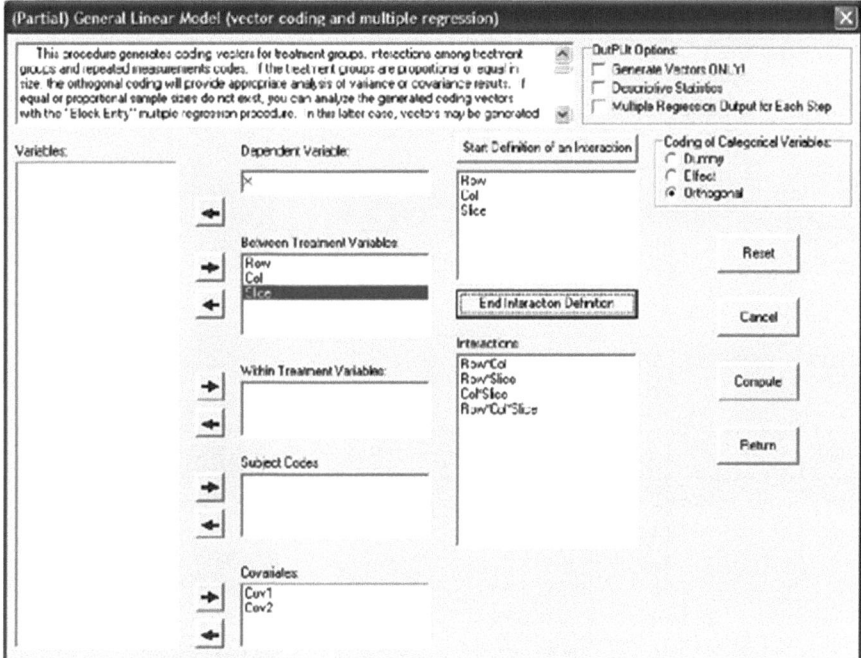

Fig. 8.39 Example 2 of sum of squares by regression

You can compare the results above with an analysis completed with the Repeated Measures procedure.

As a second example, we will complete and analysis of covariance on data that contains three treatment factors and two covariates. The file analyzed is labeled ANCOVA3.TAB. Shown above is the dialog for the analysis (Fig. 8.39) followed by the output. You can compare this output with the output obtained by analyzing the same data file with the Analysis of Covariance procedure.

```
SUMS OF SQUARES AND MEAN SQUARES BY REGRESSION

TYPE III SS - R2 = Full Model - Restricted Model

VARIABLE              SUM OF SQUARES      D.F.
        Cov1              1.275            1
        Cov2              0.783            1
        Row1             25.982            1
        Col1             71.953            1
        Slice1           13.323            1
        Slice2            0.334            1
```

```
            C1R1            21.240              1
            S1R1            11.807              1
            S2R1             0.138              1
            S1C1            13.133              1
            S2C1             0.822              1
          S1C1R1             0.081              1
          S2C1R1            47.203              1
ERROR                       46.198             58
TOTAL                      269.500             71
```

TOTAL EFFECTS SUMMARY

```
----------------------------------------------------------------
SOURCE                       SS         D.F.           MS
----------------------------------------------------------------
             Cov1          1.275          1          1.275
             Cov2          0.783          1          0.783
              Row         25.982          1         25.982
              Col         71.953          1         71.953
            Slice         13.874          2          6.937
          Row*Col         21.240          1         21.240
        Row*Slice         11.893          2          5.947
        Col*Slice         14.204          2          7.102
    Row*Col*Slice         47.247          2         23.624
----------------------------------------------------------------


----------------------------------------------------------------
SOURCE                       SS         D.F.           MS
----------------------------------------------------------------
BETWEEN SUBJECTS          208.452         13
       Covariates           2.058          2          1.029
              Row          25.982          1         25.982
              Col          71.953          1         71.953
            Slice          13.874          2          6.937
          Row*Col          21.240          1         21.240
        Row*Slice          11.893          2          5.947
        Col*Slice          14.204          2          7.102
    Row*Col*Slice          47.247          2         23.624
ERROR BETWEEN              46.198         58          0.797
----------------------------------------------------------------


----------------------------------------------------------------
TOTAL                     269.500         71
----------------------------------------------------------------

================================================================
```

The General Linear Model

We have seen in the above discussion that the multiple regression method may be used to complete an analysis of variance for a single dependent variable. The model for multiple regression is:

$$y_i = \sum_{j=1}^{k} B_j X_j + e_i$$

where the jth B value is a coefficient multiplied times the jth independent predictor score, Y is the observed dependent score and e is the error (difference between the observed and the value predicted for Y using the sum of weighted independent scores.
In some research it is desirable to determine the relationship between multiple dependent variables and multiple independent variables. Of course, one could complete a multiple regression analysis for each dependent variable but this would ignore the possible relationships among the dependent variables themselves. For example, a teacher might be interested in the relationship between the sub-scores on a standardized achievement test (independent variables) and the final examination results for several different courses (dependent variables.) Each of the final examination scores could be predicted by the sub-scores in separate analyses but most likely the interest is in knowing how well the sub-scores account for the combined variance of the achievement scores. By assigning weights to each of the dependent variables as well as the independent variables in such a way that the composite dependent score is maximally related to the composite independent score we can quantify the relationship between the two composite scores. We note that the squared product–moment correlation coefficient reflects the proportion of variance of a dependent variable predicted by the independent variable.
We can express the model for the general linear model as:

$$YM = BX + E$$

where Y is an n (the number of subjects) by m (the number of dependent variables) matrix of dependent variable values, M is a m by s (number of coefficient sets), X is a n by k (the number of independent variables) matrix, B is a k by s matrix of coefficients and E is a vector of errors for the n subjects.

Using OpenStat to Obtain Canonical Correlations

You can use the OpenStat package to obtain canonical correlations for a wide variety of applications. In production of bread, for example, a number of "dependent" quality variables may exist such as the average size of air bubbles in a slice, the density of a slice, the thickness of the crust, etc. Similarly, there are a number of "independent" variables which may be related to the dependent variables with

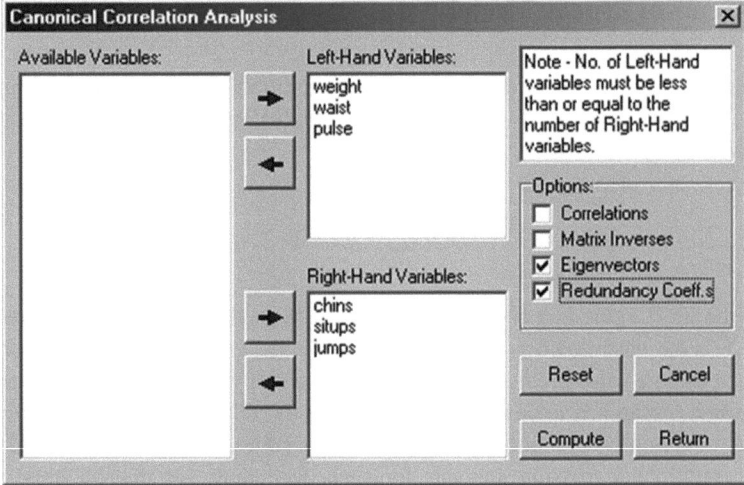

Fig. 8.40 Canonical Correlation Analysis form

examples being minutes of baking, temperature of baking, humidity in the oven, barometric pressure, time and temperature during rising of the dough, etc. The relationship between these two sets of variables might identify the "key" variables to control for maximizing the quality of the product.

To demonstrate use of OpenStat to obtain canonical correlations we will use the file labeled "cansas.txt" as an example. We will click on the Canonical Correlation option under the Correlation sub-menu of the Statistics menu. In the Figure above we show the form which appears and the data entered to initiate the analysis (Fig. 8.40):

We obtain the results as shown below:

```
CANONICAL CORRELATION ANALYSIS

Right Inverse x Right-Left Matrix with 20 valid cases.

Variables
            weight      waist       pulse
    chins   -0.102     -0.226       0.001
   situps   -0.552     -0.788       0.365
    jumps    0.193      0.448      -0.210

Left Inverse x Left-Right Matrix with 20 valid cases.

Variables
            chins      situps       jumps
   weight   0.368       0.287      -0.259
    waist  -0.882      -0.890       0.015
    pulse  -0.026       0.016      -0.055
```

Using OpenStat to Obtain Canonical Correlations

```
Canonical Function with 20 valid cases.
Variables
              Var. 1      Var. 2       Var. 3
    Var. 1    0.162       0.172        0.023
    Var. 2    0.482       0.549        0.111
    Var. 3   -0.318      -0.346       -0.032


Trace of the matrix:= 0.6785
Percent of trace extracted: 100.0000

       Canonical R    Root     % Trace    Chi-Sqr    D.F.    Prob.
    2  0.795608       0.633    93.295     16.255     9       0.062
    3  0.200556       0.040     5.928      0.718     4       0.949
    4  0.072570       0.005     0.776      0.082     1       0.775

Overall Tests of Significance:
         Statistic          Approx. Stat.  Value    D.F.   Prob.>Value
Wilk's Lambda                Chi-Squared   17.3037  9      0.0442
Hotelling-Lawley Trace       F-Test         2.4938  9 38   0.0238
Pillai Trace                 F-Test         1.5587  9 48   0.1551
Roys Largest Root            F-Test        10.9233  3 19   0.0002

Eigenvectors with 20 valid cases.

Variables
              Var. 1      Var. 2       Var. 3
    Var. 1    0.210      -0.066        0.051
    Var. 2    0.635       0.022       -0.049
    Var. 3   -0.431       0.188        0.017

Standardized Right Side Weights with 20 valid cases.

Variables
              Var. 1      Var. 2       Var. 3
   weight     0.775      -1.884        0.191
    waist    -1.579       1.181       -0.506
    pulse     0.059      -0.231       -1.051

Standardized Left Side Weights with 20 valid cases.

Variables
              Var. 1      Var. 2       Var. 3
    chins     0.349      -0.376        1.297
   situps     1.054       0.123       -1.237
    jumps    -0.716       1.062        0.419

Standardized Right Side Weights with 20 valid cases.
```

```
Variables
            Var. 1      Var. 2      Var. 3
   weight    0.775      -1.884       0.191
    waist   -1.579       1.181      -0.506
    pulse    0.059      -0.231      -1.051
```

Raw Right Side Weights with 20 valid cases.

```
Variables
            Var. 1      Var. 2      Var. 3
   weight    0.031      -0.076       0.008
    waist   -0.493       0.369      -0.158
    pulse    0.008      -0.032      -0.146
```

Raw Left Side Weights with 20 valid cases.

```
Variables
            Var. 1      Var. 2      Var. 3
    chins    0.066      -0.071       0.245
   situps    0.017       0.002      -0.020
    jumps   -0.014       0.021       0.008
```

Right Side Correlations with Function with 20 valid cases.

```
Variables
            Var. 1      Var. 2      Var. 3
   weight   -0.621      -0.772       0.135
    waist   -0.925      -0.378       0.031
    pulse    0.333       0.041      -0.942
```

Left Side Correlations with Function with 20 valid cases.

```
Variables
            Var. 1      Var. 2      Var. 3
    chins    0.728       0.237       0.644
   situps    0.818       0.573      -0.054
    jumps    0.162       0.959       0.234
```

Redundancy Analysis for Right Side Variables

	Variance Prop.	Redundancy
1	0.45080	0.28535
2	0.24698	0.00993
3	0.30222	0.00159

Redundancy Analysis for Left Side Variables

	Variance Prop.	Redundancy
1	0.40814	0.25835
2	0.43449	0.01748
3	0.15737	0.00083

Binary Logistic Regression

When this analysis is selected from the menu, the form below is used to select the dependent and independent variables (Fig. 8.41):

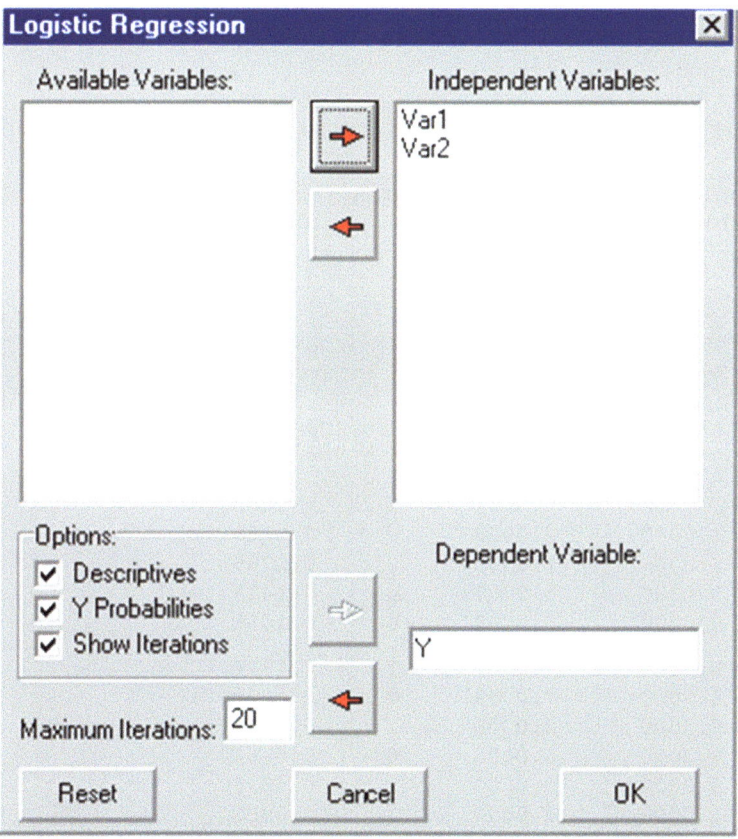

Fig. 8.41 Logistic Regression form

Output for the example analysis specified above is shown below:

```
Logistic Regression Adapted from John C. Pezzullo
Java program at http://members.aol.com/johnp71/logistic.html

Descriptive Statistics
6 cases have Y=0; 4 cases have Y=1.
Variable     Label         Average        Std.Dev.
   1          Var1          5.5000         2.8723
   2          Var2          5.5000         2.8723

Iteration History
-2 Log Likelihood =      13.4602 (Null Model)
-2 Log Likelihood =       8.7491
-2 Log Likelihood =       8.3557
-2 Log Likelihood =       8.3302
-2 Log Likelihood =       8.3300
-2 Log Likelihood =       8.3300
Converged

Overall Model Fit... Chi Square = 5.1302 with df = 2 and prob. = 0.0769

Coefficients and Standard Errors...
Variable      Label      Coeff.       StdErr         p
   1           Var1      0.3498       0.6737       0.6036
   2           Var2      0.3628       0.6801       0.5937
Intercept     -4.6669

Odds Ratios and 95% Confidence Intervals...
Variable              O.R.         Low     --    High
             Var1    1.4187       0.3788         5.3135
             Var2    1.4373       0.3790         5.4506

         X             X          Y         Prob
      1.0000        2.0000        0        0.0268
      2.0000        1.0000        0        0.0265
      3.0000        5.0000        0        0.1414
      4.0000        3.0000        0        0.1016

      5.0000        4.0000        1        0.1874
      6.0000        7.0000        0        0.4929
      7.0000        8.0000        1        0.6646
      8.0000        6.0000        0        0.5764
      9.0000       10.0000        1        0.8918
     10.0000        9.0000        1        0.8905
```

Cox Proportional Hazards Survival Regression

The specification form for this analysis is shown below with variables entered for a sample file (Fig. 8.42):

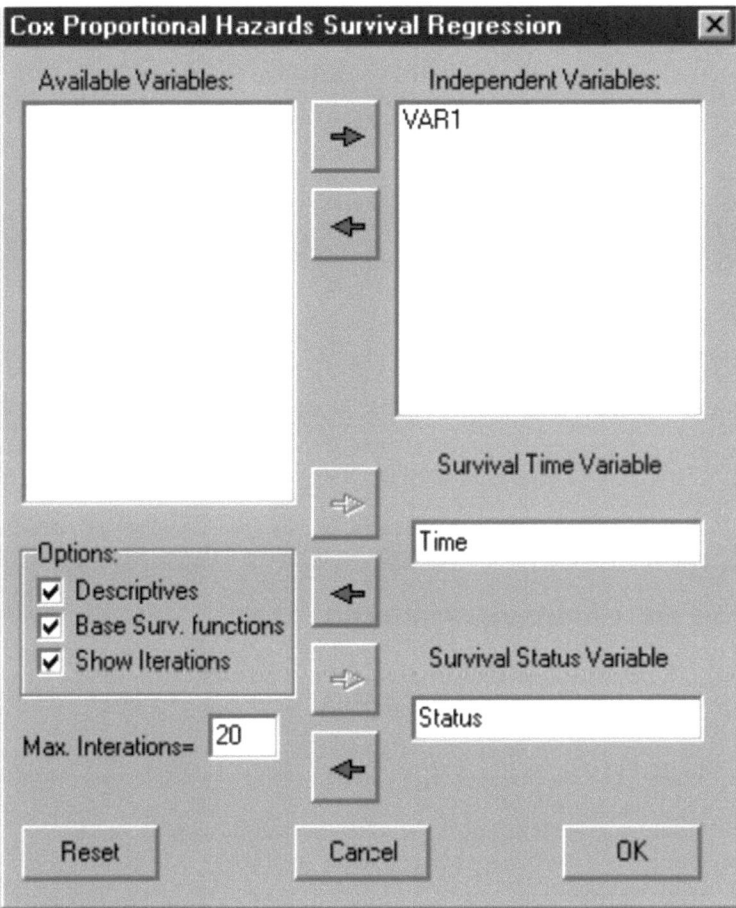

Fig. 8.42 Cox Proportional Hazards Survival Regression form

Results for the above sample are as follows:

```
Cox Proportional Hazards Survival Regression Adapted from John
C. Pezzullo's Java program at http://members.aol.com/johnp71/
prophaz.html

Descriptive Statistics
Variable      Label      Average      Std.Dev.
   1          VAR1       51.1818      10.9778
Iteration History...
-2 Log Likelihood =      11.4076 (Null Model)

-2 Log Likelihood =      6.2582
-2 Log Likelihood =      4.5390
-2 Log Likelihood =      4.1093
-2 Log Likelihood =      4.0524
-2 Log Likelihood =      4.0505
-2 Log Likelihood =      4.0505
Converged


Overall Model Fit...
Chi Square =    7.3570 with d.f. 1 and probability =    0.0067

Coefficients, Std Errs, Signif, and Confidence Intervals

Var             Coeff.    StdErr     p         Lo95%      Hi95%
      VAR1      0.3770    0.2542     0.1379    -0.1211    0.8752

Risk Ratios and Confidence Intervals

Variable       Risk Ratio    Lo95%     Hi95%
      VAR1     1.4580        0.8859    2.3993

Baseline Survivor Function (at predictor means)...
    2.0000     0.9979
    7.0000     0.9820
    9.0000     0.9525
   10.0000     0.8310
```

Weighted Least-Squares Regression

Shown below is the dialog box for the Weighted Least Squares Analysis and an analysis of the cansas.tab data file (Figs. 8.43, 8.44, 8.45).

Weighted Least-Squares Regression

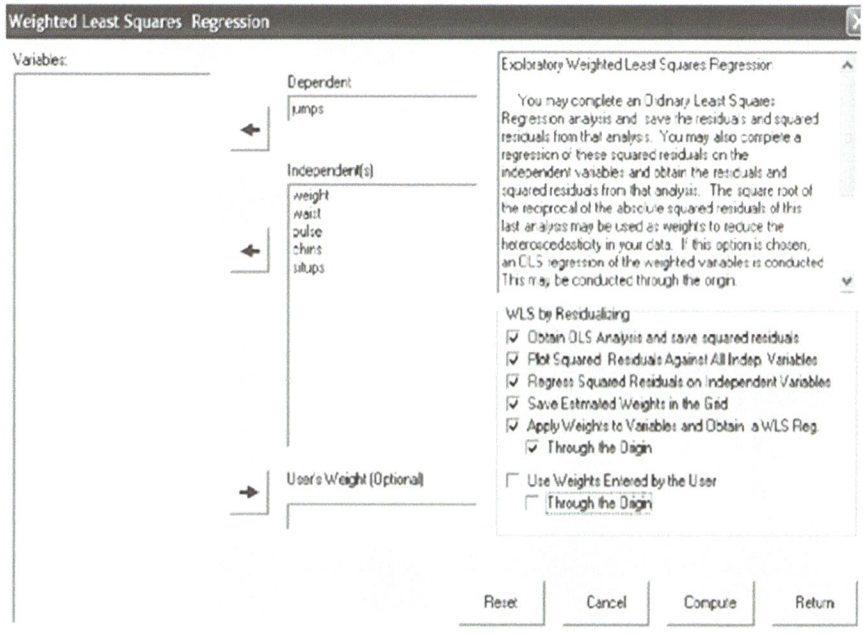

Fig. 8.43 Weighted least squares regression

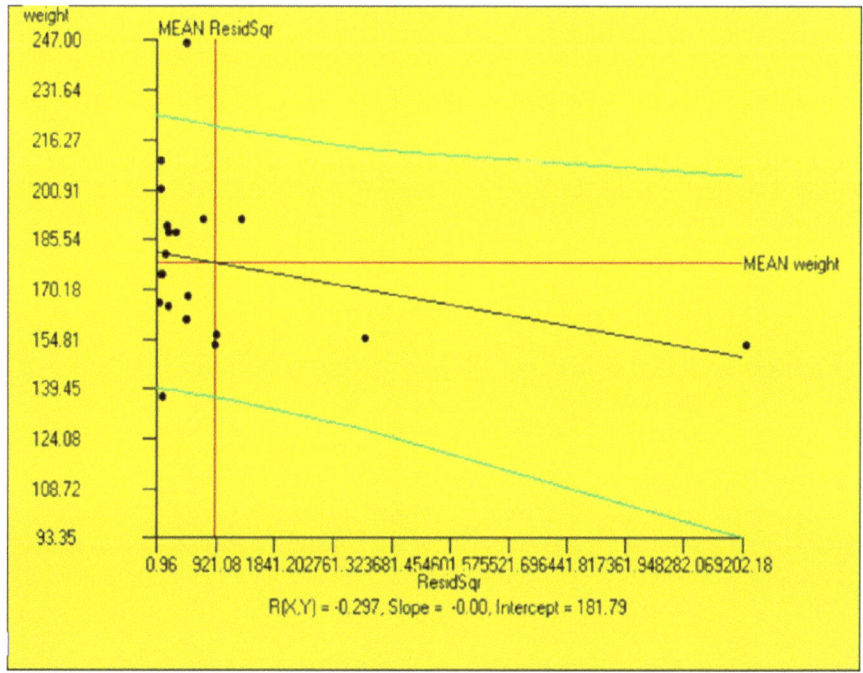

Fig. 8.44 Plot of ordinary least squares regression

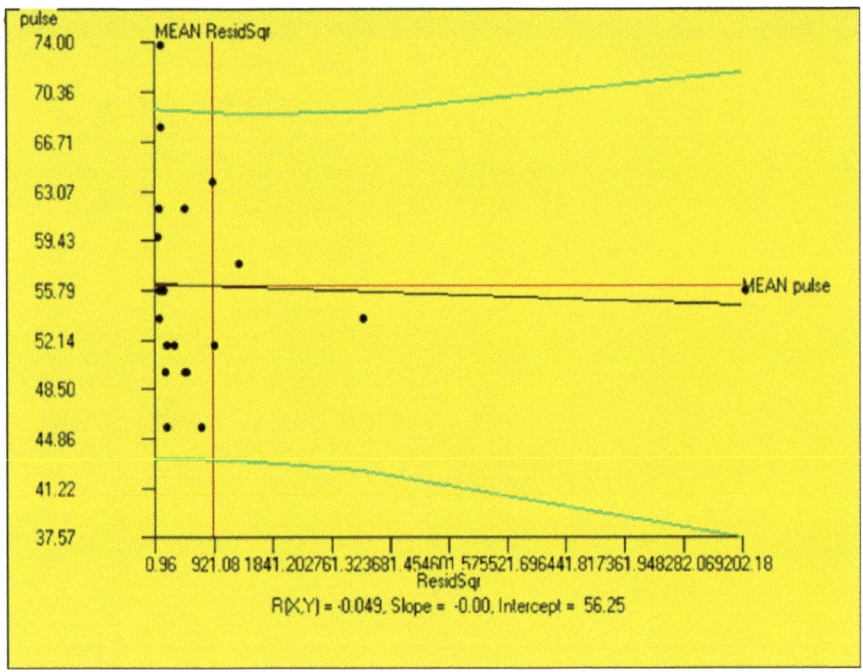

Fig. 8.45 Plot of weighted least squares regression

```
OLS REGRESSION RESULTS
Means
Variables   weight    waist    pulse    chins    situps   jumps
            178.600   35.400   56.100   9.450    145.550  70.300

Standard Deviations
Variables   weight    waist    pulse    chins    situps   jumps
            24.691    3.202    7.210    5.286    62.567   51.277

No. of valid cases = 20

CORRELATION MATRIX
            VARIABLE
            weight    waist    pulse    chins    situps   jumps
weight      1.000     0.870    -0.366   -0.390   -0.493   -0.226
waist       0.870     1.000    -0.353   -0.552   -0.646   -0.191
pulse       -0.366    -0.353   1.000    0.151    0.225    0.035
chins       -0.390    -0.552   0.151    1.000    0.696    0.496
situps      -0.493    -0.646   0.225    0.696    1.000    0.669
jumps       -0.226    -0.191   0.035    0.496    0.669    1.000

Dependent variable: jumps
```

Weighted Least-Squares Regression

```
Variable      Beta          B    Std.Err.         t  Prob.>t     VIF    TOL
weight      -0.588     -1.221       0.704    -1.734    0.105   4.424  0.226
waist        0.982     15.718       6.246     2.517    0.025   5.857  0.171
pulse       -0.064     -0.453       1.236    -0.366    0.720   1.164  0.859
chins        0.201      1.947       2.243     0.868    0.400   2.059  0.486
situps       0.888      0.728       0.205     3.546    0.003   2.413  0.414
Intercept    0.000   -366.967     183.214    -2.003    0.065

SOURCE           DF            SS           MS            F       Prob.>F
Regression        5     31793.741     6358.748        4.901        0.0084
Residual         14     18164.459     1297.461
Total            19     49958.200

R2 = 0.6364, F =      4.90, D.F. = 5 14, Prob>F = 0.0084
Adjusted R2 = 0.5066

Standard Error of Estimate = 36.02

REGRESSION OF SQUARED RESIDUALS ON INDEPENDENT VARIABLES

Means
Variables    weight      waist     pulse     chins    situps    ResidSqr
            178.600     35.400    56.100     9.450   145.550     908.196

Standard Deviations

Variables    weight      waist     pulse     chins    situps    ResidSqr
             24.691      3.202     7.210     5.286    62.567    2086.828

No. of valid cases = 20

CORRELATION MATRIX
             VARIABLE
             weight      waist     pulse     chins    situps    ResidSqr
weight        1.000      0.870    -0.366    -0.390    -0.493     -0.297
waist         0.870      1.000    -0.353    -0.552    -0.646     -0.211
pulse        -0.366     -0.353     1.000     0.151     0.225     -0.049
chins        -0.390     -0.552     0.151     1.000     0.696      0.441
situps       -0.493     -0.646     0.225     0.696     1.000      0.478
ResidSqr     -0.297     -0.211    -0.049     0.441     0.478      1.000

Dependent variable: ResidSqr

Variable      Beta           B    Std.Err.         t  Prob.>t     VIF    TOL
weight      -0.768     -64.916      36.077    -1.799    0.094   4.424  0.226
waist        0.887     578.259     320.075     1.807    0.092   5.857  0.171
pulse       -0.175     -50.564      63.367    -0.798    0.438   1.164  0.859
chins        0.316     124.826     114.955     1.086    0.296   2.059  0.486
situps       0.491      16.375      10.515     1.557    0.142   2.413  0.414
Intercept    0.000   -8694.402    9389.303    -0.926    0.370
```

```
SOURCE       DF       SS            MS          F         Prob.>F
Regression    5   35036253.363  7007250.673   2.056       0.1323
Residual     14   47705927.542  3407566.253
Total        19   82742180.905


R2 = 0.4234, F =     2.06, D.F. = 5 14, Prob>F = 0.1323
Adjusted R2 = 0.2175


Standard Error of Estimate =   1845.96
X versus Y Plot
X = ResidSqr, Y = weight from file: C:\Documents and Settings\
Owner\My Documents\Projects\Clanguage\OpenStat\cansaswls.TAB


Variable    Mean     Variance   Std.Dev.
ResidSqr    908.20   4354851.63  2086.83
weight      178.60      609.62    24.69
Correlation = -0.2973, Slope =    -0.00, Intercept =    181.79
Standard Error of Estimate =     23.57
Number of good cases = 20

WLS REGRESSION RESULTS
Means
Variables   weight    waist    pulse    chins    situps    jumps
            -0.000    0.000   -0.000    0.000   -0.000    0.000

Standard Deviations
Variables   weight    waist    pulse    chins    situps    jumps
             7.774    1.685    2.816    0.157    3.729    1.525

No. of valid cases = 20

CORRELATION MATRIX
            VARIABLE
            weight    waist    pulse    chins    situps    jumps
weight       1.000    0.994    0.936    0.442    0.742    0.697
waist        0.994    1.000    0.965    0.446    0.783    0.729
pulse        0.936    0.965    1.000    0.468    0.889    0.769
chins        0.442    0.446    0.468    1.000    0.395    0.119
situps       0.742    0.783    0.889    0.395    1.000    0.797
jumps        0.697    0.729    0.769    0.119    0.797    1.000

Dependent variable: jumps

Variable    Beta         B    Std.Err.       t  Prob.>t      VIF    TOL
weight    -2.281    -0.448    0.414    -1.082    0.298  253.984  0.004
waist      3.772     3.415    2.736     1.248    0.232  521.557  0.002
pulse     -1.409    -0.763    0.737    -1.035    0.318  105.841  0.009
chins     -0.246    -2.389    1.498    -1.594    0.133    1.363  0.734
situps     0.887     0.363    0.165     2.202    0.045    9.258  0.108
Intercept  0.000    -0.000    0.197    -0.000    1.000
```

```
SOURCE          DF         SS           MS           F         Prob.>F
Regression      5          33.376       6.675        8.624     0.0007
Residual        14         10.837       0.774
Total           19         44.212
R2 = 0.7549, F =       8.62, D.F. = 5 14, Prob>F = 0.0007
Adjusted R2 = 0.6674
Standard Error of Estimate =        0.88
```

2-Stage Least-Squares Regression

In the following example, the cansas.TAB file is analyzed. The dependent variable is the height of individual jumps. The explanatory (predictor) variables are pulse rate, no. of chinups and no. of situps the individual completes. These explanatory variables are thought to be related to the instrumental variables of weight and waist size. In the dialog box for the analysis, the option has been selected to show the regression for each of the explanatory variables that produces the predicted variables to be used in the final analysis. Results are shown below (Fig. 8.46):

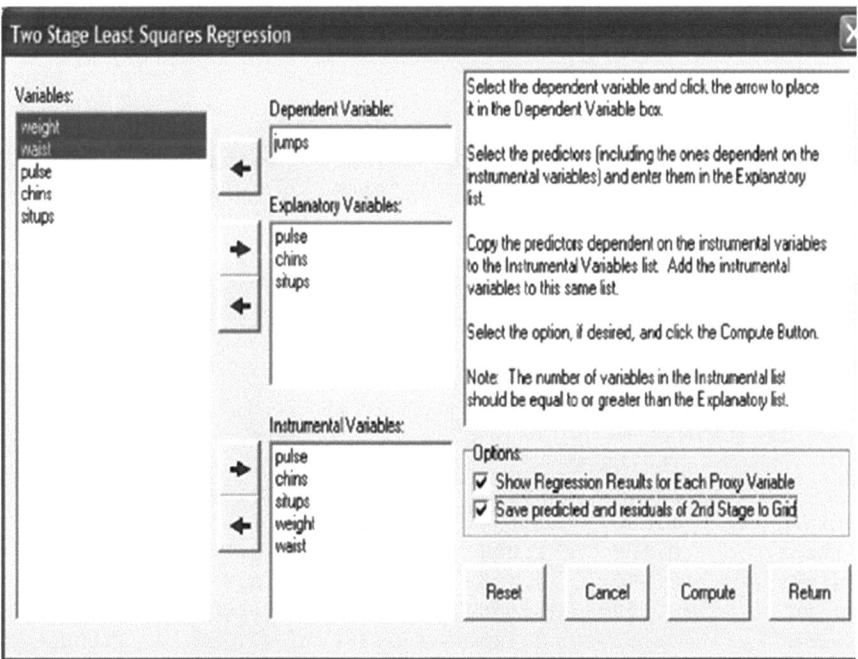

Fig. 8.46 Two Stage Least Squares Regression form

```
FILE: C:\Documents and Settings\Owner\My Documents\Projects\
Clanguage\OpenStat\cansas.TAB
Dependent = jumps
Explanatory Variables:
pulse
chins
situps
Instrumental Variables:
pulse
chins
situps
weight
waist
Proxy Variables:
P_pulse
P_chins
P_situps
```

Analysis for P_pulse

```
Dependent: pulse
Independent:
chins
situps
weight
waist
```

Means

Variables	chins	situps	weight	waist	pulse
	9.450	145.550	178.600	35.400	56.100

Standard Deviations

Variables	chins	situps	weight	waist	pulse
	5.286	62.567	24.691	3.202	7.210

No. of valid cases = 20

CORRELATION MATRIX
VARIABLE

	chins	situps	weight	waist	pulse
chins	1.000	0.696	-0.390	-0.552	0.151
situps	0.696	1.000	-0.493	-0.646	0.225
weight	-0.390	-0.493	1.000	0.870	-0.366
waist	-0.552	-0.646	0.870	1.000	-0.353
pulse	0.151	0.225	-0.366	-0.353	1.000

Dependent variable: pulse

```
Variable      Beta       B    Std.Err.        t    Prob.>t     VIF     TOL
chins       -0.062   -0.084      0.468   -0.179      0.860   2.055   0.487
situps       0.059    0.007      0.043    0.158      0.876   2.409   0.415
weight      -0.235   -0.069      0.146   -0.471      0.644   4.360   0.229
waist       -0.144   -0.325      1.301   -0.249      0.806   5.832   0.171
Intercept    0.000   79.673     32.257    2.470      0.026


SOURCE          DF          SS         MS           F       Prob.>F
Regression       4     139.176     34.794       0.615       0.6584
Residual        15     848.624     56.575
Total           19     987.800


R2 = 0.1409, F =      0.62, D.F. = 4 15, Prob>F = 0.6584
Adjusted R2 = -0.0882
Standard Error of Estimate =      7.52
```

Analysis for P_chins

```
Dependent: chins
Independent:
pulse
situps
weight
waist

Means
Variables       pulse     situps     weight      waist      chins
               56.100    145.550    178.600     35.400      9.450

Standard Deviations
Variables       pulse     situps     weight      waist      chins
                7.210     62.567     24.691      3.202      5.286

No. of valid cases = 20

CORRELATION MATRIX
           VARIABLE
            pulse     situps     weight      waist      chins
pulse       1.000      0.225     -0.366     -0.353      0.151
situps      0.225      1.000     -0.493     -0.646      0.696
weight     -0.366     -0.493      1.000      0.870     -0.390
waist      -0.353     -0.646      0.870      1.000     -0.552
chins       0.151      0.696     -0.390     -0.552      1.000

Dependent variable: chins
```

Variable	Beta	B	Std.Err.	t	Prob.>t	VIF	TOL
pulse	-0.035	-0.026	0.142	-0.179	0.860	1.162	0.861
situps	0.557	0.047	0.020	2.323	0.035	1.775	0.564
weight	0.208	0.045	0.080	0.556	0.586	4.335	0.231
waist	-0.386	-0.638	0.700	-0.911	0.377	5.549	0.180
Intercept	0.000	18.641	20.533	0.908	0.378		

SOURCE	DF	SS	MS	F	Prob.>F
Regression	4	273.089	68.272	3.971	0.0216
Residual	15	257.861	17.191		
Total	19	530.950			

R2 = 0.5143, F =   3.97, D.F. = 4 15, Prob>F = 0.0216
Adjusted R2 = 0.3848
Standard Error of Estimate =   4.15

Analysis for P_situps

Dependent: situps
Independent:
pulse
chins
weight
waist

Means
Variables	pulse	chins	weight	waist	situps
	56.100	9.450	178.600	35.400	145.550

Standard Deviations
Variables	pulse	chins	weight	waist	situps
	7.210	5.286	24.691	3.202	62.567

No. of valid cases = 20

CORRELATION MATRIX
VARIABLE
	pulse	chins	weight	waist	situps
pulse	1.000	0.151	-0.366	-0.353	0.225
chins	0.151	1.000	-0.390	-0.552	0.696
weight	-0.366	-0.390	1.000	0.870	-0.493
waist	-0.353	-0.552	0.870	1.000	-0.646
situps	0.225	0.696	-0.493	-0.646	1.000

Dependent variable: situps
Variable	Beta	B	Std.Err.	t	Prob.>t	VIF	TOL
pulse	0.028	0.246	1.555	0.158	0.876	1.162	0.861
chins	0.475	5.624	2.421	2.323	0.035	1.514	0.660
weight	0.112	0.284	0.883	0.322	0.752	4.394	0.228
waist	-0.471	-9.200	7.492	-1.228	0.238	5.322	0.188
Intercept	0.000	353.506	211.726	1.670	0.116		

2-Stage Least-Squares Regression

```
SOURCE         DF        SS          MS         F         Prob.>F
Regression      4    43556.048   10889.012    5.299       0.0073
Residual       15    30820.902    2054.727
Total          19    74376.950
```

R2 = 0.5856, F =   5.30, D.F. = 4 15, Prob>F = 0.0073
Adjusted R2 = 0.4751
Standard Error of Estimate =   45.33

Second Stage (Final) Results

Means
```
Variables     P_pulse    P_chins    P_situps    jumps
               56.100      9.450     145.550    70.300
```

Standard Deviations
```
Variables     P_pulse    P_chins    P_situps    jumps
                2.706      3.791      47.879    51.277
```

No. of valid cases = 20

CORRELATION MATRIX
```
              VARIABLE
              P_pulse    P_chins    P_situps    jumps
P_pulse        1.000      0.671      0.699      0.239
P_chins        0.671      1.000      0.847      0.555
P_situps       0.699      0.847      1.000      0.394
jumps          0.239      0.555      0.394      1.000
```

Dependent variable: jumps

```
Variable      Beta        B      Std.Err.       t     Prob.>t    VIF    TOL
P_pulse      -0.200    -3.794     5.460      -0.695    0.497   2.041  0.490
P_chins       0.841    11.381     5.249       2.168    0.046   3.701  0.270
P_situps     -0.179    -0.192     0.431      -0.445    0.662   3.979  0.251
Intercept     0.000   203.516   277.262       0.734    0.474
```

```
SOURCE         DF        SS          MS         F         Prob.>F
Regression      3    17431.811    5810.604    2.858       0.0698
Residual       16    32526.389    2032.899
Total          19    49958.200
```

R2 = 0.3489, F =   2.86, D.F. = 3 16, Prob>F = 0.0698
Adjusted R2 = 0.2269
Standard Error of Estimate =   45.09

Non-linear Regression

As an example, I have created a "parabola" function data set labeled parabola.TAB. To generate this file I used the equation y = a + b * x + c * x * x. I let a = 0, b = 5 and c = 2 for the parameters and used a sequence of x values for the independent variables in the data file that was generated. To test the non-linear fit program, I initiated the procedure and entered the values shown below (Fig. 8.47):

You can see that y is the dependent variable and x is the independent variable. Values of 1 have been entered for the initial estimates of a, b and c. The equation model was selected by clicking the parabola model from the drop-down models box. I could have entered the same equation by clicking on the equation box and typing the equation into that box or clicking parameters, math functions and variables from the drop-down boxes on the right side of the form. Notice that I selected to plot the x versus y values and also the predicted versus observed y values. I also chose to save the predicted scores and residuals (y - predicted y.) The results are as follows (Fig. 8.48):

The printed output shown below gives the model selected followed by the individual data points observed, their predicted scores, the residual, the standard error of estimate of the predicted score and the 95% confidence interval of the predicted score. These are followed by the obtained correlation coefficient and its square, root mean square of the y scores, the parameter estimates with their confidence limits and t probability for testing the significance of difference from zero (Fig. 8.49).

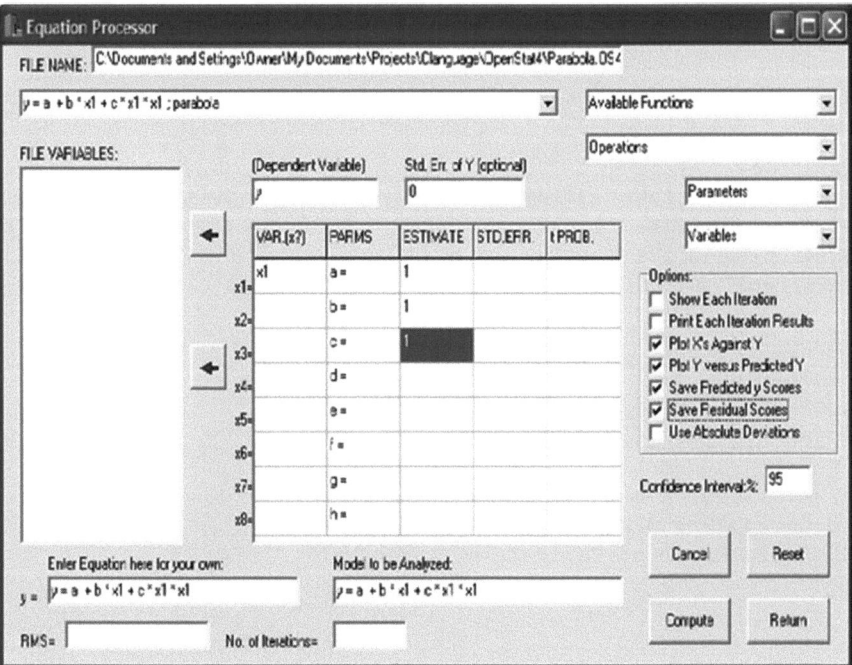

Fig. 8.47 Non-linear Regression Specifications form

Non-linear Regression

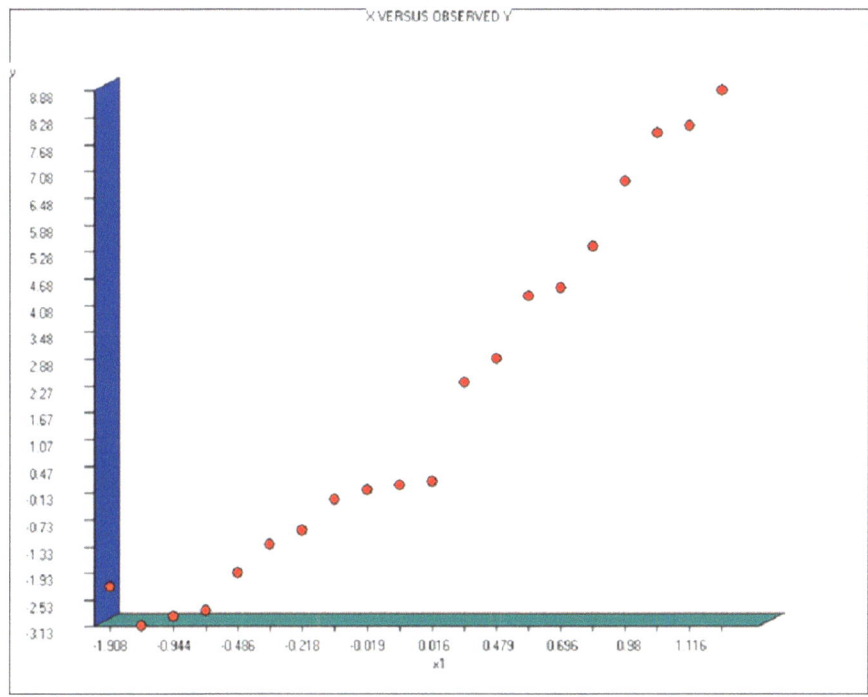

Fig. 8.48 Scores predicted by non-linear regression versus observed scores

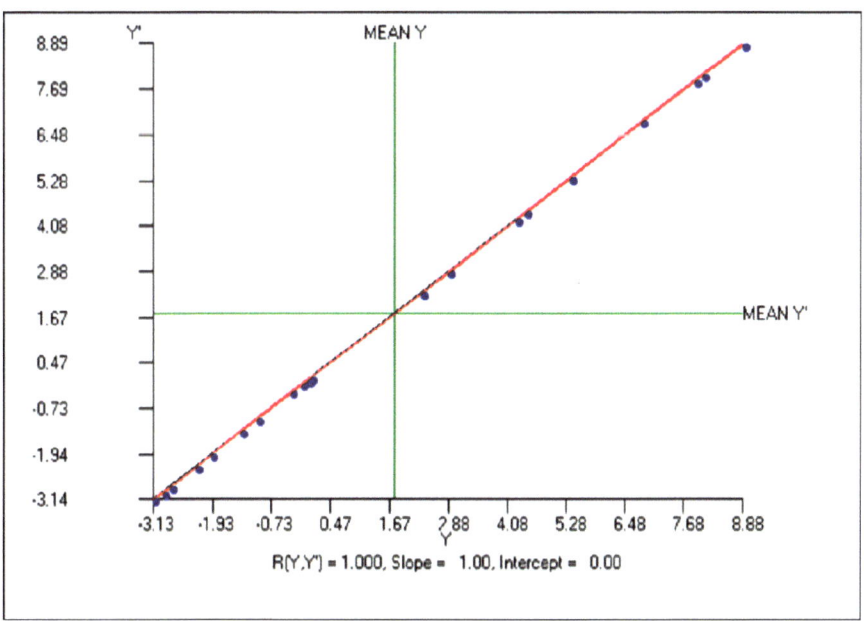

Fig. 8.49 Correlation plot between scores predicted by non-linear regression and observed scores

y = a + b * x1 + c * x1 * x1

x	y	yc	y-yc	SEest	YcLo	YcHi
0.39800	2.31000	2.30863	0.00137	0.00161	2.30582	2.31143
-1.19700	-3.13000	-3.12160	-0.00840	0.00251	-3.12597	-3.11723
-0.48600	-1.95000	-1.95878	0.00878	0.00195	-1.96218	-1.95538
-1.90800	-2.26000	-2.26113	0.00113	0.00522	-2.27020	-2.25205
-0.84100	-2.79000	-2.79228	0.00228	0.00206	-2.79586	-2.78871
-0.30100	-1.32000	-1.32450	0.00450	0.00192	-1.32784	-1.32115
0.69600	4.44000	4.45208	-0.01208	0.00168	4.44917	4.45500
1.11600	8.08000	8.07654	0.00346	0.00264	8.07195	8.08112
0.47900	2.86000	2.85607	0.00393	0.00159	2.85330	2.85884
1.09900	7.92000	7.91612	0.00388	0.00258	7.91164	7.92061
-0.94400	-2.94000	-2.93971	-0.00029	0.00214	-2.94343	-2.93600
-0.21800	-0.99000	-0.99541	0.00541	0.00190	-0.99872	-0.99211
0.81000	5.37000	5.36605	0.00395	0.00183	5.36288	5.36923
-0.06200	-0.31000	-0.30228	-0.00772	0.00185	-0.30549	-0.29907
0.67200	4.26000	4.26629	-0.00629	0.00165	4.26342	4.26917
-0.01900	-0.10000	-0.09410	-0.00590	0.00183	-0.09728	-0.09093
0.00100	0.01000	0.00525	0.00475	0.00182	0.00209	0.00841
0.01600	0.08000	0.08081	-0.00081	0.00181	0.07766	0.08396
1.19900	8.88000	8.87635	0.00365	0.00295	8.87122	8.88148
0.98000	6.82000	6.82561	-0.00561	0.00221	6.82177	6.82945

Corr. Coeff. =   1.00000 R2 =   1.00000

RMS Error =   5.99831, d.f. = 17   SSq =   611.65460

Parameter Estimates …
p1=   0.00024 +/-   0.00182 p=   0.89626
p2=   5.00349 +/-   0.00171 p=   0.00000
p3=   2.00120 +/-   0.00170 p=   0.00000

Covariance Matrix Terms and Error-Correlations…

B(1,1)=   0.00000; r=   1.00000
B(1,2)=B(2,1)=   -0.00000; r= -0.28318
B(1,3)=B(3,1)=   -0.00000; r= -0.67166
B(2,2)=   0.00000; r=   1.00000
B(2,3)=B(3,2)=   0.00000; r=   0.32845
B(3,3)=   0.00000; r=   1.00000

X versus Y Plot

Non-linear Regression

```
X = Y, Y = Y' from file: C:\Documents and Settings\Owner\My
Documents\Projects\Clanguage\OpenStat\Parabola.TAB

Variable     Mean    Variance  Std.Dev.
Y            1.76    16.29     4.04
Y'           1.76    16.29     4.04

Correlation = 1.0000, Slope =       1.00, Intercept =       0.00

Standard Error of Estimate =       0.01

Number of good cases = 20
```

You can see that the fit is quite good between the observed and predicted scores. Once you have obtained the results you will notice that the parameters, their standard errors and the t probabilities are also entered in the dialog form. Had you elected to proceed in a step-fashion, these results would be updated at each step so you can observe the convergence to the best fit (the root mean square shown in the lower left corner.) (Fig. 8.50).

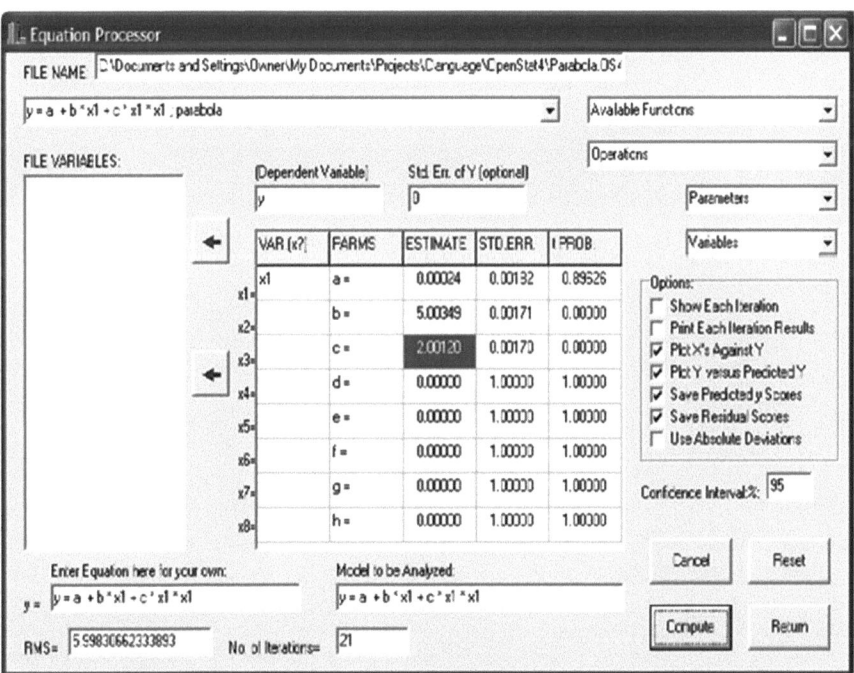

Fig. 8.50 Completed non-linear regression parameter estimates of regression coefficients

Chapter 9
Multivariate

Discriminant Function / MANOVA

An Example

We will use the file labeled ManoDiscrim.txt for our example. A file of the same name (or a .tab file) should be in your directory. Load the file and then click on the Statistics / Multivariate / Discriminant Function option. You should see the form below completed for a discriminant function analysis (Fig. 9.1):

You will notice we have asked for all options and have specified that classification use the a priori (sample) sizes for classification. When you click the Compute button, the following results are obtained (Fig. 9.2):

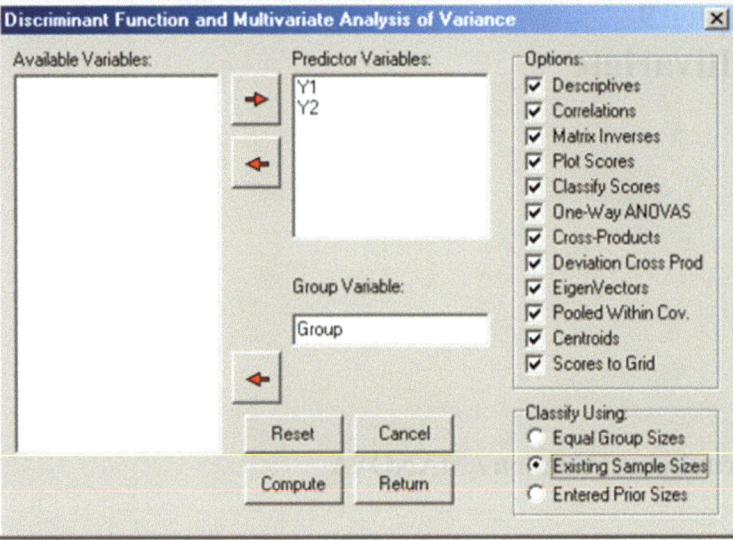

Fig. 9.1 Specifications for a discriminant function analysis

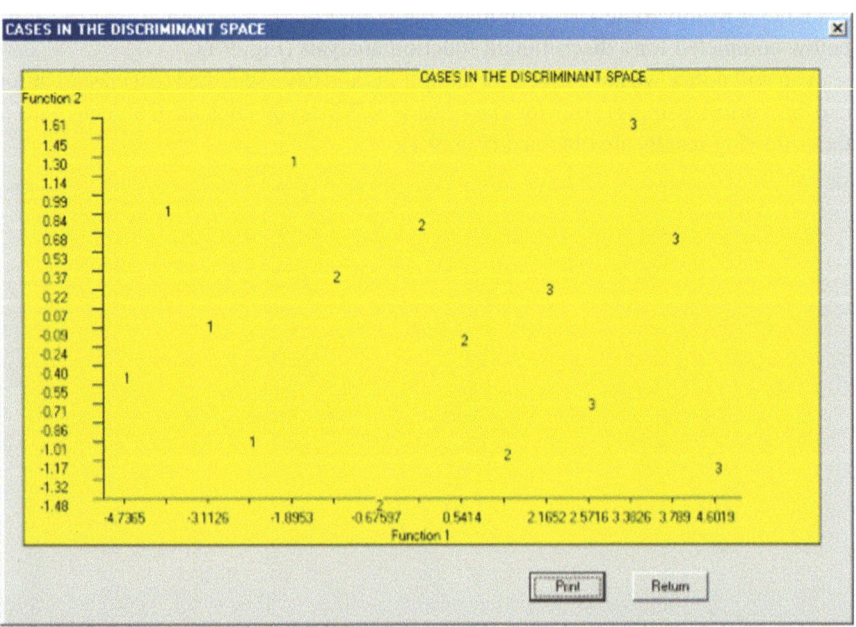

Fig. 9.2 Plot of cases in the discriminant space

Discriminant Function / MANOVA

```
MULTIVARIATE ANOVA / DISCRIMINANT FUNCTION
Reference: Multiple Regression in Behavioral Research
Elazar J. Pedhazur, 1997, Chapters 20-21
Harcourt Brace College Publishers


Total Cases := 15, Number of Groups := 3

SUM OF CROSS-PRODUCTS forGroup 1, N = 5 with 5 valid cases.

Variables
                 Y1          Y2
      Y1     111.000     194.000
      Y2     194.000     343.000

WITHIN GROUP SUM OF DEVIATION CROSS-PROD with 5 valid cases.

Variables
                 Y1          Y2
      Y1       5.200       5.400
      Y2       5.400       6.800

MEANS FOR GROUP 1, N := 5 with 5 valid cases.

Variables        Y1          Y2
               4.600       8.200

VARIANCES FOR GROUP 1 with 5 valid cases.

Variables        Y1          Y2
               1.300       1.700

STANDARD DEVIATIONS FOR GROUP 1 with 5 valid cases.

Variables        Y1          Y2
               1.140       1.304

SUM OF CROSS-PRODUCTS forGroup 2, N = 5 with 5 valid cases.

Variables
                 Y1          Y2
      Y1     129.000     169.000
      Y2     169.000     223.000

WITHIN GROUP SUM OF DEVIATION CROSS-PROD with 5 valid cases.
```

```
Variables
                Y1          Y2
        Y1      4.000       4.000
        Y2      4.000       5.200
```

MEANS FOR GROUP 2, N := 5 with 5 valid cases.

```
Variables       Y1          Y2
                5.000       6.600
```

VARIANCES FOR GROUP 2 with 5 valid cases.

```
Variables       Y1          Y2
                1.000       1.300
```

STANDARD DEVIATIONS FOR GROUP 2 with 5 valid cases.

```
Variables       Y1          Y2
                1.000       1.140
```

SUM OF CROSS-PRODUCTS forGroup 3, N = 5 with 5 valid cases.

```
Variables
                Y1          Y2
        Y1      195.000     196.000
        Y2      196.000     199.000
```

WITHIN GROUP SUM OF DEVIATION CROSS-PROD with 5 valid cases.

```
Variables
        Y1      Y2
        Y1      2.800       3.800
        Y2      3.800       6.800
```

MEANS FOR GROUP 3, N := 5 with 5 valid cases.

```
Variables       Y1          Y2
                6.200       6.200
```

VARIANCES FOR GROUP 3 with 5 valid cases.

```
Variables       Y1          Y2
                0.700       1.700
```

STANDARD DEVIATIONS FOR GROUP 3 with 5 valid cases.

Discriminant Function / MANOVA

```
Variables            Y1         Y2
                   0.837      1.304
```

TOTAL SUM OF CROSS-PRODUCTS with 15 valid cases.

```
Variables
                   Y1         Y2
       Y1      435.000    559.000
       Y2      559.000    765.000
```

TOTAL SUM OF DEVIATION CROSS-PRODUCTS with 15 valid cases.

```
Variables
                   Y1         Y2
       Y1       18.933      6.000
       Y2        6.000     30.000
```

MEANS with 15 valid cases.

```
Variables            Y1         Y2
                   5.267      7.000
```

VARIANCES with 15 valid cases.

```
Variables            Y1         Y2
                   1.352      2.143
```

STANDARD DEVIATIONS with 15 valid cases.

```
Variables            Y1         Y2
                   1.163      1.464
```

BETWEEN GROUPS SUM OF DEV. CPs with 15 valid cases.

```
Variables
                   Y1         Y2
       Y1        6.933     -7.200
       Y2       -7.200     11.200
```

UNIVARIATE ANOVA FOR VARIABLE Y1
```
SOURCE      DF      SS         MS        F         PROB > F
BETWEEN      2    6.933      3.467     3.467       0.065
ERROR       12   12.000      1.000
TOTAL       14   18.933
```

UNIVARIATE ANOVA FOR VARIABLE Y2
```
SOURCE      DF      SS        MS        F         PROB > F
BETWEEN     2       11.200    5.600     3.574     0.061
ERROR       12      18.800    1.567
TOTAL       14      30.000
```

Inv. of Pooled Within Dev. CPs Matrix with 15 valid cases.

Variables
```
            Y1         Y2
    Y1      0.366     -0.257
    Y2     -0.257      0.234
```

Number of roots extracted := 2
Percent of trace extracted := 100.0000
Roots of the W inverse time B Matrix

```
No.  Root     Proportion   Canonical R   Chi-Squared   D.F.   Prob.
1    8.7985   0.9935       0.9476        25.7156       4      0.000
2    0.0571   0.0065       0.2325         0.6111       1      0.434
```

Eigenvectors of the W inverse x B Matrix with 15 valid cases.

Variables
```
               1          2
    Y1     -2.316      0.188
    Y2      1.853      0.148
```

Pooled Within-Groups Covariance Matrix with 15 valid cases.

Variables
```
            Y1         Y2
    Y1      1.000      1.100
    Y2      1.100      1.567
```

Total Covariance Matrix with 15 valid cases.

Variables
```
            Y1         Y2
    Y1      1.352      0.429
    Y2      0.429      2.143
```

Raw Function Coeff.s from Pooled Cov. with 15 valid cases.

Variables
```
               1          2
    Y1     -2.030      0.520
    Y2      1.624      0.409
```

Raw Discriminant Function Constants with 15 valid cases.

Discriminant Function / MANOVA

```
Variables         1         2
               -0.674    -5.601

Fisher Discriminant Functions
Group 1 Constant := -24.402
Variable    Coefficient
    1         -5.084
    2          8.804

Group 2 Constant := -14.196
Variable    Coefficient
    1          1.607
    2          3.084

Group 3 Constant := -19.759
Variable    Coefficient
    1          8.112
    2         -1.738
```

CLASSIFICATION OF CASES

SUBJECT ID NO.	ACTUAL GROUP	HIGH IN	PROBABILITY GROUP P(G/D)	SEC.D GROUP	HIGH P(G/D)	DISCRIM SCORE
1	1	1	0.9999	2	0.0001	4.6019
						-1.1792
2	1	1	0.9554	2	0.0446	2.5716
						-0.6590
3	1	1	0.8903	2	0.1097	2.1652
						0.2699
4	1	1	0.9996	2	0.0004	3.7890
						0.6786
5	1	1	0.9989	2	0.0011	3.3826
						1.6075
6	2	2	0.9746	3	0.0252	-0.6760
						-1.4763
7	2	2	0.9341	1	0.0657	0.9478
						-1.0676
8	2	2	0.9730	1	0.0259	0.5414
						-0.1387
9	2	2	0.5724	3	0.4276	-1.4888
						0.3815
10	2	2	0.9842	1	0.0099	0.1350
						0.7902
11	3	3	0.9452	2	0.0548	-2.7062
						-0.9560
12	3	3	0.9999	2	0.0001	-4.7365
						-0.4358
13	3	3	0.9893	2	0.0107	-3.1126
						-0.0271
14	3	3	0.9980	2	0.0020	-3.5191
						0.9018
15	3	3	0.8007	2	0.1993	-1.8953
						1.3104

CLASSIFICATION TABLE

```
              PREDICTED GROUP
Variables
              1      2      3      TOTAL
      1       5      0      0        5
      2       0      5      0        5
      3       0      0      5        5
   TOTAL      5      5      5       15
```

Standardized Coeff. from Pooled Cov. with 15 valid cases.

```
Variables
              1          2
     Y1    -2.030      0.520
     Y2     2.032      0.511
```

Centroids with 15 valid cases.

```
Variables
              1          2
     1      3.302      0.144
     2     -0.108     -0.302
     3     -3.194      0.159
```

Raw Coefficients from Total Cov. with 15 valid cases.

```
Variables
              1          2
     Y1    -0.701      0.547
     Y2     0.560      0.429
```

Raw Discriminant Function Constants with 15 valid cases.

```
Variables     1          2
           -0.674     -5.601
```

Standardized Coeff.s from Total Cov. with 15 valid cases.

```
Variables
              1          2
     Y1    -0.815      0.636
     Y2     0.820      0.628
```

Total Correlation Matrix with 15 valid cases.

Discriminant Function / MANOVA

```
Variables

                 Y1           Y2
       Y1      1.000        0.252
       Y2      0.252        1.000

Corr.s Between Variables and Functions with 15 valid cases.

Variables

                  1            2
       Y1      -0.608        0.794
       Y2       0.615        0.788

Wilk's Lambda = 0.0965.
F = 12.2013 with D.F. 4 and 22 . Prob > F = 0.0000
Bartlett Chi-Squared = 26.8845 with 4 D.F. and prob. = 0.0000
Pillai Trace = 0.9520
```

You will notice that we have obtained cross-products and deviation cross-products for each group as well as the combined between and within groups as well as descriptive statistics (means, variances, standard deviations.) Two roots were obtained, the first significant at the 0.05 level using a chi-square test. The one-way analyses of variances completed for each continuous variable were not significant at the 0.05 level which demonstrates that a multivariate analysis may identify group differences not caught by individual variable analysis. The discriminant functions can be used to plot the group subjects in the (orthogonal) space of the functions. If you examine the plot you can see that the individuals in the three groups analyzed are easily separated using just the first discriminant function (the horizontal axis.) Raw and standardized coefficients for the discriminant functions are presented as well as Fisher's discriminant functions for each group. The latter are used to classify the subjects and the classifications are shown along with a table which summarizes the classifications. Note that in this example, all cases are correctly classified. Certainly, a cross-validation of the functions for classification would likely encounter some errors of classification. Since we asked that the discriminant scores be placed in the data grid, the data grid will now contain two new variables the Fisher discriminant scores.

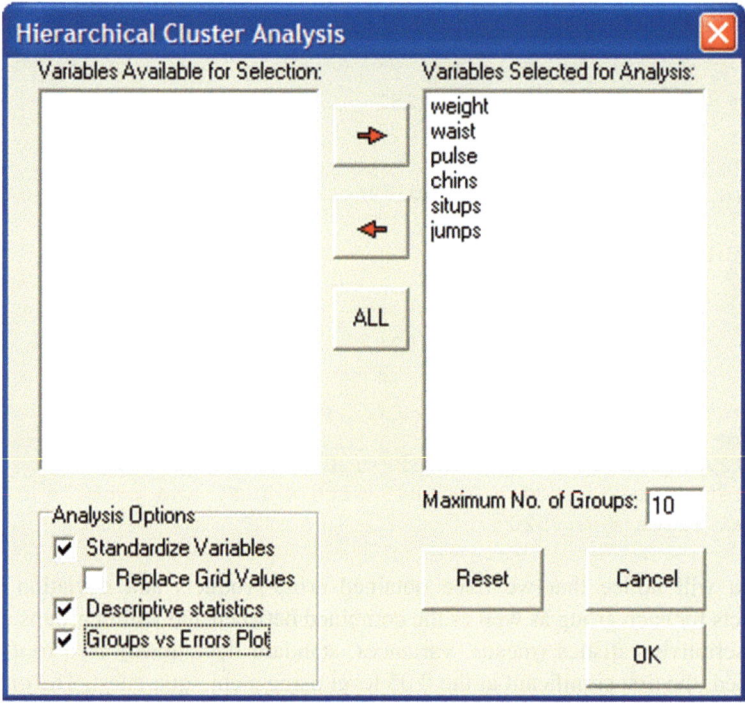

Fig. 9.3 Hierarchical Cluster Analysis form

Cluster Analyses

Hierarchical Cluster Analysis

To demonstrate the Hierarchical Clustering program, the data to be analyzed is the one labeled cansas.TAB. You will see the form above with specifications for the grouping (Fig. 9.3):

Results for the hierarchical analysis that you would obtain after clicking the Compute button are presented below (Fig. 9.4):

Cluster Analyses

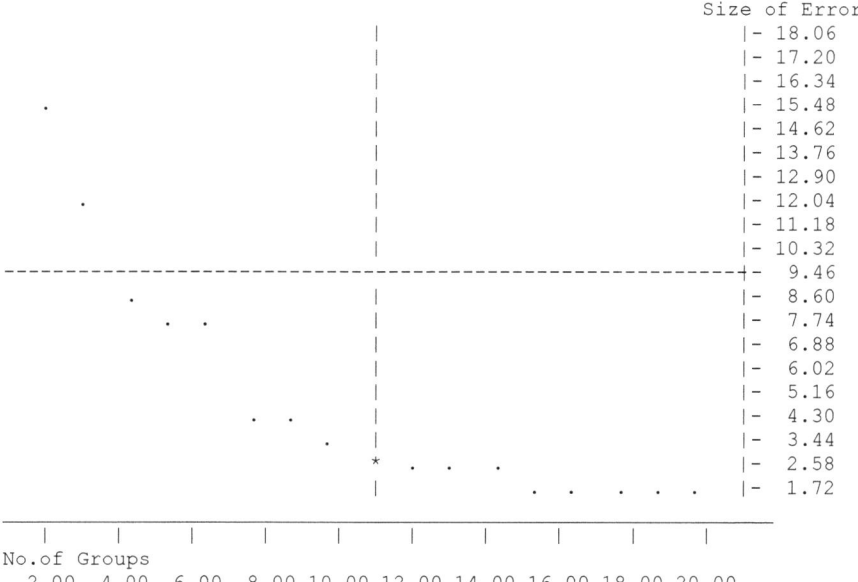

Fig. 9.4 Plot of grouping errors in the discriminant analysis

```
Hierarchical Cluster Analysis

Number of object to cluster = 20 on 6 variables.

Variable Means

Variables    weight     waist      pulse      chins      situps     jumps
             178.600    35.400     56.100     9.450      145.550    70.300

Variable Variances

Variables    weight     waist      pulse      chins      situps     jumps
             609.621    10.253     51.989     27.945     3914.576   2629.379

Variable Standard Deviations

Variables    weight     waist      pulse      chins      situps     jumps
             24.691     3.202      7.210      5.286      62.567     51.277
```

```
19 groups after combining group 1 (n = 1 ) and group 5 (n = 1)
error = 0.386
18 groups after combining group 17 (n = 1 ) and group 18 (n = 1)
error = 0.387
17 groups after combining group 11 (n = 1 ) and group 17 (n = 2)
error = 0.556
16 groups after combining group 1 (n = 2 ) and group 16 (n = 1)
error = 0.663
15 groups after combining group 3 (n = 1 ) and group 7 (n = 1)
error = 0.805
14 groups after combining group 4 (n = 1 ) and group 10 (n = 1)
error = 1.050
13 groups after combining group 2 (n = 1 ) and group 6 (n = 1)
error = 1.345
12 groups after combining group 1 (n = 3 ) and group 14 (n = 1)
error = 1.402
11 groups after combining group 0 (n = 1 ) and group 1 (n = 4)
error = 1.489
10 groups after combining group 11 (n = 3 ) and group 12 (n = 1)
error = 2.128
Group 1 (n= 5)
     Object = CASE 1
     Object = CASE 2
     Object = CASE 6
     Object = CASE 15
     Object = CASE 17
Group 3 (n= 2)
     Object = CASE 3
     Object = CASE 7
Group 4 (n= 2)
     Object = CASE 4
     Object = CASE 8
Group 5 (n= 2)
     Object = CASE 5
     Object = CASE 11
Group 9 (n= 1)
     Object = CASE 9
Group 10 (n= 1)
     Object = CASE 10
Group 12 (n= 4)
     Object = CASE 12
     Object = CASE 13
     Object = CASE 18
     Object = CASE 19
Group 14 (n= 1)
     Object = CASE 14
Group 16 (n= 1)
     Object = CASE 16
```

```
Group 20 (n= 1)
      Object = CASE 20

(…. for 9 groups, 8 groups, etc. down to 2 groups)

4 groups after combining group 4 (n = 6 ) and group 9 (n = 1)
error = 11.027
Group 1 (n= 8)
     Object = CASE 1
     Object = CASE 2
     Object = CASE 3
     Object = CASE 6
     Object = CASE 7
     Object = CASE 15
     Object = CASE 16
     Object = CASE 17
Group 4 (n= 4)
     Object = CASE 4
     Object = CASE 8
     Object = CASE 9
     Object = CASE 20
Group 5 (n= 7)
     Object = CASE 5
     Object = CASE 10
     Object = CASE 11
     Object = CASE 12
     Object = CASE 13
     Object = CASE 18
     Object = CASE 19
Group 14 (n= 1)
     Object = CASE 14

3 groups after combining group 0 (n = 8 ) and group 13 (n = 1)
error = 13.897
Group 1 (n= 9)
       Object = CASE 1
       Object = CASE 2
       Object = CASE 3
       Object = CASE 6
       Object = CASE 7
       Object = CASE 14
       Object = CASE 15
       Object = CASE 16
       Object = CASE 17
Group 4 (n= 4)
       Object = CASE 4
       Object = CASE 8
       Object = CASE 9
       Object = CASE 20
```

```
Group 5 (n= 7)
       Object = CASE 5
       Object = CASE 10
       Object = CASE 11
       Object = CASE 12
       Object = CASE 13
       Object = CASE 18
       Object = CASE 19

2 groups after combining group 3 (n = 4 ) and group 4 (n = 7)
error = 17.198
Group 1 (n= 9)
       Object = CASE 1
       Object = CASE 2
       Object = CASE 3
       Object = CASE 6
       Object = CASE 7
       Object = CASE 14
       Object = CASE 15
       Object = CASE 16
       Object = CASE 17
Group 4 (n= 11)
       Object = CASE 4
       Object = CASE 5
       Object = CASE 8
       Object = CASE 9
       Object = CASE 10
       Object = CASE 11
       Object = CASE 12
       Object = CASE 13
       Object = CASE 18
       Object = CASE 19
       Object = CASE 20
```

If you compare the results above with a discriminant analysis analysis on the same data, you will see that the clustering procedure does not necessarily replicate the original groups. Clearly, "nearest neighbor" grouping in Euclidean space does not necessarily result in the same a priori groups from the discriminant analysis.

By examining the increase in error (variance of subjects within the groups) as a function of the number of groups, one can often make some decision about the number of groups they wish to interpret. There is a large increase in error when going from 8 groups down to 7 in this analysis which suggests there are possibly 7 or 8 groups which might be examined. If we had more information on the objects of those groups, we might see a pattern or commonality shared by objects of those groups.

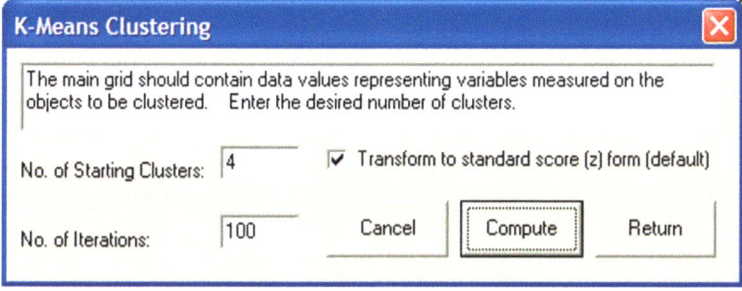

Fig. 9.5 The K Means Clustering form

K-Means Clustering Analysis

With this procedure, one first specifies the number of groups to be formed among the objects. The procedure uses a procedure to load each of the k groups with one object in a somewhat random manner. The procedure then iteratively adds or subtracts objects from each group based on an error measure of the distance between the objects in the group. The procedure ends when subsequent iterations do not produce a lower value or the number of iterations has been exceeded.

In this example, we loaded the cansas.TAB file to group the 20 subjects into four groups. The results may be compared with the other cluster methods of this chapter (Fig. 9.5).

Results are:

```
K-Means Clustering. Adapted from AS 136 APPL. STATIST. (1979)
VOL.28, NO.1

File = C:\Documents and Settings\Owner\My Documents\Projects\
Clanguage\OpenStat\cansas.TAB
No. Cases = 20, No. Variables = 6, No. Clusters = 4

NUMBER OF SUBJECTS IN EACH CLUSTER
Cluster = 1 with 1 cases.
Cluster = 2 with 7 cases.
Cluster = 3 with 9 cases.
Cluster = 4 with 3 cases.

PLACEMENT OF SUBJECTS IN CLUSTERS
CLUSTER         SUBJECT
      1            14
      2             2
      2             6
      2             8
      2             1
      2            15
```

```
        2           17
        2           20
        3           11
        3           12
        3           13
        3            4
        3            5
        3            9
        3           18
        3           19
        3           10
        4            7
        4           16
        4            3
```

```
AVERAGE VARIABLE VALUES BY CLUSTER
         VARIABLES
CLUSTER     1        2        3        4        5        6
   1       0.11     1.03    -0.12    -0.30    -0.02    -0.01
   2      -0.00     0.02    -0.02    -0.19    -0.01    -0.01
   3      -0.02    -0.20     0.01     0.17     0.01     0.01
   4       0.04     0.22     0.05     0.04    -0.00     0.01

WITHIN CLUSTER SUMS OF SQUARES
Cluster 1 = 0.000
Cluster 2 = 0.274
Cluster 3 = 0.406
Cluster 4 = 0.028
```

Average Linkage Hierarchical Cluster Analysis

This cluster procedure clusters objects based on their similarity (or dissimilarity) as recorded in a data matrix. The correlation among objects is often used as a measure of similarity. In this example, we first loaded the file labeled "cansas.TAB". We then "rotated" the data using the rotate function in the Edit menu so that columns represent subjects and rows represent variables. We then used the Correlation procedure (with the option to save the correlation matrix) to obtain the correlation among the 20 subjects as a measure of similarity. We then closed the file. Next, we opened the matrix file we had just saved using the File / Open a Matrix File option. We then clicked on the Analyses / Multivariate / Cluster / Average Linkage option. Shown below is the dialogue box for the analysis (Fig. 9.6):

Cluster Analyses

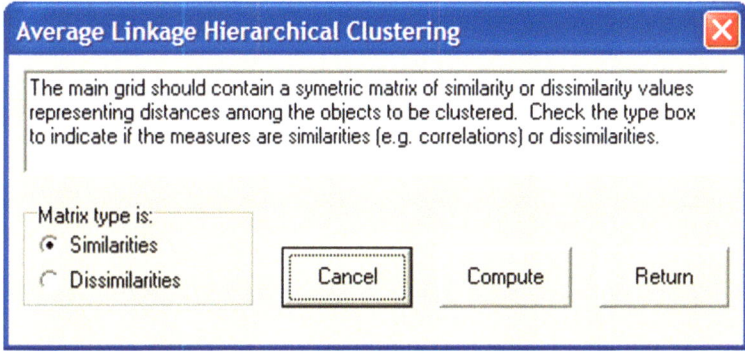

Fig. 9.6 Average Linkage dialog form

Output of the analysis includes a listing of which objects (groups) are combined at each step followed by a dendogram of the combinations. You can compare this method of clustering subjects with that obtained in the previous analysis.

```
Average Linkage Cluster Analysis. Adopted from ClusBas by John
S. Uebersax
Group  18 is joined by group  19. N is   2 ITER =   1 SIM =   0.999
Group   1 is joined by group   5. N is   2 ITER =   2 SIM =   0.998
Group   6 is joined by group   7. N is   2 ITER =   3 SIM =   0.995
Group  15 is joined by group  17. N is   2 ITER =   4 SIM =   0.995
Group  12 is joined by group  13. N is   2 ITER =   5 SIM =   0.994
Group   8 is joined by group  11. N is   2 ITER =   6 SIM =   0.993
Group   4 is joined by group   8. N is   3 ITER =   7 SIM =   0.992
Group   2 is joined by group   6. N is   3 ITER =   8 SIM =   0.988
Group  12 is joined by group  16. N is   3 ITER =   9 SIM =   0.981
Group  14 is joined by group  15. N is   3 ITER =  10 SIM =   0.980
Group   2 is joined by group   4. N is   6 ITER =  11 SIM =   0.978
Group  12 is joined by group  18. N is   5 ITER =  12 SIM =   0.972
Group   2 is joined by group  20. N is   7 ITER =  13 SIM =   0.964
Group   1 is joined by group   2. N is   9 ITER =  14 SIM =   0.962
Group   9 is joined by group  12. N is   6 ITER =  15 SIM =   0.933
Group   1 is joined by group   3. N is  10 ITER =  16 SIM =   0.911
Group   1 is joined by group  14. N is  13 ITER =  17 SIM =   0.900
Group   1 is joined by group   9. N is  19 ITER =  18 SIM =   0.783
Group   1 is joined by group  10. N is  20 ITER =  19 SIM =   0.558

No. of objects = 20
Matrix defined similarities among objects.
```

180 9 Multivariate

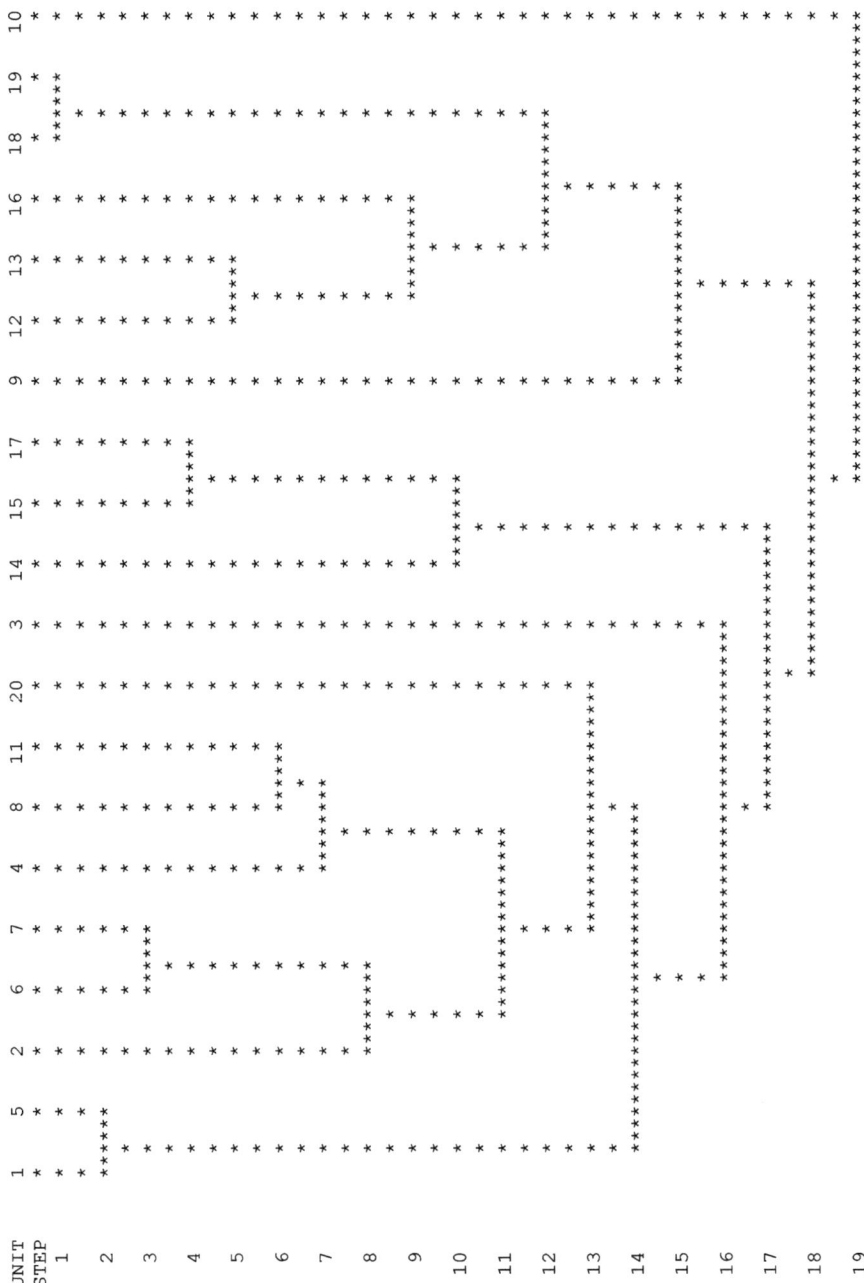

Path Analysis

To illustrate path analysis, you could utilize an example from page 788 of the book by Elazar J. Pedhazur (Multiple Regression in Behavioral Science, 1997.) Four variables in the study are labeled SES (Socio-Economic Status), IQ (Intelligence Quotient), AM (Achievement Motivation) and GPA (Grade Point Average.) Our theoretical speculations lead us to believe that AM is "caused" by SES and IQ and that GPA is "caused" by AM as well as SES and IQ. You would enter the correlations among these variables into the data grid of OpenStat then analyze the matrix with the path analysis procedure.

Example of a Path Analysis

In this example we will use the file CANSAS.TXT. The user begins by selecting the Path Analysis option of the Statistics / Multivariate menu. In the figure below (Fig. 9.7) we have selected all variables to analyze and have entered our first path indicating that waist size is "caused" by weight:

We will also hypothesize that pulse rate is "caused" by weight, chin-ups are "caused" by weight, waist and pulse, that the number of sit-ups is "caused" by weight, waist and pulse and that jumps are "caused" by weight, waist and pulse. Each time we enter a new causal relationship we click the scroll bar to move to a new model number prior to entering the "caused" and "causing" variables. Once we have entered each model, we then click on the Compute button. Note we have elected to print descriptive statistics, each models correlation matrix, and the reproduced correlation matrix which will be our measure of how well the models "fit" the data. The results are shown below:

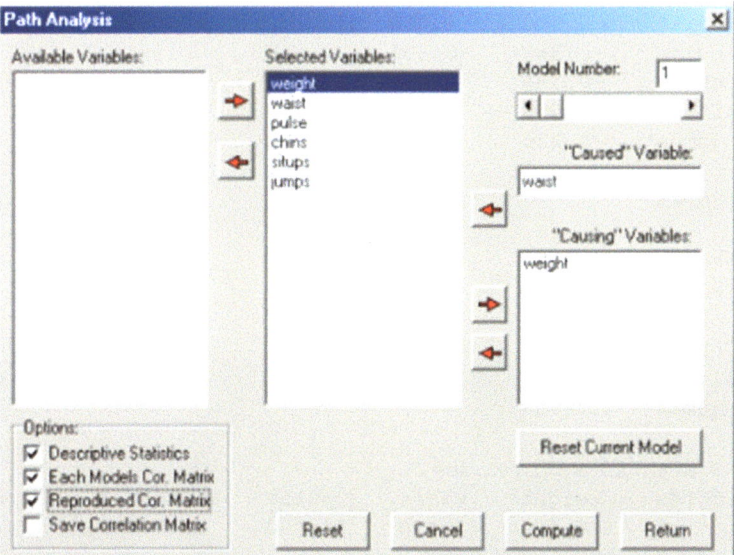

Fig. 9.7 Path Analysis dialog form

```
PATH ANALYSIS RESULTS

CAUSED VARIABLE: waist
     Causing Variables:
     weight
CAUSED VARIABLE: pulse
     Causing Variables:
     weight
CAUSED VARIABLE: chins
     Causing Variables:
     weight
     waist
     pulse
CAUSED VARIABLE: situps
     Causing Variables:
     weight
     waist
     pulse
CAUSED VARIABLE: jumps
     Causing Variables:
     weight
     waist
     pulse
```

```
Correlation Matrix with 20 valid cases.

Variables
           weight    waist    pulse    chins    situps
  weight    1.000    0.870   -0.366   -0.390   -0.493
   waist    0.870    1.000   -0.353   -0.552   -0.646
   pulse   -0.366   -0.353    1.000    0.151    0.225
   chins   -0.390   -0.552    0.151    1.000    0.696
  situps   -0.493   -0.646    0.225    0.696    1.000
   jumps   -0.226   -0.191    0.035    0.496    0.669

Variables
           jumps
  weight   -0.226
   waist   -0.191
   pulse    0.035
   chins    0.496
  situps    0.669
   jumps    1.000

MEANS with 20 valid cases.
Variables    weight    waist    pulse    chins    situps
            178.600   35.400   56.100    9.450   145.550
```

```
Variables      jumps
               70.300
VARIANCES      with 20 valid cases.
Variables      weight      waist      pulse      chins      situps
               609.621     10.253     51.989     27.945     3914.576

Variables      jumps
               2629.379

STANDARD DEVIATIONS with 20 valid cases.

Variables      weight      waist      pulse      chins      situps
               24.691      3.202      7.210      5.286      62.567

Variables      jumps
               51.277

Dependent Variable = waist

Correlation Matrix with 20 valid cases.

Variables
          weight      waist
weight    1.000       0.870
waist     0.870       1.000

MEANS with 20 valid cases.
Variables      weight      waist
               178.600     35.400

VARIANCES with 20 valid cases.
Variables      weight      waist
               6 09.621    10.253

STANDARD DEVIATIONS with 20 valid cases.
Variables      weight      waist
               24.691      3.202
```

```
Dependent Variable = waist

       R          R2          F       Prob.>F   DF1   DF2
    0.870       0.757      56.173      0.000     1    18
Adjusted R Squared = 0.744

Std. Error of Estimate =        1.621

Variable        Beta        B       Std.Error   t        Prob.>t
weight          0.870     0.113      0.015     7.495      0.000

Constant = 15.244
```

```
Dependent Variable = pulse

Correlation Matrix with 20 valid cases.

Variables
              weight        pulse
weight        1.000        -0.366
pulse        -0.366         1.000

MEANS with 20 valid cases.
Variables     weight        pulse
              178.600       56.100

VARIANCES with 20 valid cases.

Variables     weight        pulse
              609.621       51.989

STANDARD DEVIATIONS with 20 valid cases.

Variables     weight        pulse
              24.691        7.210

Dependent Variable = pulse

       R          R2          F       Prob.>F   DF1   DF2
    0.366       0.134       2.780     0.113      1    18
Adjusted R Squared = 0.086

Std. Error of Estimate = 6.895

Variable        Beta        B       Std.Error   t        Prob.>t
  weight       -0.366     -0.107     0.064     -1.667     0.113

Constant = 75.177
```

Path Analysis

```
Dependent Variable = chins

Correlation Matrix with 20 valid cases.

Variables
            weight      waist       pulse       chins
   weight    1.000      0.870      -0.366      -0.390
    waist    0.870      1.000      -0.353      -0.552
    pulse   -0.366     -0.353       1.000       0.151
    chins   -0.390     -0.552       0.151       1.000


MEANS with 20 valid cases.

Variables    weight      waist       pulse       chins
            178.600     35.400      56.100       9.450

VARIANCES with 20 valid cases.

Variables    weight      waist       pulse       chins
            609.621     10.253      51.989      27.945

STANDARD DEVIATIONS with 20 valid cases.

Variables    weight      waist       pulse       chins
             24.691      3.202       7.210       5.286
```

```
Dependent Variable = chins
       R         R2          F      Prob.>F   DF1   DF2
    0.583      0.340      2.742      0.077     3    16
Adjusted R Squared = 0.216
Std. Error of Estimate =      4.681

Variable      Beta       B       Std.Error    t        Prob.>t
   weight    0.368     0.079      0.089      0.886      0.389
    waist   -0.882    -1.456      0.683     -2.132      0.049
    pulse   -0.026    -0.019      0.160     -0.118      0.907
Constant =   47.968
```

```
Dependent Variable = situps

Correlation Matrix with 20 valid cases.

Variables
             weight      waist      pulse     situps
   weight    1.000       0.870     -0.366    -0.493
    waist    0.870       1.000     -0.353    -0.646
    pulse   -0.366      -0.353      1.000     0.225
   situps   -0.493      -0.646      0.225     1.000

MEANS with 20 valid cases.

Variables    weight      waist      pulse     situps
            178.600     35.400     56.100    145.550

VARIANCES with 20 valid cases.

Variables    weight      waist      pulse     situps
            609.621     10.253     51.989   3914.576

STANDARD DEVIATIONS with 20 valid cases.

Variables    weight      waist      pulse     situps
             24.691      3.202      7.210     62.567
```

```
Dependent Variable = situps

      R          R2          F       Prob.>F   DF1   DF2
   0.661      0.436       4.131      0.024      3    16
Adjusted R Squared = 0.331

Std. Error of Estimate = 51.181

Variable     Beta       B        Std.Error    t         Prob.>t
   weight    0.287     0.728      0.973      0.748       0.466
    waist   -0.890   -17.387      7.465     -2.329       0.033
    pulse    0.016     0.139      1.755      0.079       0.938

Constant = 623.282
```

Path Analysis

```
Dependent Variable = jumps

Correlation Matrix with 20 valid cases.

Variables
            weight       waist       pulse       jumps
   weight    1.000       0.870      -0.366      -0.226
    waist    0.870       1.000      -0.353      -0.191
    pulse   -0.366      -0.353       1.000       0.035
    jumps   -0.226      -0.191       0.035       1.000

MEANS with 20 valid cases.

Variables    weight       waist       pulse       jumps
            178.600      35.400      56.100      70.300

VARIANCES with 20 valid cases.

Variables    weight       waist       pulse       jumps
            609.621      10.253      51.989    2629.379

STANDARD DEVIATIONS with 20 valid cases.

Variables    weight       waist       pulse       jumps
             24.691       3.202       7.210      51.277
```

```
Dependent Variable = jumps
      R           R2           F      Prob.>F   DF1   DF2
   0.232        0.054       0.304      0.822     3    16
Adjusted R Squared = -0.123

Std. Error of Estimate = 54.351

Variable     Beta          B       Std.Error      t      Prob.>t
  weight    -0.259      -0.538       1.034      -0.520    0.610
   waist     0.015       0.234       7.928       0.029    0.977
   pulse    -0.055      -0.389       1.863      -0.209    0.837

Constant =    179.887
```

```
Matrix of Path Coefficients with 20 valid cases.

Variables

            weight      waist      pulse      chins     situps
  weight     0.000      0.870     -0.366      0.368      0.287
   waist     0.870      0.000      0.000     -0.882     -0.890
   pulse    -0.366      0.000      0.000     -0.026      0.016
   chins     0.368     -0.882     -0.026      0.000      0.000
  situps     0.287     -0.890      0.016      0.000      0.000
   jumps    -0.259      0.015     -0.055      0.000      0.000

Variables

             jumps
  weight    -0.259
   waist     0.015
   pulse    -0.055
   chins     0.000
  situps     0.000
   jumps     0.000
```

```
SUMMARY OF CAUSAL MODELS
Var. Caused     Causing Var.     Path Coefficient
      waist         weight             0.870
      pulse         weight            -0.366
      chins         weight             0.368
      chins         waist             -0.882
      chins         pulse             -0.026
     situps         weight             0.287
     situps         waist             -0.890
     situps         pulse              0.016
      jumps         weight            -0.259
      jumps         waist              0.015
      jumps         pulse             -0.055
```

Factor Analysis

```
Reproduced Correlation Matrix with 20 valid cases.

Variables

             weight      waist      pulse      chins     situps
    weight    1.000      0.870     -0.366     -0.390    -0.493
     waist    0.870      1.000     -0.318     -0.553    -0.645
     pulse   -0.366     -0.318      1.000      0.120     0.194
     chins   -0.390     -0.553      0.120      1.000     0.382
    situps   -0.493     -0.645      0.194      0.382     1.000
     jumps   -0.226     -0.193      0.035      0.086     0.108

Variables
              jumps
    weight   -0.226
     waist   -0.193
     pulse    0.035
     chins    0.086
    situps    0.108
     jumps    1.000

Average absolute difference between observed and reproduced
coefficients := 0.077
Maximum difference found := 0.562
```

We note that pulse is not a particularly important predictor of chin-ups or sit-ups. The largest discrepancy of 0.562 between an original correlation and a correlation reproduced using the path coefficients indicates our model of causation may have been inadequate.

Factor Analysis

The sample factor analysis completed below utilizes a data set labeled CANSAS. TXT as used in the previous path analysis example . The canonical factor analysis method was used and the varimax rotation method was used.

Shown below is the factor analysis form selected by choosing the factor analysis option under the Statistics / Multivariate menu (Fig. 9.8):

Note the options elected in the above form. The results obtained are shown below (Fig. 9.9):

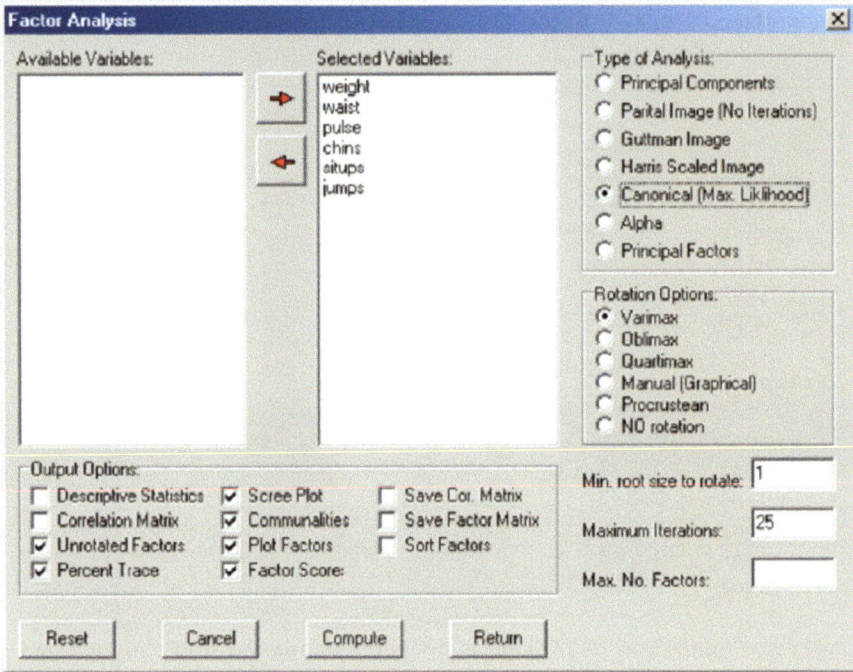

Fig. 9.8 Factor Analysis dialog form

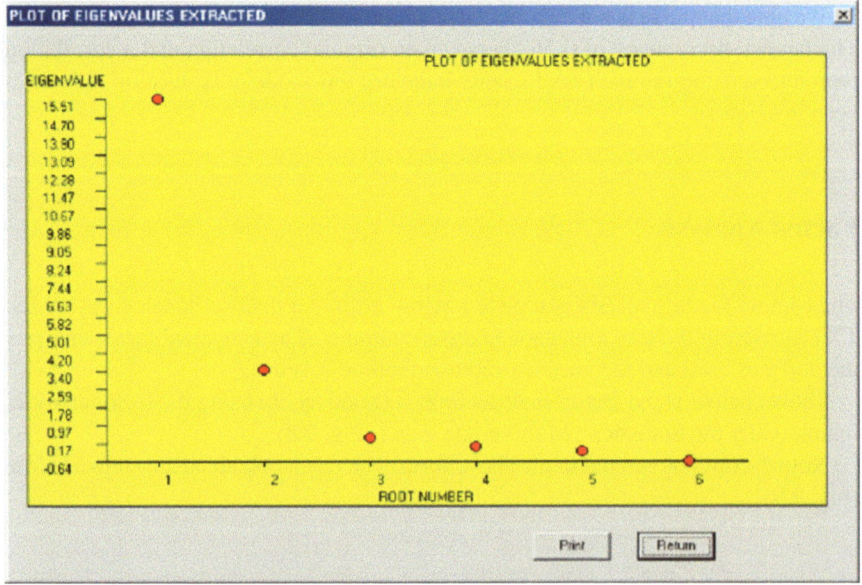

Fig. 9.9 Screen plot of eigenvalues

Factor Analysis

```
Factor Analysis
See Rummel, R.J., Applied Factor Analysis
Northwestern University Press, 1970

Canonical Factor Analysis
Original matrix trace = 18.56
Roots (Eigenvalues) Extracted:
   1  15.512
   2   3.455
   3   0.405
   4   0.010
   5  -0.185
   6  -0.641
```

```
Unrotated Factor Loadings

FACTORS with 20 valid cases.

Variables
          Factor 1   Factor 2   Factor 3   Factor 4   Factor 5
  weight    0.858     -0.286      0.157     -0.006      0.000
   waist    0.928     -0.201     -0.066     -0.003      0.000
   pulse   -0.360      0.149     -0.044     -0.089      0.000
   chins   -0.644     -0.382      0.195      0.009      0.000
  situps   -0.770     -0.472      0.057     -0.009      0.000
   jumps   -0.409     -0.689     -0.222      0.005      0.000
```

```
Variables

          Factor 6
  weight    0.000
   waist    0.000
   pulse    0.000
   chins    0.000
  situps    0.000
   jumps    0.000
```

```
Percent of Trace In Each Root:
  1 Root := 15.512 Trace := 18.557 Percent :=  83.593
  2 Root :=  3.455 Trace := 18.557 Percent :=  18.621
  3 Root :=  0.405 Trace := 18.557 Percent :=   2.180
  4 Root :=  0.010 Trace := 18.557 Percent :=   0.055
  5 Root := -0.185 Trace := 18.557 Percent :=  -0.995
  6 Root := -0.641 Trace := 18.557 Percent :=  -3.455

COMMUNALITY ESTIMATES
  1 weight    0.844
  2 waist     0.906
  3 pulse     0.162
  4 chins     0.598
  5 situps    0.819
  6 jumps     0.692
```

```
Proportion of variance in unrotated factors
  1    48.364
  2    16.475

Communality Estimates as percentages:
  1    81.893
  2    90.153
  3    15.165
  4    56.003
  5    81.607
  6    64.217

Varimax Rotated Loadings with 20 valid cases.

Variables
               Factor 1      Factor 2
  weight        -0.882        -0.201
   waist        -0.898        -0.310
   pulse         0.385         0.059
   chins         0.352         0.660
  situps         0.413         0.803
   jumps        -0.009         0.801
```

Factor Analysis

```
Percent of Variation in Rotated Factors
Factor   1 33.776
Factor   2 31.064
Total Percent of Variance in Factors : 64.840

Communalities as Percentages
1 for    weight    81.893
2 for    waist     90.153
3 for    pulse     15.165
4 for    chins     56.003
5 for    situps    81.607
6 for    jumps     64.217
```

```
              SCATTERPLOT - FACTOR PLOT
Factor 2
 |                         |                       |-  0.95-  1.00
 |                         |                       |-  0.90-  0.95
 |                         |                       |-  0.85-  0.90
 |                         |      2        1       |-  0.80-  0.85
 |                         |                       |-  0.75-  0.80
 |                         |                       |-  0.70-  0.75
 |                         |              3        |-  0.65-  0.70
 |                         |                       |-  0.60-  0.65
 |                         |                       |-  0.55-  0.60
 |                         |                       |-  0.50-  0.55
 |                         |                       |-  0.45-  0.50
 |                         |                       |-  0.40-  0.45
 |                         |                       |-  0.35-  0.40
 |                         |                       |-  0.30-  0.35
 |                         |                       |-  0.25-  0.30
 |                         |                       |-  0.20-  0.25
 |                         |                       |-  0.15-  0.20
 |                         |                       |-  0.10-  0.15
 |                         |              4        |-  0.05-  0.10
 |-------------------------------------------------|-  0.00-  0.05
 |                         |                       |- -0.05-  0.00
 |                         |                       |- -0.10- -0.05
 |                         |                       |- -0.15- -0.10
 |                         |                       |- -0.20- -0.15
 |    5                    |                       |- -0.25- -0.20
 |                         |                       |- -0.30- -0.25
 |  6                      |                       |- -0.35- -0.30
 |                         |                       |- -0.40- -0.35
 |                         |                       |- -0.45- -0.40
 |                         |                       |- -0.50- -0.45
 |                         |                       |- -0.55- -0.50
 |                         |                       |- -0.60- -0.55
 |                         |                       |- -0.65- -0.60
 |                         |                       |- -0.70- -0.65
 |                         |                       |- -0.75- -0.70
 |                         |                       |- -0.80- -0.75
 |                         |                       |- -0.85- -0.80
 |                         |                       |- -0.90- -0.85
 |                         |                       |- -0.95- -0.90
 |                         |                       |- -1.00- -0.95
 -------------------------------------------------
  |   |   |   |   |   |   |   |   |   |   |   |   | Factor 1
 -1.0-0.9-0.7-0.6-0.5-0.3-0.2-0.1 0.1 0.2 0.3 0.5 0.6 0.7 0.9 1.0
```

```
Labels:
 1 = situps
 2 = jumps
 3 = chins
 4 = pulse
 5 = weight
 6 = waist
```

```
SUBJECT FACTOR SCORE RESULTS:

Regression Coefficients with 20 valid cases.

Variables
              Factor 1        Factor 2
  weight      -0.418           0.150
   waist      -0.608           0.080
   pulse       0.042          -0.020
   chins      -0.024           0.203
  situps      -0.069           0.526
   jumps      -0.163           0.399
Standard Error of Factor Scores:
Factor 1    0.946
Factor 2    0.905
```

We note that two factors were extracted with eigenvalues greater than 1.0 and when rotated indicate that the three body measurements appear to load on one factor and that the performance measures load on the second factor. The data grid also now contains the "least-squares" factor scores for each subject. Hummm! I wonder what a hierarchical grouping of these subjects on the two factor scores would produce!

General Linear Model (Sums of Squares by Regression)

Two examples will be provided in this section. The first example demonstrates the use of the GLM procedure for completing a three-way analysis of variance. The second will demonstrate the use of the GLM procedure a repeated measures analysis of variance. Alternative procedures will also be presented to aid in the interpretation of the results.

General Linear Model (Sums of Squares by Regression) 195

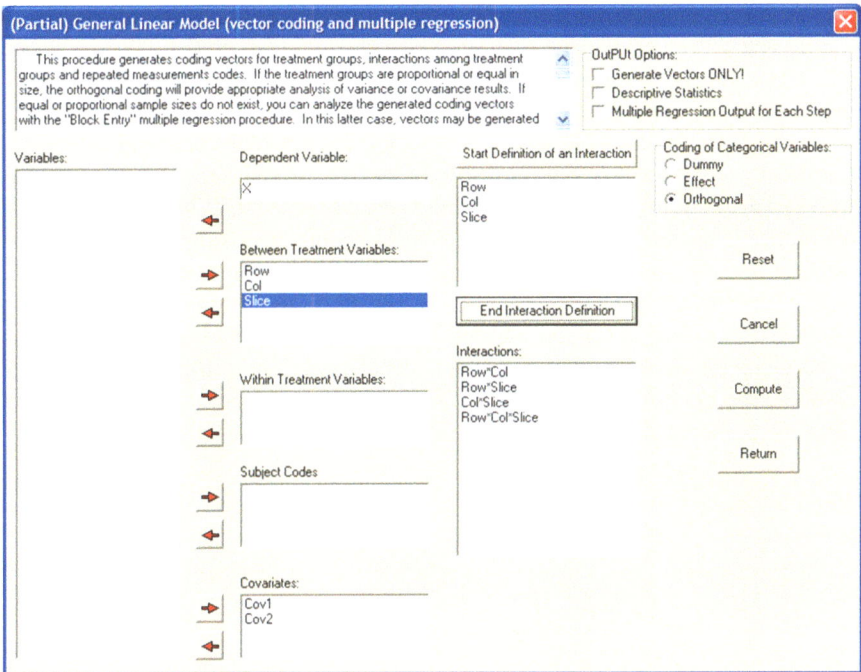

Fig. 9.10 The GLM dialog form

Example 1

The file labeled Ancova3.tab is loaded. Next, select the Analyses / Multivariate / Sums of Squares by Regression option from the menu. Shown below is the form for specifying a three-way, analysis of covariance. The dependent variable X has been entered in the continuous dependent variable list. The independent variables Row, Column, Slice have been entered in the fixed effects dependent list box. The two covariates have been entered in the covariates box. The coding method elected for creating vectors representing the categories of the independent variables is the orthogonal coding method. To specify the interactions for the analysis model, the button "begin definition of an interaction" is clicked followed by clicking of each term to be included in the interaction. The specification of the interaction is ended by clicking the "end definition of an interaction" button. This procedure was repeated for each of the interactions desired: row by column, row by slice, column by slice and row by column by slice. You will note that these interaction definitions are summarized using abbreviations in the list of defined interactions. You may also select the output options desired before clicking the "Compute" button. It is suggested that you select the option for all multiple regression results only if you wish to fully understand how the analysis is completed since the output is voluminous. The output shown below is the result of NOT selecting any of the options (Fig. 9.10).

The results obtained are shown below. Each predictor (coded vector) is entered one-by-one with the increment in variance (squared multiple correlation). This is then followed by computing the full model (the model with all variables entered) minus each independent variable to obtain the decrement in variance associated with each specific independent variable. Again, for brevity, this part of the output is not shown. A summary table then provides the results of the incremental and decrement effect of each variable. The final table summarizes the results for the analysis of variance. You will notice that, through the use of orthogonal coding, we can verify the independence of the row, column and slice effect variables. The inter-correlation among the coding vectors for a balanced design will be zero (0.0). Attempting to do a three-way analysis of variance using the traditional "partitioning of variance" method may result in a program error when a design is unbalanced, that is, the cell sizes are not equal or proportional across the factors. The unique contributions of each factor can, however, be assessed using multiple regression as in the general linear model.

```
----------------------------------------------------------
SUMS OF SQUARES AND MEAN SQUARES BY REGRESSION
TYPE III SS - R2 = Full Model - Restricted Model

VARIABLE         SUM OF SQUARES    D.F.
          Cov1         1.275         1
          Cov2         0.783         1
          Row1        25.982         1
          Col1        71.953         1
         Slice1       13.323         1
         Slice2        0.334         1
          C1R1        21.240         1
          S1R1        11.807         1
          S2R1         0.138         1
          S1C1        13.133         1
          S2C1         0.822         1
         S1C1R1        0.081         1
         S2C1R1       47.203         1
ERROR                 46.198        58
TOTAL                269.500        71


TOTAL EFFECTS SUMMARY
----------------------------------------------------------
SOURCE                  SS     D.F.        MS
----------------------------------------------------------
         Cov1         1.275     1         1.275
         Cov2         0.783     1         0.783
          Row        25.982     1        25.982
          Col        71.953     1        71.953
        Slice        13.874     2         6.937
      Row*Col        21.240     1        21.240
    Row*Slice        11.893     2         5.947
    Col*Slice        14.204     2         7.102
Row*Col*Slice        47.247     2        23.624
----------------------------------------------------------
```

General Linear Model (Sums of Squares by Regression)

```
-----------------------------------------------------------------
SOURCE                       SS    D.F.        MS
-----------------------------------------------------------------
BETWEEN SUBJECTS         208.452    13
        Covariates        2.058     2        1.029
               Row       25.982     1       25.982
               Col       71.953     1       71.953
             Slice       13.874     2        6.937
           Row*Col       21.240     1       21.240
         Row*Slice       11.893     2        5.947
         Col*Slice       14.204     2        7.102
     Row*Col*Slice       47.247     2       23.624
ERROR BETWEEN            46.198    58        0.797
-----------------------------------------------------------------

-----------------------------------------------------------------
TOTAL                   269.500    71
-----------------------------------------------------------------
```

The output above may be compared with the results obtained using the analysis of covariance procedure under the Analysis of Variance menu. The results from that analysis are shown next. You can see that the results are essentially identical although the ANCOVA procedure also includes some tests of the assumptions of homogeneity.

```
Test for Homogeneity of Group Regression Coefficients
Change in R2 = 0.1629. F = 31.437 Prob.> F = 0.0000 with d.f. 22 and 36

Unadjusted Group Means for Group Variables Row
Means
Variables     Group 1      Group 2
               3.500        4.667

Intercepts for Each Group Regression Equation for Variable: Row
Intercepts
Variables     Group 1      Group 2
               4.156        5.404

Adjusted Group Means for Group Variables Row
Means
Variables     Group 1      Group 2
               3.459        4.707

Unadjusted Group Means for Group Variables Col
Means
Variables     Group 1      Group 2
               3.000        5.167
```

```
Intercepts for Each Group Regression Equation for Variable: Col
Intercepts
Variables     Group 1      Group 2
              4.156        5.404

Adjusted Group Means for Group Variables Col
Means
Variables     Group 1      Group 2
              2.979        5.187

Unadjusted Group Means for Group Variables Slice
Means
Variables     Group 1      Group 2      Group 3
              3.500        4.500        4.250

Intercepts for Each Group Regression Equation for Variable: Slice
Intercepts
Variables     Group 1      Group 2      Group 3
              4.156        3.676        6.508
Adjusted Group Means for Group Variables Slice
Means
Variables     Group 1      Group 2      Group 3
              3.493        4.572        4.185
```

Test for Each Source of Variance Obtained by Eliminating from the Regression Model for ANCOVA the Vectors Associated with Each Fixed Effect.

SOURCE	Deg.F.	SS	MS	F	Prob>F
Cov1	1	1.27	1.27	1.600	0.2109
Cov2	1	0.78	0.78	0.983	0.3255
A	1	25.98	25.98	32.620	0.0000
B	1	71.95	71.95	90.335	0.0000
C	2	13.87	6.94	8.709	0.0005
AxB	1	21.24	21.24	26.666	0.0000
AxC	2	11.89	5.95	7.466	0.0013
BxC	2	14.20	7.10	8.916	0.0004
AxBxC	2	47.25	23.62	29.659	0.0000
ERROR 58	46.20	0.80			
TOTAL 71	269.50				

ANALYSIS FOR COVARIATES ONLY

Covariates	2	6.99	3.49	0.918	0.4041

General Linear Model (Sums of Squares by Regression)

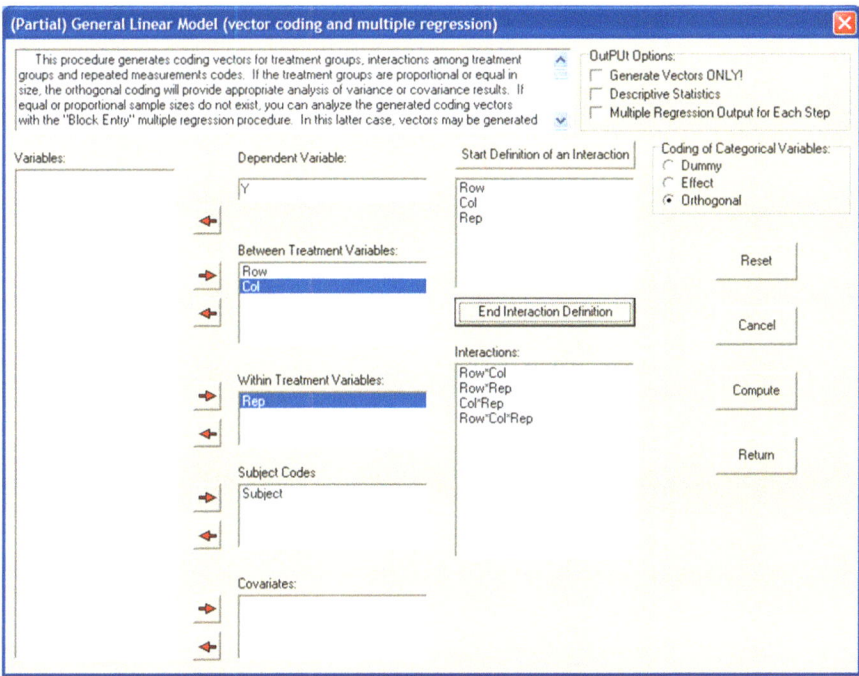

Fig. 9.11 GLM Specifications for a repeated measures ANOVA

Example Two

The second example of the GLM procedure involves a repeated measures analysis of variance similar to that you might complete with the "two between and one within anova" procedure. In this example, we have used the file labeled REGSS2.TAB. The data include a dependent variable, row and column variables, a repeated measures variable and a subject code for each of the row and column combinations. There are three subjects within each of the row and column combinations and four repeated measures within each row-column combination. The specification for the analysis is shown above (Fig. 9.11):

The results of the analysis are as follows:

```
SUMS OF SQUARES AND MEAN SQUARES BY REGRESSION
TYPE III SS - R2 = Full Model - Restricted Model
VARIABLE        SUM OF SQUARES   D.F.
         Row1        10.083        1
         Col1         8.333        1
         Rep1       150.000        1
         Rep2       312.500        1
         Rep3       529.000        1
         C1R1        80.083        1
         R1R1         0.167        1
         R2R1         2.000        1
         R3R1         6.250        1
         R1C1         4.167        1
         R2C1         0.889        1
         R3C1         7.111        1
       R1C1R1         6.000        1
       R2C1R1         0.500        1
       R3C1R1         6.250        1
ERROR                134.667       32
TOTAL               1258.000       47
```

```
TOTAL EFFECTS SUMMARY
-----------------------------------------------------------------
SOURCE                       SS    D.F.        MS
-----------------------------------------------------------------
            Row          10.083     1       10.083
            Col           8.333     1        8.333
            Rep         991.500     3      330.500
        Row*Col          80.083     1       80.083
        Row*Rep           8.417     3        2.806
        Col*Rep          12.167     3        4.056
    Row*Col*Rep          12.750     3        4.250
-----------------------------------------------------------------


-----------------------------------------------------------------
SOURCE                               SS   D.F.        MS
-----------------------------------------------------------------
BETWEEN SUBJECTS                181.000    11
             Row                 10.083     1       10.083
             Col                  8.333     1        8.333
         Row*Col                 80.083     1       80.083
ERROR BETWEEN                    82.500     8       10.312
-----------------------------------------------------------------
```

General Linear Model (Sums of Squares by Regression)

```
WITHIN SUBJECTS           1077.000   36
              Rep           991.500    3       330.500
          Row*Rep             8.417    3         2.806
          Col*Rep            12.167    3         4.056
      Row*Col*Rep            12.750    3         4.250
ERROR WITHIN                 52.167   24         2.174
-------------------------------------------------------------
TOTAL                      1258.000   47
-------------------------------------------------------------
```

A comparable analysis may be performed using the file labeled ABRData.tab. In this file, the repeated measures for each subject are entered along with the row and column codes on the same line. In the previously analyzed file, we had to code the repeated dependent values on separate lines and include a code for the subject and a code for the repeated measure. Here are the results for this analysis (Fig. 9.12):

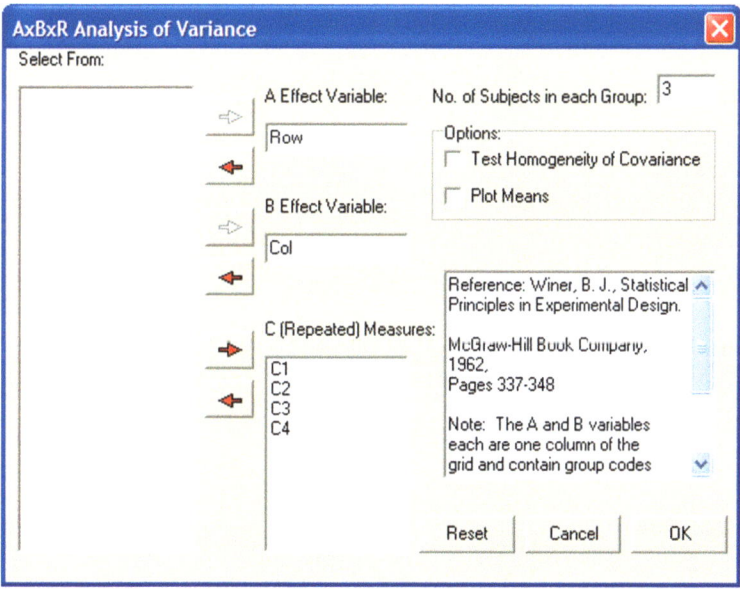

Fig. 9.12 A × B × R ANOVA dialog form

```
SOURCE              DF      SS        MS        F        PROB.
Between Subjects    11    181.000
  A Effects          1     10.083    10.083    0.978     0.352
  B Effects          1      8.333     8.333    0.808     0.395
  AB Effects         1     80.083    80.083    7.766     0.024
  Error Between      8     82.500    10.312
Within Subjects     36   1077.000
  C Replications     3    991.500   330.500  152.051     0.000
  AC Effects         3      8.417     2.806    1.291     0.300
  BC Effects         3     12.167     4.056    1.866     0.162
  ABC Effects        3     12.750     4.250    1.955     0.148
  Error Within      24     52.167     2.174
Total               47   1258.000
```

ABR Means Table

	Repeated Measures			
	C1	C2	C3	C4
A1 B1	17.000	12.000	8.667	4.000
A1 B2	15.333	10.000	7.000	2.333
A2 B1	16.667	10.000	6.000	2.333
A2 B2	17.000	14.000	9.333	8.333

AB Means Table

	B Levels	
	B1	B2
A1	10.417	8.667
A2	8.750	12.167

AC Means Table

	C Levels			
	C1	C2	C3	C4
A1	16.167	11.000	7.833	3.167
A2	16.833	12.000	7.667	5.333

BC Means Table

	C Levels			
	C1	C2	C3	C4
B1	16.833	11.000	7.333	3.167
B2	16.167	12.000	8.167	5.333

It may be observed that the sums of squares and mean squares for the two analyses above are identical. The analysis of variance procedure (second analysis) does give the F tests as well as means (and plots if elected) for the various variance components. What is demonstrated however is that the analysis of variance model may be completely defined using multiple regression methods. It might also be noted that one can choose NOT to include all interaction terms in the GLM procedure if there is an adequate basis for expecting such interactions to be zero. Notice that we might also have included covariates in the GLM procedure. That is, one can complete a repeated measures analysis of covariance which is not an option in the regular anova procedures!

Median Polish Analysis

Our example uses the file labeled "GeneChips.TEX" which is an array of cells with one observation per cell. The dialogue for the analysis appears as (Fig. 9.13):

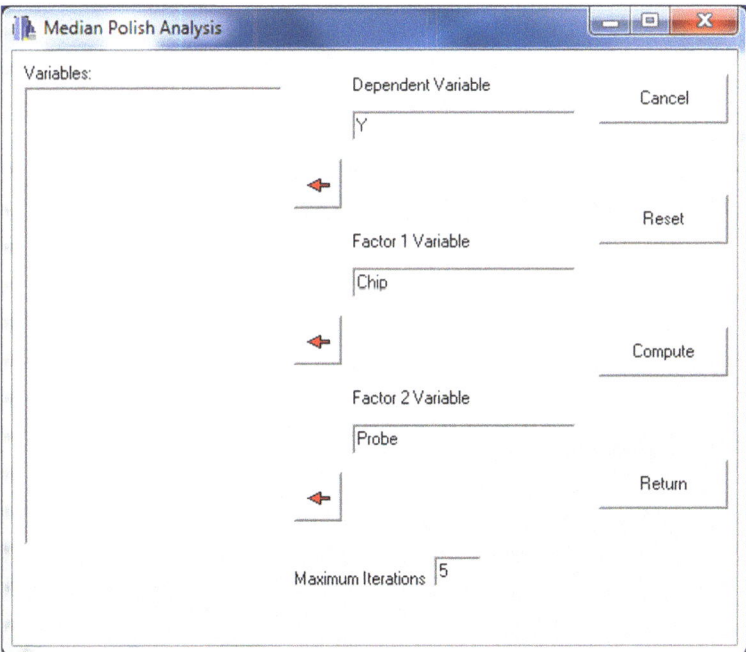

Fig. 9.13 Dialog for the Median Polish analysis

The results obtained are:

```
Observed Data
ROW             COLUMNS
            1           2           3           4           5
1       18.000      11.000       8.000      21.000       4.000
2       13.000       7.000       5.000      16.000       7.000
3       15.000       6.000       7.000      16.000       6.000
4       19.000      15.000      12.000      18.000       5.000

Adjusted Data
MEDIAN      1           2           3           4           5       Residuals
-----------------------------------------------------------------
 0.000    0.500       0.000      -1.250       1.750      -2.250       0.000
 0.000   -0.500       0.000      -0.250       0.750       4.750       0.000
 0.000    0.000      -2.500       0.250      -0.750       2.250       0.000
 0.000    0.000       2.500       1.250      -2.750      -2.750       0.000
-----------------------------------------------------------------
Col.Resid.    0.000      0.000       0.000       0.000       0.000
Col.Median    0.000      0.000       0.000       0.000       0.000

Cumulative absolute value of Row Residuals
Row = 1   Cum.Residuals =    10.250
Row = 2   Cum.Residuals =    21.750
Row = 3   Cum.Residuals =    17.250
Row = 4   Cum.Residuals =    10.250

Cumulative absolute value of Column Residuals
Column = 1   Cum.Residuals =    1.000
Column = 2   Cum.Residuals =    1.000
Column = 3   Cum.Residuals =    2.000
Column = 4   Cum.Residuals =    7.000
Column = 5   Cum.Residuals =    6.000
```

Bartlett Test of Sphericity

This test is often used to determine the degree of sphericity in a matrix. A chi-squared test is used to determine the probability of the degree of sphericity found. As an example, the "cansas.TEX" file provides a significant degree of sphericity as shown in the analysis below (Fig. 9.14):

Bartlett Test of Sphericity

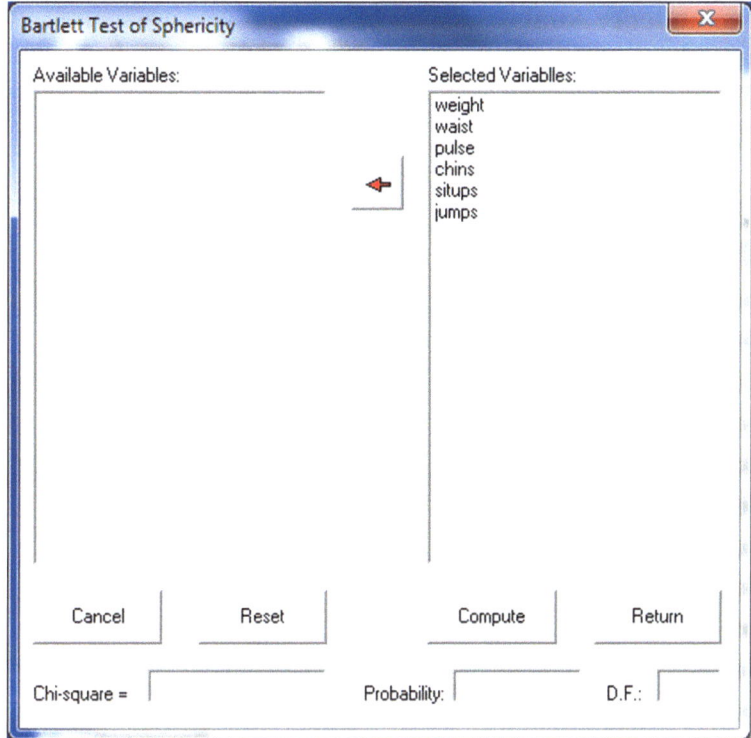

Fig. 9.14 Dialog for the Bartlett Test of Sphericity

```
CORRELATION MATRIX
Variables    weight    waist    pulse    chins    situps    jumps
   weight    1.000    0.870    -0.366   -0.390   -0.493    -0.226
    waist    0.870    1.000    -0.353   -0.552   -0.646    -0.191
    pulse   -0.366   -0.353     1.000    0.151    0.225     0.035
    chins   -0.390   -0.552     0.151    1.000    0.696     0.496
   situps   -0.493   -0.646     0.225    0.696    1.000     0.669
    jumps   -0.226   -0.191     0.035    0.496    0.669     1.000

Determinant = -3.873, log of determinant = 0.000

Chi-square = 69.067, D.F. = 15, Probability greater value = 0.0000
```

Correspondence Analysis

This procedure analyzes data such as that found in the "smokers.TEX" file and shown below:

```
CASES FOR FILE C:\Users\wgmiller\Projects\Data\Smokers.TEX

         UNITS          Group    None    Light    Medium    Heavy
        CASE 1    Senior_Mgr.       4        2         3        2
        CASE 2    Junior_Mgr.       4        3         7        4
        CASE 3    Senior_Emp.      25       10        12        4
        CASE 4    Junior_Emp.      18       24        33       13
        CASE 5    Secretaries      10        6         7        2
```

The dialog for the analysis appears as (Fig. 9.15):
The results obtained are (Figs. 9.16, 9.17, 9.18):

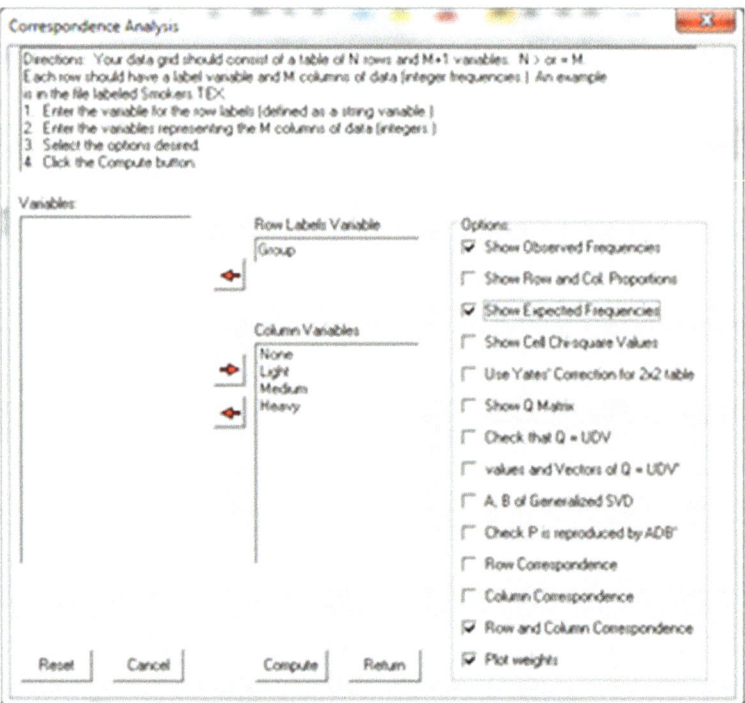

Fig. 9.15 Dialog for Correspondence Analysis

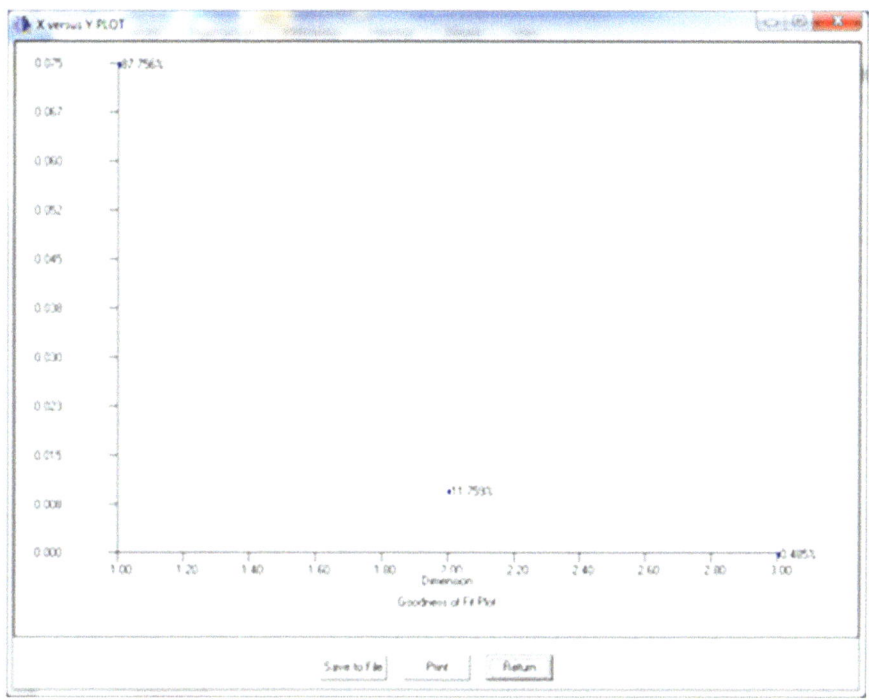

Fig. 9.16 Correspondence Analysis plot 1

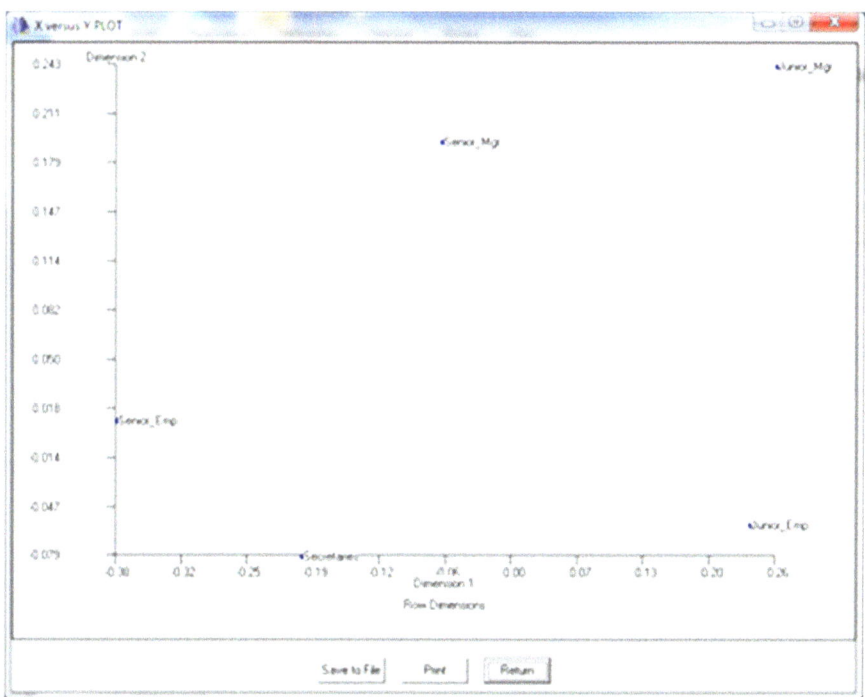

Fig. 9.17 Correspondence Analysis plot 2

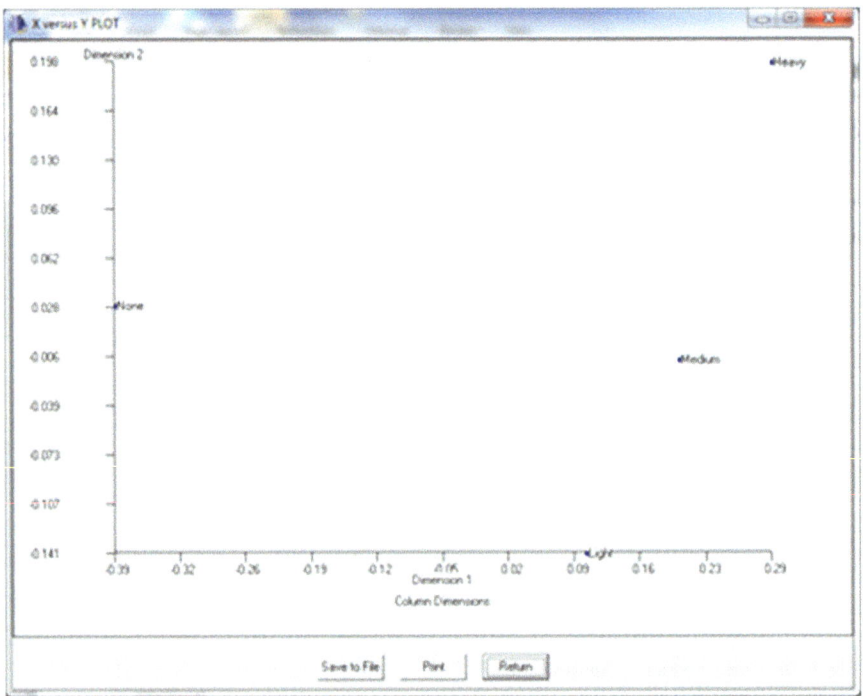

Fig. 9.18 Correspondence Analysis plot 3

```
CORRESPONDENCE ANALYSIS
Based on formulations of Bee-Leng Lee
Chapter 11 Correspondence Analysis for ViSta
Results are based on the Generalized Singular Value Decomposition
of P = A x D x B' where P is the relative frequencies observed,
A are the left generalized singular vectors,
D is a diagonal matrix of generalized singular values, and
B' is the transpose of the right generalized singular vectors.
NOTE: The first value and corresponding vectors are 1 and are
to be ignored.
An intermediate step is the regular SVD of the matrix Q = UDV'
where Q = Dr^-1/2 x P x Dc^-1/2 where Dr is a diagonal matrix
of total row relative frequencies and Dc is a diagonal matrix
of total column relative frequencies.
Chi-square Analysis Results
No. of Cases = 193
```

Correspondence Analysis

```
OBSERVED FREQUENCIES
            Frequencies
               None      Light     Medium     Heavy      Total
Senior_Mgr.     4          2         3          2         11
Junior_Mgr.     4          3         7          4         18
Senior_Emp.    25         10        12          4         51
Junior_Emp.    18         24        33         13         88
Secretaries    10          6         7          2         25
Total          61         45        62         25        193

EXPECTED FREQUENCIES

            Expected Values
               None      Light     Medium     Heavy
Senior_Mgr.    3.477     2.565     3.534      1.425
Junior_Mgr.    5.689     4.197     5.782      2.332
Senior_Emp.   16.119    11.891    16.383      6.606
Junior_Emp.   27.813    20.518    28.269     11.399
Secretaries    7.902     5.829     8.031      3.238

PROPORTIONS OF TOTAL N

            Proportions
               None      Light     Medium     Heavy      Total
Senior_Mgr.    0.021     0.010     0.016      0.010      0.057
Junior_Mgr.    0.021     0.016     0.036      0.021      0.093
Senior_Emp.    0.130     0.052     0.062      0.021      0.264
Junior_Emp.    0.093     0.124     0.171      0.067      0.456
Secretaries    0.052     0.031     0.036      0.010      0.130
Total          0.316     0.233     0.321      0.130      1.000
```

Chi-square = 16.442 with D.F. = 12. Prob. > value = 0.172

Liklihood Ratio = 16.348 with prob. > value = 0.1758

phi correlation = 0.2919

Pearson Correlation r = 0.0005

Mantel-Haenszel Test of Linear Association = 0.000 with probability > value = 0.9999

The coefficient of contingency = 0.280

Cramer's V = 0.169

Inertia = 0.0852

```
Row Dimensions
             (Ignore Column 1)
              None      Light    Medium     Heavy
Senior_Mgr.   1.000    -0.066     0.194     0.071
Junior_Mgr.   1.000     0.259     0.243    -0.034
Senior_Emp.   1.000    -0.381     0.011    -0.005
Junior_Emp.   1.000     0.233    -0.058     0.003
Secretaries   1.000    -0.201    -0.079    -0.008

Column Dimensions
             (Ignore Column 1)
              None      Light    Medium     Heavy
None          1.000    -0.393     0.030    -0.001
Light         1.000     0.099    -0.141     0.022
Medium        1.000     0.196    -0.007    -0.026
Heavy         1.000     0.294     0.198     0.026
```

Log Linear Screening, A×B and A×B×C Analyses

The chi-squared test is often used for testing the independence of observed frequencies in a two-way table. However, there may be three classifications in which objects counted. Moreover, one may be interested in the model that best describes the observed values. OpenStat contains three procedures to analyzed cross-classified data. The first is an "over-all" screening, the second is for analyzing a two-way classification table and the third is to analyze a three-way classification table. To demonstrate these procedures, we will use a file labeled "ABCLogLinData.TEX" from the sample data files (Figs. 9.19, 9.20, 9.21).

Log Linear Screening, A×B and A×B×C Analyses

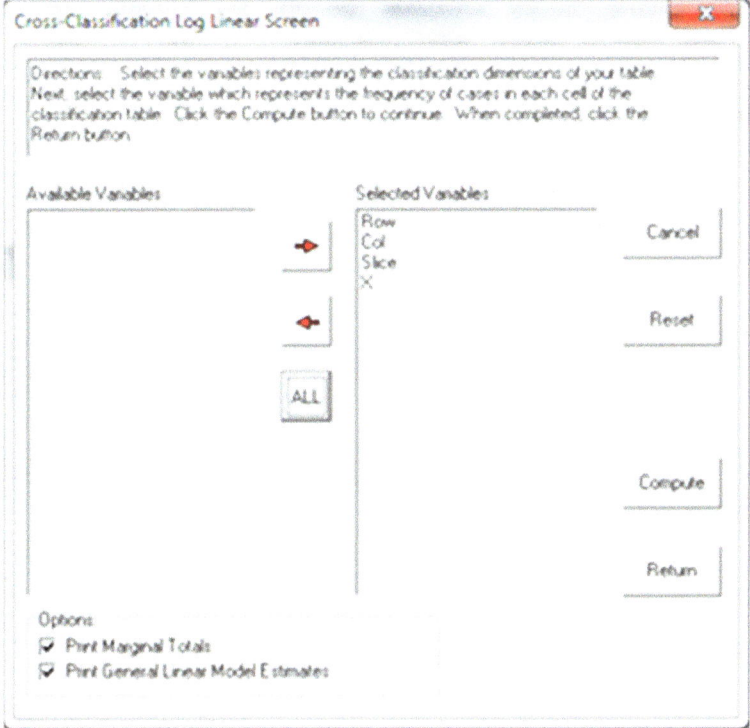

Fig. 9.19 Dialog for Log Linear Screening

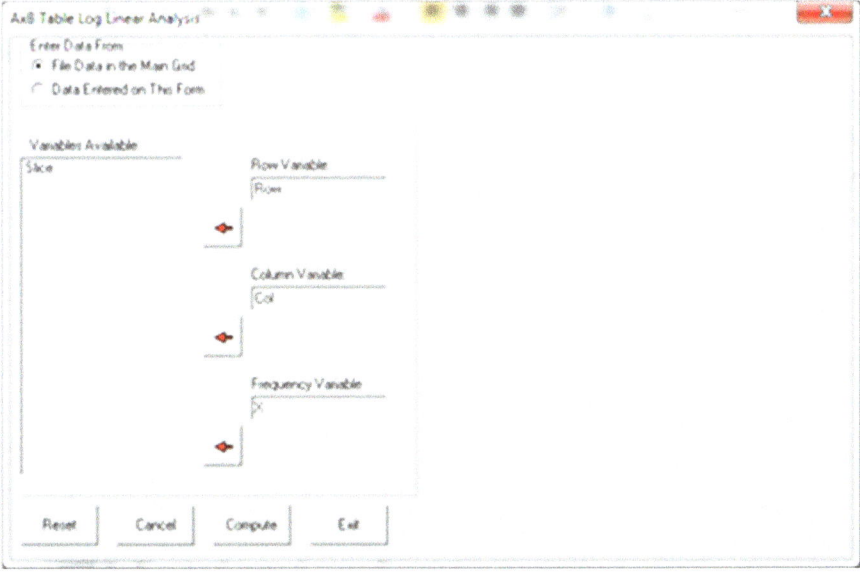

Fig. 9.20 Dialog for the A × B Log Linear Analysis

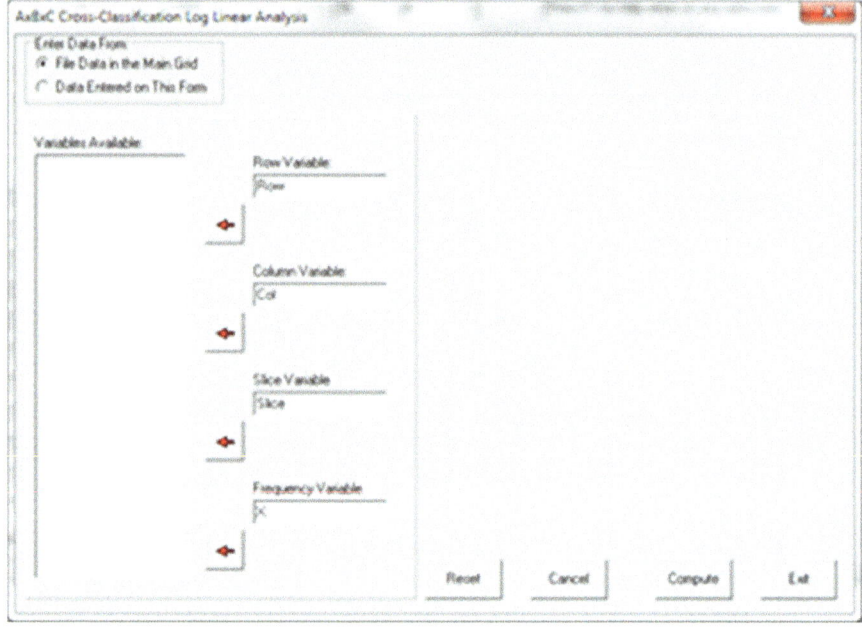

Fig. 9.21 Dialog for the A × B × C Log Linear Analysis

The Screening Procedure

```
FILE: C:\Users\wgmiller\Projects\Data\ABCLogLinData.tex

Marginal Totals for Row
Level Frequency
    1         63
    2         84

Marginal Totals for Col
Level Frequency
    1         54
    2         93

Marginal Totals for Slice
Level Frequency
    1         42
    2         54
    3         51

Total Frequencies = 147
```

```
FILE: C:\Users\wgmiller\Projects\Data\ABCLogLinData.tex

EXPECTED CELL VALUES FOR MODEL OF COMPLETE INDEPENDENCE

Cell              Observed      Expected     Log Expected
  1   1   1          6            6.61          1.889
  2   1   1          6            8.82          2.177
  1   2   1         15           11.39          2.433
  2   2   1         15           15.18          2.720
  1   1   2          9            8.50          2.140
  2   1   2         15           11.34          2.428
  1   2   2         12           14.64          2.684
  2   2   2         18           19.52          2.972
  1   1   3         12            8.03          2.083
  2   1   3          6           10.71          2.371
  1   2   3          9           13.83          2.627
  2   2   3         24           18.44          2.914
Chisquare =      11.310 with probability = 0.004      (DF = 2)
G squared =      11.471 with probability = 0.003      (DF = 2)

U (mu) for general loglinear model = 2.45

First Order LogLinear Model Factors and N of Cells in Each
CELL          U1    N Cells     U2    N Cells    U3    N Cells
  1  1  1  -0.144      6      -0.272     6     -0.148     4
  2  1  1   0.144      6      -0.272     6     -0.148     4
  1  2  1  -0.144      6       0.272     6     -0.148     4
  2  2  1   0.144      6       0.272     6     -0.148     4
  1  1  2  -0.144      6      -0.272     6      0.103     4
  2  1  2   0.144      6      -0.272     6      0.103     4
  1  2  2  -0.144      6       0.272     6      0.103     4
  2  2  2   0.144      6       0.272     6      0.103     4
  1  1  3  -0.144      6      -0.272     6      0.046     4
  2  1  3   0.144      6      -0.272     6      0.046     4
  1  2  3  -0.144      6       0.272     6      0.046     4
  2  2  3   0.144      6       0.272     6      0.046     4

Second Order Loglinear Model Terms and N of Cells in Each
CELL         U12   N Cells    U13   N Cells    U23   N Cells
  1  1  1  -0.416     3      -0.292     2     -0.420     2
  2  1  1  -0.128     3      -0.005     2     -0.420     2
  1  2  1   0.128     3      -0.292     2      0.123     2
  2  2  1   0.416     3      -0.005     2      0.123     2
  1  1  2  -0.416     3      -0.041     2     -0.169     2
  2  1  2  -0.128     3       0.247     2     -0.169     2
  1  2  2   0.128     3      -0.041     2      0.375     2
  2  2  2   0.416     3       0.247     2      0.375     2
  1  1  3  -0.416     3      -0.098     2     -0.226     2
  2  1  3  -0.128     3       0.190     2     -0.226     2
  1  2  3   0.128     3      -0.098     2      0.317     2
  2  2  3   0.416     3       0.190     2      0.317     2
```

```
SCREEN FOR INTERACTIONS AMONG THE VARIABLES
Adapted from the Fortran program by Lustbader and Stodola printed in
Applied Statistics, Volume 30, Issue 1, 1981, pages 97-105 as Algorithm
AS 160 Partial and Marginal Association in Multidimensional Contingency
Tables
Statistics for tests that the interactions of a given order are zero
ORDER     STATISTIC    D.F.     PROB.
  1         15.108       4       0.004
  2          6.143       5       0.293
  3          5.328       2       0.070

Statistics for Marginal Association Tests
VARIABLE    ASSOC.    PART ASSOC.    MARGINAL ASSOC.    D.F.    PROB
   1          1          3.010            3.010          1      0.083
   1          2         10.472           10.472          1      0.001
   1          3          1.626            1.626          2      0.444
   2          1          2.224            1.773          1      0.183
   2          2          1.726            1.275          2      0.529
   2          3          3.095            2.644          2      0.267
```

The A × B Log Linear Analysis

```
ANALYSES FOR AN I BY J CLASSIFICATION TABLE

Reference: G.J.G. Upton, The Analysis of Cross-tabulated Data, 1980

Cross-Products Odds Ratio =  1.583
Log odds of the cross-products ratio =  0.460

Saturated Model Results

Observed Frequencies
ROW/COL        1        2       TOTAL
    1        27.00    36.00     63.00
    2        27.00    57.00     84.00
TOTAL        54.00    93.00    147.00

Log frequencies, row average and column average of log frequencies
ROW/COL        1        2       TOTAL
    1         3.30     3.58      3.44
    2         3.30     4.04      3.67
TOTAL         3.30     3.81      3.55

Expected Frequencies
ROW/COL        1        2       TOTAL
    1        27.00    36.00     63.00
    2        27.00    57.00     84.00
TOTAL        54.00    93.00    147.00
```

```
Cell Parameters
ROW    COL    MU      LAMBDA ROW    LAMBDA COL    LAMBDA ROW x COL
 1      1    3.555     -0.115         -0.259          0.115
 1      2    3.555     -0.115          0.259         -0.115
 2      1    3.555      0.115         -0.259         -0.115
 2      2    3.555      0.115          0.259          0.115

Y squared statistic for model fit = -0.000 D.F. = 0

Independent Effects Model Results

Expected Frequencies
ROW/COL        1         2       TOTAL
     1       23.14     39.86     63.00
     2       30.86     53.14     84.00
TOTAL        54.00     93.00    147.00

Cell Parameters
ROW    COL    MU      LAMBDA ROW    LAMBDA COL    LAMBDA ROW x COL
 1      1    3.557     -0.144         -0.272          0.000
 1      2    3.557     -0.144          0.272          0.000
 2      1    3.557      0.144         -0.272          0.000
 2      2    3.557      0.144          0.272          0.000

Y squared statistic for model fit = 1.773 D.F. = 1
Chi-squared = 1.778 with 1 D.F.

No Column Effects Model Results

Expected Frequencies
ROW/COL        1         2       TOTAL
     1       31.50     31.50     63.00
     2       42.00     42.00     84.00
TOTAL        73.50     73.50    147.00
```

```
Cell Parameters
ROW    COL    MU      LAMBDA ROW    LAMBDA COL    LAMBDA ROW x COL
 1      1    3.594     -0.144         0.000          -0.000
 1      2    3.594     -0.144         0.000          -0.000
 2      1    3.594      0.144         0.000          -0.000
 2      2    3.594      0.144         0.000          -0.000

Y squared statistic for model fit = 12.245 D.F. = 2

No Row Effects Model Results

Expected Frequencies
ROW/COL        1         2       TOTAL
     1       27.00     46.50     73.50
     2       27.00     46.50     73.50
TOTAL        54.00     93.00    147.00

Cell Parameters
ROW    COL    MU      LAMBDA ROW    LAMBDA COL    LAMBDA ROW x COL
 1      1    3.568      0.000        -0.272          0.000
 1      2    3.568      0.000         0.272          0.000
 2      1    3.568      0.000        -0.272          0.000
 2      2    3.568      0.000         0.272          0.000

Y squared statistic for model fit = 4.783 D.F. = 2

Equiprobability Effects Model Results

Expected Frequencies
ROW/COL        1         2       TOTAL
     1       36.75     36.75     36.75
     2       36.75     36.75     36.75
TOTAL        36.75     36.75    147.00

Cell Parameters
ROW    COL    MU      LAMBDA ROW    LAMBDA COL    LAMBDA ROW x COL
 1      1    3.604      0.000         0.000          0.000
 1      2    3.604      0.000         0.000          0.000
 2      1    3.604      0.000         0.000          0.000
 2      2    3.604      0.000         0.000          0.000

Y squared statistic for model fit = 15.255 D.F. = 3
```

The A × B × C Log Linear Analysis

```
Log-Linear Analysis of a Three Dimension Table
Observed    Frequencies
  1   1   1      6.000
  1   1   2      9.000
  1   1   3     12.000
  1   2   1     15.000
  1   2   2     12.000
  1   2   3      9.000
  2   1   1      6.000
  2   1   2     15.000
  2   1   3      6.000
  2   2   1     15.000
  2   2   2     18.000
  2   2   3     24.000
Totals for Dimension A
Row 1    63.000
Row 2    84.000
Totals for Dimension B
Col 1    54.000
Col 2    93.000
Totals for Dimension C
Slice 1   42.000
Slice 2   54.000
Slice 3   51.000

Sub-matrix AB
ROW/COL        1         2
     1    27.000    36.000
     2    27.000    57.000

Sub-matrix AC
ROW/COL        1         2         3
     1    21.000    21.000    21.000
     2    21.000    33.000    30.000

Sub-matrix BC
ROW/COL        1         2         3
     1    12.000    24.000    18.000
     2    30.000    30.000    33.000
```

```
Saturated Model
Expected    Frequencies
   1    1    1      6.000
   1    1    2      9.000
   1    1    3     12.000
   1    2    1     15.000
   1    2    2     12.000
   1    2    3      9.000
   2    1    1      6.000
   2    1    2     15.000
   2    1    3      6.000
   2    2    1     15.000
   2    2    2     18.000
   2    2    3     24.000
Totals for Dimension A
Row 1      63.000
Row 2      84.000
Totals for Dimension B
Col 1      54.000
Col 2      93.000
Totals for Dimension C
Slice 1    42.000
Slice 2    54.000
Slice 3    51.000

Log Frequencies
   1    1    1      1.792
   1    1    2      2.197
   1    1    3      2.485
   1    2    1      2.708
   1    2    2      2.485
   1    2    3      2.197
   2    1    1      1.792
   2    1    2      2.708
   2    1    3      1.792
   2    2    1      2.708
   2    2    2      2.890
   2    2    3      3.178
Totals for Dimension A
Row 1      2.311
Row 2      2.511
Totals for Dimension B
Col 1      2.128
Col 2      2.694
Totals for Dimension C
Slice 1    2.250
Slice 2    2.570
Slice 3    2.413
```

Log Linear Screening, A×B and A×B×C Analyses

```
Cell Parameters
ROW   COL   SLICE    MU       LAMBDA A   LAMBDA B    LAMBDA C
                   LAMBDA AB  LAMBDA AC  LAMBDA BC   LAMBDA ABC
 1     1     1      2.411     -0.100     -0.283      -0.161
                    0.131      0.100     -0.175      -0.131
 1     1     2      2.411     -0.100     -0.283       0.159
                    0.131     -0.129      0.166      -0.157
 1     1     3      2.411     -0.100     -0.283       0.002
                    0.131      0.028      0.009       0.288
 1     2     1      2.411     -0.100      0.283      -0.161
                   -0.131      0.100      0.175       0.131
 1     2     2      2.411     -0.100      0.283       0.159
                   -0.131     -0.129     -0.166       0.157
 1     2     3      2.411     -0.100      0.283       0.002
                   -0.131      0.028     -0.009      -0.288
 2     1     1      2.411      0.100     -0.283      -0.161
                   -0.131     -0.100     -0.175       0.131
 2     1     2      2.411      0.100     -0.283       0.159
                   -0.131      0.129      0.166       0.157
 2     1     3      2.411      0.100     -0.283       0.002
                   -0.131     -0.028      0.009      -0.288
 2     2     1      2.411      0.100      0.283      -0.161
                    0.131     -0.100      0.175      -0.131
 2     2     2      2.411      0.100      0.283       0.159
                    0.131      0.129     -0.166      -0.157
 2     2     3      2.411      0.100      0.283       0.002
                    0.131     -0.028     -0.009       0.288

G squared statistic for model fit = 0.000  D.F. = 0

Model of Independence

Expected Frequencies
    1    1    1     6.612
    1    1    2     8.501
    1    1    3     8.029
    1    2    1    11.388
    1    2    2    14.641
    1    2    3    13.828
    2    1    1     8.816
    2    1    2    11.335
    2    1    3    10.706
    2    2    1    15.184
    2    2    2    19.522
    2    2    3    18.437
```

```
Totals for Dimension A
Row 1     63.000
Row 2     84.000
Totals for Dimension B
Col 1     54.000
Col 2     93.000
Totals for Dimension C
Slice 1   42.000
Slice 2   54.000
Slice 3   51.000

Log Frequencies
    1   1   1    1.889
    1   1   2    2.140
    1   1   3    2.083
    1   2   1    2.433
    1   2   2    2.684
    1   2   3    2.627
    2   1   1    2.177
    2   1   2    2.428
    2   1   3    2.371
    2   2   1    2.720
    2   2   2    2.972
    2   2   3    2.914
Totals for Dimension A
Row 1     2.309
Row 2     2.597
Totals for Dimension B
Col 1     2.181
Col 2     2.725
Totals for Dimension C
Slice 1   2.305
Slice 2   2.556
Slice 3   2.499
```

Log Linear Screening, A×B and A×B×C Analyses   221

```
Cell Parameters
ROW   COL   SLICE      MU        LAMBDA A    LAMBDA B    LAMBDA C
                    LAMBDA AB    LAMBDA AC   LAMBDA BC   LAMBDA ABC
 1     1     1        2.453       -0.144      -0.272      -0.148
                      0.000        0.000       0.000      -0.000
 1     1     2        2.453       -0.144      -0.272       0.103
                      0.000       -0.000       0.000       0.000
 1     1     3        2.453       -0.144      -0.272       0.046
                      0.000        0.000       0.000       0.000
 1     2     1        2.453       -0.144       0.272      -0.148
                      0.000        0.000       0.000       0.000
 1     2     2        2.453       -0.144       0.272       0.103
                      0.000       -0.000      -0.000       0.000
 1     2     3        2.453       -0.144       0.272       0.046
                      0.000        0.000      -0.000       0.000
 2     1     1        2.453        0.144      -0.272      -0.148
                      0.000        0.000       0.000      -0.000
 2     1     2        2.453        0.144      -0.272       0.103
                      0.000       -0.000       0.000       0.000
 2     1     3        2.453        0.144      -0.272       0.046
                      0.000        0.000       0.000      -0.000
 2     2     1        2.453        0.144       0.272      -0.148
                     -0.000        0.000       0.000       0.000
 2     2     2        2.453        0.144       0.272       0.103
                     -0.000       -0.000      -0.000       0.000
 2     2     3        2.453        0.144       0.272       0.046
                     -0.000        0.000      -0.000       0.000

G squared statistic for model fit = 11.471 D.F. = 7

No AB Effect

Expected Frequencies
     1   1   1      6.000
     1   1   2      9.333
     1   1   3      7.412
     1   2   1     15.000
     1   2   2     11.667
     1   2   3     13.588
     2   1   1      6.000
     2   1   2     14.667
     2   1   3     10.588
     2   2   1     15.000
     2   2   2     18.333
     2   2   3     19.412
```

```
Totals for Dimension A
Row 1     63.000
Row 2     84.000
Totals for Dimension B
Col 1     54.000
Col 2     93.000
Totals for Dimension C
Slice 1   42.000
Slice 2   54.000
Slice 3   51.000

Log Frequencies
   1   1   1    1.792
   1   1   2    2.234
   1   1   3    2.003
   1   2   1    2.708
   1   2   2    2.457
   1   2   3    2.609
   2   1   1    1.792
   2   1   2    2.686
   2   1   3    2.360
   2   2   1    2.708
   2   2   2    2.909
   2   2   3    2.966
Totals for Dimension A
Row 1     2.300
Row 2     2.570
Totals for Dimension B
Col 1     2.144
Col 2     2.726
Totals for Dimension C
Slice 1   2.250
Slice 2   2.571
Slice 3   2.484
```

```
Cell Parameters
ROW   COL   SLICE     MU        LAMBDA A    LAMBDA B    LAMBDA C
                      LAMBDA AB LAMBDA AC   LAMBDA BC   LAMBDA ABC
 1     1     1        2.435     -0.135      -0.291      -0.185
                      0.000      0.135      -0.167       0.000
 1     1     2        2.435     -0.135      -0.291       0.136
                      0.000     -0.091       0.179       0.000
 1     1     3        2.435     -0.135      -0.291       0.049
                      0.000     -0.044      -0.012       0.000
 1     2     1        2.435     -0.135       0.291      -0.185
                      0.000      0.135       0.167       0.000
 1     2     2        2.435     -0.135       0.291       0.136
                      0.000     -0.091      -0.179       0.000
 1     2     3        2.435     -0.135       0.291       0.049
                      0.000     -0.044       0.012       0.000
 2     1     1        2.435      0.135      -0.291      -0.185
                      0.000     -0.135      -0.167      -0.000
 2     1     2        2.435      0.135      -0.291       0.136
                      0.000      0.091       0.179      -0.000
 2     1     3        2.435      0.135      -0.291       0.049
                      0.000      0.044      -0.012      -0.000
 2     2     1        2.435      0.135       0.291      -0.185
                      0.000     -0.135       0.167       0.000
 2     2     2        2.435      0.135       0.291       0.136
                      0.000      0.091      -0.179       0.000
 2     2     3        2.435      0.135       0.291       0.049
                      0.000      0.044       0.012       0.000

G squared statistic for model fit = 7.552 D.F. = 3

No AC Effect

Expected Frequencies
     1     1     1     6.000
     1     1     2    12.000
     1     1     3     9.000
     1     2     1    11.613
     1     2     2    11.613
     1     2     3    12.774
     2     1     1     6.000
     2     1     2    12.000
     2     1     3     9.000
     2     2     1    18.387
     2     2     2    18.387
     2     2     3    20.226
```

Totals for Dimension A
Row 1   63.000
Row 2   84.000
Totals for Dimension B
Col 1   54.000
Col 2   93.000
Totals for Dimension C
Slice 1   42.000
Slice 2   54.000
Slice 3   51.000

Log Frequencies
    1   1   1   1.792
    1   1   2   2.485
    1   1   3   2.197
    1   2   1   2.452
    1   2   2   2.452
    1   2   3   2.547
    2   1   1   1.792
    2   1   2   2.485
    2   1   3   2.197
    2   2   1   2.912
    2   2   2   2.912
    2   2   3   3.007
Totals for Dimension A
Row 1   2.321
Row 2   2.551
Totals for Dimension B
Col 1   2.158
Col 2   2.714
Totals for Dimension C
Slice 1   2.237
Slice 2   2.583
Slice 3   2.487

Log Linear Screening, A×B and A×B×C Analyses

```
Cell Parameters
ROW   COL   SLICE     MU       LAMBDA A    LAMBDA B    LAMBDA C
                   LAMBDA AB   LAMBDA AC   LAMBDA BC   LAMBDA ABC
 1     1     1      2.436      -0.115      -0.278      -0.199
                    0.115       0.000      -0.167       0.000
 1     1     2      2.436      -0.115      -0.278       0.148
                    0.115       0.000       0.179       0.000
 1     1     3      2.436      -0.115      -0.278       0.051
                    0.115      -0.000      -0.012       0.000
 1     2     1      2.436      -0.115       0.278      -0.199
                   -0.115       0.000       0.167       0.000
 1     2     2      2.436      -0.115       0.278       0.148
                   -0.115       0.000      -0.179       0.000
 1     2     3      2.436      -0.115       0.278       0.051
                   -0.115      -0.000       0.012       0.000
 2     1     1      2.436       0.115      -0.278      -0.199
                   -0.115       0.000      -0.167      -0.000
 2     1     2      2.436       0.115      -0.278       0.148
                   -0.115       0.000       0.179      -0.000
 2     1     3      2.436       0.115      -0.278       0.051
                   -0.115       0.000      -0.012      -0.000
 2     2     1      2.436       0.115       0.278      -0.199
                    0.115       0.000       0.167      -0.000
 2     2     2      2.436       0.115       0.278       0.148
                    0.115       0.000      -0.179      -0.000
 2     2     3      2.436       0.115       0.278       0.051
                    0.115       0.000       0.012      -0.000

G squared statistic for model fit = 7.055  D.F. = 4

No BC Effect

Expected Frequencies
      1    1    1     9.000
      1    1    2     9.000
      1    1    3     9.000
      1    2    1    12.000
      1    2    2    12.000
      1    2    3    12.000
      2    1    1     6.750
      2    1    2    10.607
      2    1    3     9.643
      2    2    1    14.250
      2    2    2    22.393
      2    2    3    20.357
```

```
Totals for Dimension A
Row 1     63.000
Row 2     84.000
Totals for Dimension B
Col 1     54.000
Col 2     93.000
Totals for Dimension C
Slice 1   42.000
Slice 2   54.000
Slice 3   51.000

Log Frequencies
   1   1   1    2.197
   1   1   2    2.197
   1   1   3    2.197
   1   2   1    2.485
   1   2   2    2.485
   1   2   3    2.485
   2   1   1    1.910
   2   1   2    2.362
   2   1   3    2.266
   2   2   1    2.657
   2   2   2    3.109
   2   2   3    3.013
Totals for Dimension A
Row 1     2.341
Row 2     2.553
Totals for Dimension B
Col 1     2.188
Col 2     2.706
Totals for Dimension C
Slice 1   2.312
Slice 2   2.538
Slice 3   2.490
```

Log Linear Screening, A×B and A×B×C Analyses   227

```
Cell Parameters
ROW    COL    SLICE      MU        LAMBDA A    LAMBDA B    LAMBDA C
                      LAMBDA AB    LAMBDA AC   LAMBDA BC   LAMBDA ABC
  1     1      1       2.447       -0.106      -0.259      -0.135
                       0.115        0.135       0.000      -0.000
  1     1      2       2.447       -0.106      -0.259       0.091
                       0.115       -0.091       0.000      -0.000
  1     1      3       2.447       -0.106      -0.259       0.044
                       0.115       -0.044      -0.000       0.000
  1     2      1       2.447       -0.106       0.259      -0.135
                      -0.115        0.135      -0.000       0.000
  1     2      2       2.447       -0.106       0.259       0.091
                      -0.115       -0.091      -0.000       0.000
  1     2      3       2.447       -0.106       0.259       0.044
                      -0.115       -0.044      -0.000       0.000
  2     1      1       2.447        0.106      -0.259      -0.135
                      -0.115       -0.135       0.000       0.000
  2     1      2       2.447        0.106      -0.259       0.091
                      -0.115        0.091       0.000       0.000
  2     1      3       2.447        0.106      -0.259       0.044
                      -0.115        0.044      -0.000       0.000
  2     2      1       2.447        0.106       0.259      -0.135
                       0.115       -0.135      -0.000       0.000
  2     2      2       2.447        0.106       0.259       0.091
                       0.115        0.091      -0.000       0.000
  2     2      3       2.447        0.106       0.259       0.044
                       0.115        0.044      -0.000       0.000

G squared statistic for model fit = 8.423 D.F. = 4

Model of No Slice (C) effect

Expected Frequencies
    1    1    1      7.714
    1    1    2      7.714
    1    1    3      7.714
    1    2    1     13.286
    1    2    2     13.286
    1    2    3     13.286
    2    1    1     10.286
    2    1    2     10.286
    2    1    3     10.286
    2    2    1     17.714
    2    2    2     17.714
    2    2    3     17.714
```

```
Totals for Dimension A
Row 1     63.000
Row 2     84.000
Totals for Dimension B
Col 1     54.000
Col 2     93.000
Totals for Dimension C
Slice 1   49.000
Slice 2   49.000
Slice 3   49.000

Log Frequencies
   1    1    1    2.043
   1    1    2    2.043
   1    1    3    2.043
   1    2    1    2.587
   1    2    2    2.587
   1    2    3    2.587
   2    1    1    2.331
   2    1    2    2.331
   2    1    3    2.331
   2    2    1    2.874
   2    2    2    2.874
   2    2    3    2.874
Totals for Dimension A
Row 1     2.315
Row 2     2.603
Totals for Dimension B
Col 1     2.187
Col 2     2.731
Totals for Dimension C
Slice 1   2.459
Slice 2   2.459
Slice 3   2.459
```

Log Linear Screening, A×B and A×B×C Analyses

```
Cell Parameters
ROW   COL   SLICE      MU       LAMBDA A    LAMBDA B    LAMBDA C
                    LAMBDA AB   LAMBDA AC   LAMBDA BC   LAMBDA ABC
 1     1     1      2.459       -0.144      -0.272       0.000
                    0.000        0.000       0.000      -0.000
 1     1     2      2.459       -0.144      -0.272       0.000
                    0.000        0.000       0.000      -0.000
 1     1     3      2.459       -0.144      -0.272       0.000
                    0.000        0.000       0.000      -0.000
 1     2     1      2.459       -0.144       0.272       0.000
                    0.000        0.000       0.000       0.000
 1     2     2      2.459       -0.144       0.272       0.000
                    0.000        0.000       0.000       0.000
 1     2     3      2.459       -0.144       0.272       0.000
                    0.000        0.000       0.000       0.000
 2     1     1      2.459        0.144      -0.272       0.000
                    0.000        0.000       0.000      -0.000
 2     1     2      2.459        0.144      -0.272       0.000
                    0.000        0.000       0.000      -0.000
 2     1     3      2.459        0.144      -0.272       0.000
                    0.000        0.000       0.000      -0.000
 2     2     1      2.459        0.144       0.272       0.000
                   -0.000        0.000       0.000       0.000
 2     2     2      2.459        0.144       0.272       0.000
                   -0.000        0.000       0.000       0.000
 2     2     3      2.459        0.144       0.272       0.000
                   -0.000        0.000       0.000       0.000

G squared statistic for model fit = 13.097  D.F. = 9

Model of no Column (B) effect

Expected Frequencies
     1     1     1      9.000
     1     1     2     11.571
     1     1     3     10.929
     1     2     1      9.000
     1     2     2     11.571
     1     2     3     10.929
     2     1     1     12.000
     2     1     2     15.429
     2     1     3     14.571
     2     2     1     12.000
     2     2     2     15.429
     2     2     3     14.571
```

```
Totals for Dimension A
Row 1    63.000
Row 2    84.000
Totals for Dimension B
Col 1    73.500
Col 2    73.500
Totals for Dimension C
Slice 1    42.000
Slice 2    54.000
Slice 3    51.000

Log Frequencies
   1   1   1    2.197
   1   1   2    2.449
   1   1   3    2.391
   1   2   1    2.197
   1   2   2    2.449
   1   2   3    2.391
   2   1   1    2.485
   2   1   2    2.736
   2   1   3    2.679
   2   2   1    2.485
   2   2   2    2.736
   2   2   3    2.679
Totals for Dimension A
Row 1   2.346
Row 2   2.633
Totals for Dimension B
Col 1   2.490
Col 2   2.490
Totals for Dimension C
Slice 1    2.341
Slice 2    2.592
Slice 3    2.535
```

Log Linear Screening, A×B and A×B×C Analyses

```
Cell Parameters
ROW   COL   SLICE     MU         LAMBDA A     LAMBDA B     LAMBDA C
                      LAMBDA AB  LAMBDA AC    LAMBDA BC    LAMBDA ABC
 1     1     1        2.490      -0.144       -0.000       -0.148
                      0.000       0.000        0.000       -0.000
 1     1     2        2.490      -0.144       -0.000        0.103
                      0.000       0.000        0.000       -0.000
 1     1     3        2.490      -0.144       -0.000        0.046
                      0.000       0.000        0.000       -0.000
 1     2     1        2.490      -0.144       -0.000       -0.148
                      0.000       0.000        0.000       -0.000
 1     2     2        2.490      -0.144       -0.000        0.103
                      0.000       0.000        0.000       -0.000
 1     2     3        2.490      -0.144       -0.000        0.046
                      0.000       0.000        0.000       -0.000
 2     1     1        2.490       0.144       -0.000       -0.148
                      0.000       0.000        0.000       -0.000
 2     1     2        2.490       0.144       -0.000        0.103
                      0.000       0.000        0.000       -0.000
 2     1     3        2.490       0.144       -0.000        0.046
                      0.000       0.000        0.000       -0.000
 2     2     1        2.490       0.144       -0.000       -0.148
                      0.000       0.000        0.000       -0.000
 2     2     2        2.490       0.144       -0.000        0.103
                      0.000       0.000        0.000       -0.000
 2     2     3        2.490       0.144       -0.000        0.046
                      0.000       0.000        0.000       -0.000

G squared statistic for model fit = 21.943 D.F. = 8

Model of no Row (A) effect

Expected Frequencies
     1    1    1       7.714
     1    1    2       9.918
     1    1    3       9.367
     1    2    1      13.286
     1    2    2      17.082
     1    2    3      16.133
     2    1    1       7.714
     2    1    2       9.918
     2    1    3       9.367
     2    2    1      13.286
     2    2    2      17.082
     2    2    3      16.133
```

```
Totals for Dimension A
Row 1    73.500
Row 2    73.500
Totals for Dimension B
Col 1    54.000
Col 2    93.000
Totals for Dimension C
Slice 1    42.000
Slice 2    54.000
Slice 3    51.000

Log Frequencies
    1   1   1    2.043
    1   1   2    2.294
    1   1   3    2.237
    1   2   1    2.587
    1   2   2    2.838
    1   2   3    2.781
    2   1   1    2.043
    2   1   2    2.294
    2   1   3    2.237
    2   2   1    2.587
    2   2   2    2.838
    2   2   3    2.781
Totals for Dimension A
Row 1    2.463
Row 2    2.463
Totals for Dimension B
Col 1    2.192
Col 2    2.735
Totals for Dimension C
Slice 1    2.315
Slice 2    2.566
Slice 3    2.509
```

```
Cell Parameters
ROW   COL   SLICE      MU        LAMBDA A    LAMBDA B    LAMBDA C
                   LAMBDA AB    LAMBDA AC   LAMBDA BC   LAMBDA ABC
 1     1     1       2.463        0.000      -0.272      -0.148
                     0.000       -0.000       0.000       0.000
 1     1     2       2.463        0.000      -0.272       0.103
                     0.000       -0.000       0.000       0.000
 1     1     3       2.463        0.000      -0.272       0.046
                     0.000       -0.000       0.000       0.000
 1     2     1       2.463        0.000       0.272      -0.148
                    -0.000       -0.000       0.000       0.000
 1     2     2       2.463        0.000       0.272       0.103
                    -0.000       -0.000       0.000       0.000
 1     2     3       2.463        0.000       0.272       0.046
                    -0.000       -0.000       0.000       0.000
 2     1     1       2.463        0.000      -0.272      -0.148
                     0.000       -0.000       0.000       0.000
 2     1     2       2.463        0.000      -0.272       0.103
                     0.000       -0.000       0.000       0.000
 2     1     3       2.463        0.000      -0.272       0.046
                     0.000       -0.000       0.000       0.000
 2     2     1       2.463        0.000       0.272      -0.148
                    -0.000       -0.000       0.000       0.000
 2     2     2       2.463        0.000       0.272       0.103
                    -0.000       -0.000       0.000       0.000
 2     2     3       2.463        0.000       0.272       0.046
                    -0.000       -0.000       0.000       0.000

G squared statistic for model fit = 14.481 D.F. = 8

Equi-probability Model

Expected Frequencies
      1     1     1     12.250
      1     1     2     12.250
      1     1     3     12.250
      1     2     1     12.250
      1     2     2     12.250
      1     2     3     12.250
      2     1     1     12.250
      2     1     2     12.250
      2     1     3     12.250
      2     2     1     12.250
      2     2     2     12.250
      2     2     3     12.250
```

```
Totals for Dimension A
Row 1     73.500
Row 2     73.500
Totals for Dimension B
Col 1     73.500
Col 2     73.500
Totals for Dimension C
Slice     1 49.000
Slice     2 49.000
Slice     3 49.000

Log Frequencies
   1    1    1      2.506
   1    1    2      2.506
   1    1    3      2.506
   1    2    1      2.506
   1    2    2      2.506
   1    2    3      2.506
   2    1    1      2.506
   2    1    2      2.506
   2    1    3      2.506
   2    2    1      2.506
   2    2    2      2.506
   2    2    3      2.506
Totals for Dimension A
Row 1     2.506
Row 2     2.506
Totals for Dimension B
Col 1     2.506
Col 2     2.506
Totals for Dimension C
Slice 1   2.506
Slice 2   2.506
Slice 3   2.506
```

```
Cell Parameters
ROW    COL    SLICE     MU        LAMBDA A    LAMBDA B    LAMBDA C
                        LAMBDA AB LAMBDA AC   LAMBDA BC   LAMBDA ABC
 1      1      1        2.506     0.000       0.000       0.000
                        0.000     0.000       0.000       0.000
 1      1      2        2.506     0.000       0.000       0.000
                        0.000     0.000       0.000       0.000
 1      1      3        2.506     0.000       0.000       0.000
                        0.000     0.000       0.000       0.000
 1      2      1        2.506     0.000       0.000       0.000
                        0.000     0.000       0.000       0.000
 1      2      2        2.506     0.000       0.000       0.000
                        0.000     0.000       0.000       0.000
 1      2      3        2.506     0.000       0.000       0.000
                        0.000     0.000       0.000       0.000
 2      1      1        2.506     0.000       0.000       0.000
                        0.000     0.000       0.000       0.000
 2      1      2        2.506     0.000       0.000       0.000
                        0.000     0.000       0.000       0.000
 2      1      3        2.506     0.000       0.000       0.000
                        0.000     0.000       0.000       0.000
 2      2      1        2.506     0.000       0.000       0.000
                        0.000     0.000       0.000       0.000
 2      2      2        2.506     0.000       0.000       0.000
                        0.000     0.000       0.000       0.000
 2      2      3        2.506     0.000       0.000       0.000
                        0.000     0.000       0.000       0.000

G squared statistic for model fit = 26.579 D.F. = 11
```

Chapter 10
Non-parametric

Contingency Chi-Square

Example Contingency Chi Square

In this example we will use the data file ChiData.txt which consists of two columns of data representing the row and column of a three by three contingency table. The rows represent each observation with the row and column of that observation recorded in columns one and two. We begin by selecting the Statistics/Non Parametric / Contingency Chi Square option of the menu. The following figure (Fig. 10.1) demonstrates that the row and column labels have been selected for the option of reading a data file containing individual cases. We have also elected all options except saving the frequency file.

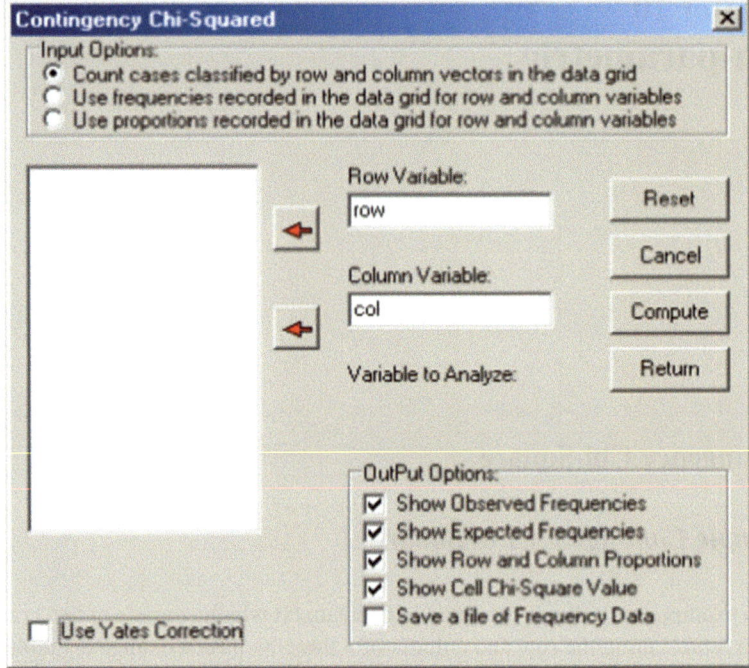

Fig. 10.1 Contingency Chi-Square Dialog form

When we click the compute button, we obtain the results shown below:

```
Chi-square Analysis Results
OBSERVED FREQUENCIES
                        Rows
Variables
                COL.1       COL.2       COL.3       COL.4       Total
    Row 1         5           5           5           5            20
    Row 2        10           4           7           3            24
    Row 3         5          10          10           2            27
    Total        20          19          22          10            71
EXPECTED FREQUENCIES with 71 valid cases.

Variables
                COL.1       COL.2       COL.3       COL.4
    Row 1       5.634       5.352       6.197       2.817
    Row 2       6.761       6.423       7.437       3.380
    Row 3       7.606       7.225       8.366       3.803
ROW PROPORTIONS with 71 valid cases.
```

```
Variables
            COL.1       COL.2       COL.3       COL.4       Total
    Row 1   0.250       0.250       0.250       0.250       1.000
    Row 2   0.417       0.167       0.292       0.125       1.000
    Row 3   0.185       0.370       0.370       0.074       1.000
    Total   0.282       0.268       0.310       0.141       1.000
COLUMN PROPORTIONS with 71 valid cases.

Variables
            COL.1   COL.2   COL.3   COL.4   Total
    Row 1   0.250       0.263       0.227       0.500       0.282
    Row 2   0.500       0.211       0.318       0.300       0.338
    Row 3   0.250       0.526       0.455       0.200       0.380
    Total   1.000       1.000       1.000       1.000       1.000
PROPORTIONS OF TOTAL N with 71 valid cases.

Variables
            COL.1       COL.2       COL.3       COL.4       Total
    Row 1   0.070       0.070       0.070       0.070       0.282
    Row 2   0.141       0.056       0.099       0.042       0.338
    Row 3   0.070       0.141       0.141       0.028       0.380
    Total   0.282       0.268       0.310       0.141       1.000
CHI-SQUARED VALUE FOR CELLS with 71 valid cases.

Variables
            COL.1       COL.2       COL.3       COL.4
    Row 1   0.071       0.023       0.231       1.692
    Row 2   1.552       0.914       0.026       0.043
    Row 3   0.893       1.066       0.319       0.855
Chi-square = 7.684 with D.F. = 6. Prob. > value = 0.262
```

It should be noted that the user has the option of reading data in three different formats. We have shown the first format where individual cases are classified by row and column. It is sometimes more convenient to record the actual frequencies in each row and cell combination. Examine the file labeled ChiSquareOne.TXT for such an example. Sometimes the investigator may only know the cell proportions and the total number of observations. In this case the third file format may be used where the proportion in each row and column combination are recorded. See the example file labeled ChiSquareTwo.TXT.

Spearman Rank Correlation

Example Spearman Rank Correlation

We will use the file labeled Spearman.txt for our example. The third variable represents rank data with ties. Select the Statistics/Non Parametric/Spearman Rank Correlation option from the menu. Shown below is the specification form for the analysis (Fig. 10.2):

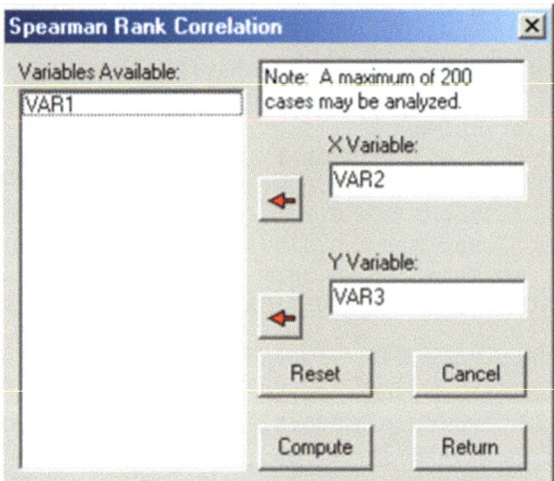

Fig. 10.2 The Spearman rank correlation dialog

When we click the Compute button we obtain:

```
Spearman Rank Correlation Between VAR2 & VAR3
Observed scores, their ranks and differences between ranks
     VAR2        Ranks        VAR3       Ranks      Rank Difference
     42.00       3.00         0.00       1.50            1.50
     46.00       4.00         0.00       1.50            2.50
     39.00       2.00         1.00       3.50           -1.50
     37.00       1.00         1.00       3.50           -2.50
     65.00       8.00         3.00       5.00            3.00
     88.00      11.00         4.00       6.00            5.00
     86.00      10.00         5.00       7.00            3.00
     56.00       6.00         6.00       8.00           -2.00
     62.00       7.00         7.00       9.00           -2.00
     92.00      12.00         8.00      10.50            1.50
     54.00       5.00         8.00      10.50           -5.50
     81.00       9.00        12.00      12.00           -3.00
Spearman Rank Correlation =   0.615
t-test value for hypothesis r = 0 is 2.467
Probability > t = 0.0333
```

Notice that the original scores have been converted to ranks and where ties exist they have been averaged.

Mann-Whitney U Test

As an example, load the file labeled MannWhitU.txt and then select the option Statistics/Non Parametric/Mann-Whitney U Test from the menu. Shown below is the specification form in which we have indicated the analysis to perform (Fig. 10.3):

Fig. 10.3 The Mann-Whitney U Test dialog form

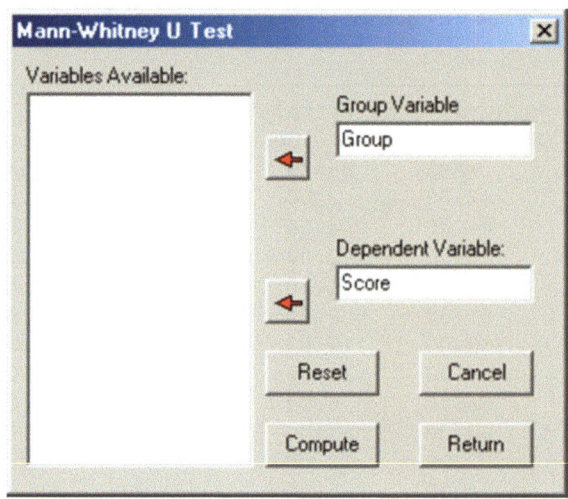

Upon clicking the Compute button you obtain:

```
Mann-Whitney U Test
See pages 116-127 in S. Siegel: Nonparametric Statistics for the
Behavioral Sciences

        Score       Rank        Group

         6.00       1.50           1
         6.00       1.50           2
         7.00       5.00           1
         7.00       5.00           1
         7.00       5.00           1
         7.00       5.00           1
         7.00       5.00           1
         8.00       9.50           1
         8.00       9.50           2
         8.00       9.50           2
         8.00       9.50           1
         9.00      12.00           1
        10.00      16.00           1
        10.00      16.00           2
        10.00      16.00           2
        10.00      16.00           2
        10.00      16.00           1
```

Fisher's Exact Test

```
        10.00     16.00           1
        10.00     16.00           1
        11.00     20.50           2
        11.00     20.50           2
        12.00     24.50           2
        12.00     24.50           2
        12.00     24.50           2
        12.00     24.50           2
        12.00     24.50           1
        12.00     24.50           1
        13.00     29.50           1
        13.00     29.50           2
        13.00     29.50           2
        13.00     29.50           2
        14.00     33.00           2
        14.00     33.00           2
        14.00     33.00           2
        15.00     36.00           2
        15.00     36.00           2
        15.00     36.00           2
        16.00     38.00           2
        17.00     39.00           2

Sum of Ranks in each Group
Group    Sum    No. in Gr
  1       200.00    16
  2       580.00    23
No. of tied rank groups =   9
Statistic U = 304.0000
z Statistic (corrected for ties) =   3.4262, Prob. > z = 0.0003
```

Fisher's Exact Test

When you elect the Statistics/NonParametric / Fisher's Exact Test option from the menu, you are shown a specification form which provides for four different formats for entering data. We have elected the last format (entry of frequencies on the form itself) (Fig. 10.4):

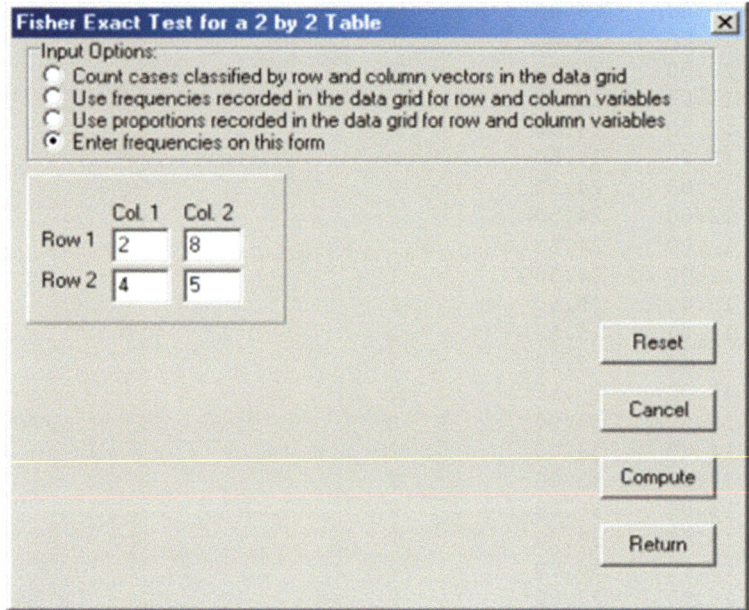

Fig. 10.4 Fisher's Exact Test dialog form

When we click the Compute button we obtain:

```
Fisher Exact Probability Test
Contingency Table for Fisher Exact Test
                Column
Row              1              2
 1               2              8
 2               4              5
Probability := 0.2090


Cumulative Probability := 0.2090

Contingency Table for Fisher Exact Test
                Column
Row              1              2
 1               1              9
 2               5              4
Probability := 0.0464


Cumulative Probability := 0.2554
```

```
Contingency Table for Fisher Exact Test
                Column
Row             1               2
 1              0               10
 2              6               3
Probability := 0.0031


Cumulative Probability := 0.2585

Tocher ratio computed: 0.002
A random value of 0.893 selected was greater than the Tocher value.
Conclusion: Accept the null Hypothesis
```

Notice that the probability of each combination of cell values as extreme or more extreme than that observed is computed and the probabilities summed.

Alternative formats for data files are the same as for the Contingency Chi Square analysis discussed in the previous section.

Kendall's Coefficient of Concordance

Our example analysis will use the file labeled Concord2.txt . Load the file and select the Statistics / NonParametric/Coefficient of Concordance option. Shown below is the form completed for the analysis (Fig. 10.5):

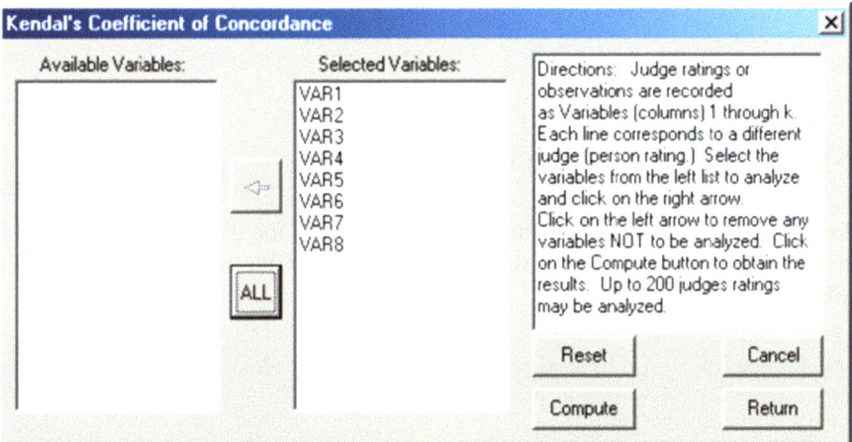

Fig. 10.5 Kendal's coefficient of concordance

Clicking the Compute button results in the following output:

If you are observing competition in the Olympics or other athletic competitions, it is fun to record the judge's scores and examine the degree to which there is agreement among them!

```
Kendall Coefficient of Concordance Analysis

Ranks Assigned to Judge Ratings of Objects

Judge 1        Objects
VAR1   VAR2    VAR3    VAR4    VAR5    VAR6    VAR7    VAR8
12.0   1.5000  3.5000  3.5000  5.5000  5.5000  7.0000  8.0000

Judge 2        Objects
VAR1   VAR2    VAR3    VAR4    VAR5    VAR6    VAR7    VAR8
12.0   2.0000  3.0000  4.0000  5.0000  6.0000  7.0000  8.0000

Judge 3        Objects
VAR1   VAR2    VAR3    VAR4    VAR5    VAR6    VAR7    VAR8
12.0   2.5000  2.5000  2.5000  6.5000  6.5000  6.5000  6.5000

Sum of Ranks for Each Object Judged
      Objects
VAR1   VAR2    VAR3    VAR4     VAR5     VAR6     VAR7     VAR8
12.0   6.0000  9.0000  10.0000  17.0000  18.0000  20.5000  22.5000

Coefficient of concordance :=        0.942
Average Spearman Rank Correlation :=        0.913
Chi-Square Statistic :=     19.777
Probability of a larger Chi-Square := 0.0061
```

Kruskal-Wallis One-Way ANOVA

As an example, load the file labeled kwanova.txt into the data grid and select the menu option for the analysis. Below is the form and the results of the analysis (Fig. 10.6):

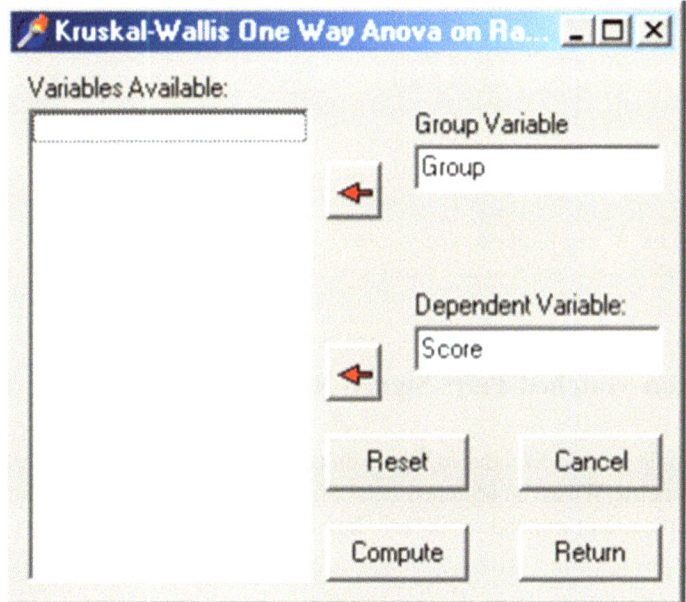

Fig. 10.6 Kruskal-Wallis one way ANOVA on ranks dialog

```
Kruskal - Wallis One-Way Analysis of Variance
See pages 184-194 in S. Siegel: Nonparametric Statistics for the
Behavioral Sciences
     Score      Rank      Group

     61.00      1.00         1
     82.00      2.00         2
     83.00      3.00         1
     96.00      4.00         1
    101.00      5.00         1
    109.00      6.00         2
    115.00      7.00         3
    124.00      8.00         2
    128.00      9.00         1
    132.00     10.00         2
    135.00     11.00         2
    147.00     12.00         3
    149.00     13.00         3
    166.00     14.00         3
```

```
Sum of Ranks in each Group
Group    Sum     No. in Group
  1      22.00       5
  2      37.00       5
  3      46.00       4
No. of tied rank groups = 0
Statisic H uncorrected for ties = 6.4057
Correction for Ties = 1.0000
Statistic H corrected for ties = 6.4057
Corrected H is approx. chi-square with 2 D.F. and probability = 0.0406
```

Wilcoxon Matched-Pairs Signed Ranks Test

Our example uses the file labeled Wilcoxon.txt. Load this file and select the Statistics/NonParametric/Wilcoxon Matched-Pairs Signed Ranks Test option from the menu. The specification form and results are shown below (Fig. 10.7):

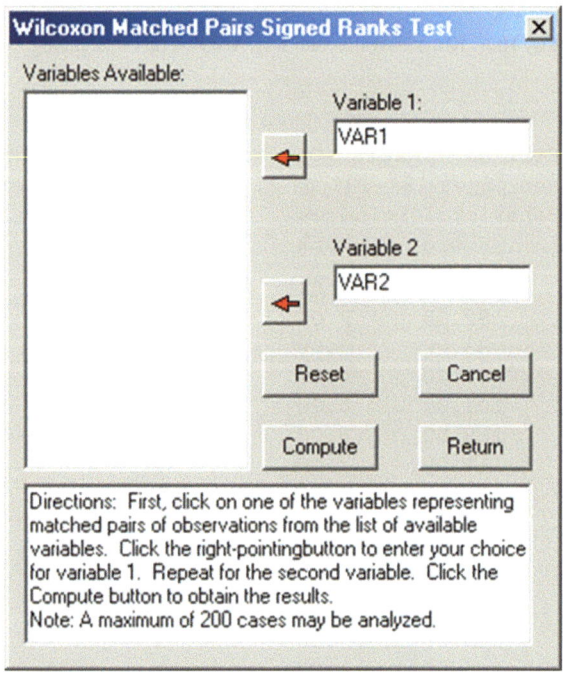

Fig. 10.7 Wilcoxon matched pairs signed ranks test dialog

The Wilcoxon Matched-Pairs Signed-Ranks Test

See pages 75-83 in S. Seigel: Nonparametric Statistics for the Social Sciences

```
Ordered Cases with cases having 0 differences eliminated:
Number of cases with absolute differences greater than 0 = 10
CASE       VAR1        VAR2       Difference     Signed Rank
  3       73.00       74.00         -1.00           -1.00
  8       65.00       62.00          3.00            2.00
  7       76.00       80.00         -4.00           -3.00
  4       43.00       37.00          6.00            4.00
  5       58.00       51.00          7.00            5.00
  6       56.00       43.00         13.00            6.50
 10       56.00       43.00         13.00            6.50
  9       82.00       63.00         19.00            8.50
  1       82.00       63.00         19.00            8.50
  2       69.00       42.00         27.00           10.00

Smaller sum of ranks (T) = 4.00
Approximately normal z for test statistic T = 2.395
Probability (1-tailed) of greater z = 0.0083
NOTE: For N < 25 use tabled values for Wilcoxon Test
```

Cochran Q Test

Load the file labeled Qtest.txt and select the Statistics/NonParametric/Cochran Q Test option from the menu. Shown below is the specification form completed for the analysis of the file data and the results obtained when you click the Compute button (Fig. 10.8):

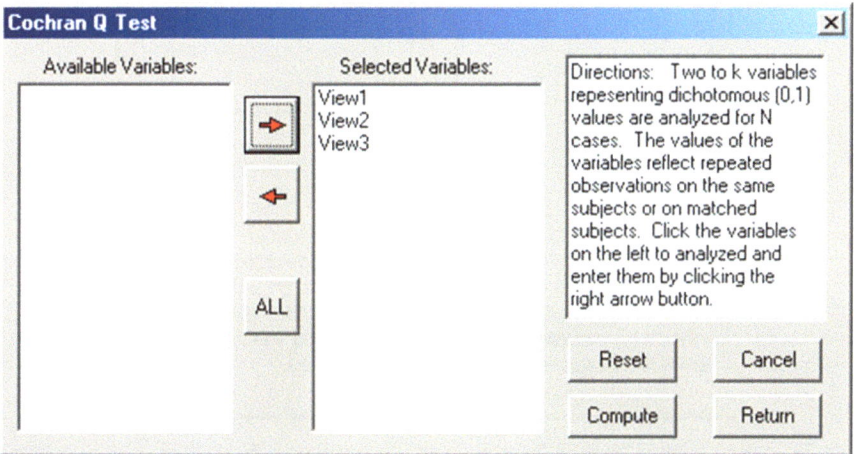

Fig. 10.8 Cochran Q Test Dialog form

Cochran Q Test for Related Samples
See pages 161-166 in S. Siegel: Nonparametric Statistics for the
Behavioral Sciences
McGraw-Hill Book Company, New York, 1956

Cochran Q Statistic = 16.667
which is distributed as chi-square with 2 D.F. and probability = 0.0002

Sign Test

The file labeled SignTest.txt contains male and female cases in which have been matched on relevant criteria and observations have been made on a 5-point Likert-type instrument. The program will compare the two scores for each pair and assign a positive or negative difference indicator. Load the file into the data grid and select the Statistics/NonParametric/Sign Test option. Shown below is the specification form which appears and the results obtained when clicking the Compute button (Fig. 10.9):

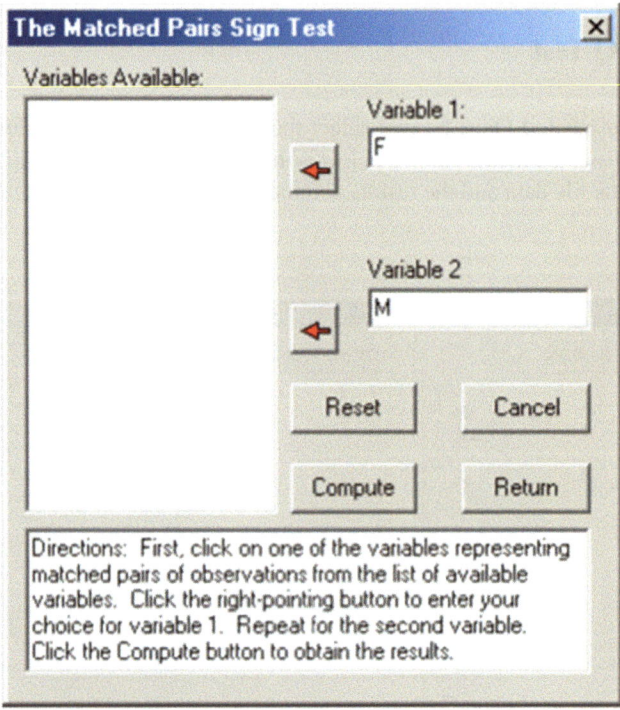

Fig. 10.9 The matched pairs sign test dialog

Friedman Two Way ANOVA

Results for the Sign Test

Frequency of 11 out of 17 observed + sign differences.
Frequency of 3 out of 17 observed - sign differences.
Frequency of 3 out of 17 observed no differences.
The theoretical proportion expected for +'s or -'s is 0.5
The test is for the probability of the +'s or -'s (which ever is fewer)
as small or smaller than that observed given the expected proportion.

Binary Probability of 0 = 0.0001
Binary Probability of 1 = 0.0009
Binary Probability of 2 = 0.0056
Binary Probability of 3 = 0.0222
Binomial Probability of 3 or smaller out of 14 = 0.0287

Friedman Two Way ANOVA

For an example analysis, load the file labeled Friedman.txt and select Statistics / NonParametric / Friedman Two Way ANOVA from the menu. The data represent four treatments or repeated measures for three groups, each containing one subject. Shown below is the specification form and the results following a click of the Compute button (Fig. 10.10):

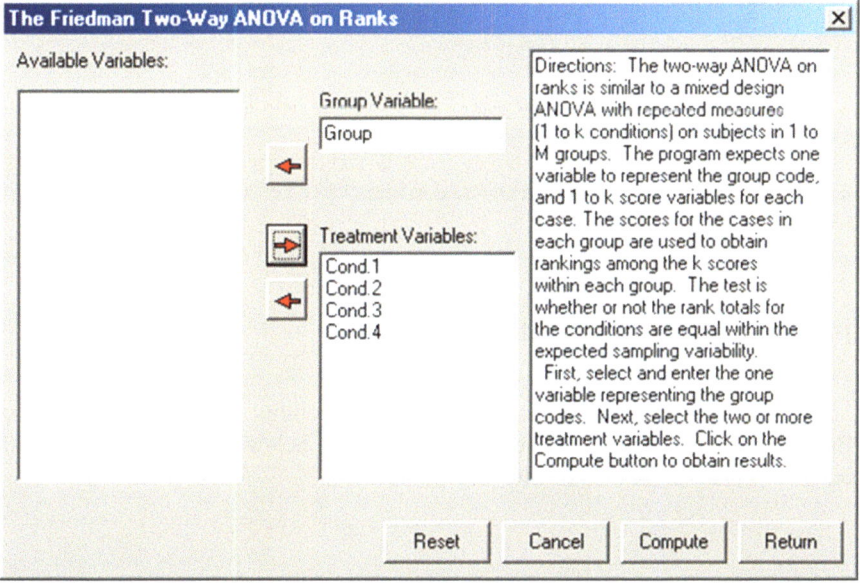

Fig. 10.10 The Friedman Two-Way ANOVA dialog

```
FRIEDMAN TWO-WAY ANOVA ON RANKS
See pages 166-173 in S. Siegel's Nonparametric Statistics
for the Behavioral Sciences, McGraw-Hill Book Co., New York, 1956

Treatment means - values to be ranked. with 3 valid cases.

Variables
                Cond.1      Cond.2      Cond.3      Cond.4
    Group 1     9.000       4.000       1.000       7.000
    Group 2     6.000       5.000       2.000       8.000
    Group 3     9.000       1.000       2.000       6.000

Number in each group's treatment.
                        GROUP
Variables
                Cond.1      Cond.2      Cond.3      Cond.4
    Group 1       1           1           1           1
    Group 2       1           1           1           1
    Group 3       1           1           1           1

Score Rankings Within Groups with 3 valid cases.
Variables
                Cond.1      Cond.2      Cond.3      Cond.4
    Group 1     4.000       2.000       1.000       3.000
    Group 2     3.000       2.000       1.000       4.000
    Group 3     4.000       1.000       2.000       3.000

TOTAL RANKS with 3 valid cases.
Variables       Cond.1      Cond.2      Cond.3      Cond.4
                11.000      5.000       4.000       10.000

Chi-square with 3 D.F. :=    7.400 with probability := 0.0602
Chi-square too approximate-use exact table (TABLE N)
page 280-281 in Siegel
```

Probability of a Binomial Event

Select the Statistics/NonParametric/Binomial Probability option from the menu. Enter the values as shown in the specification form below and press the Compute button to obtain the shown results (Fig. 10.11).

Fig. 10.11 The binomial probability dialog

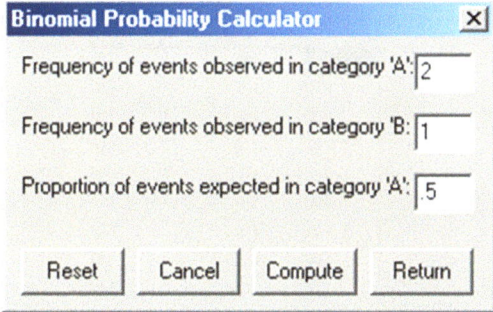

```
Binomial Probability Test

Frequency of 2 out of 3 observed
The theoretical proportion expected in category A is 0.500
The test is for the probability of a value in category A as small
or smaller
than that observed given the expected proportion.
Probability of 0 = 0.1250
Probability of 1 = 0.3750
Probability of 2 = 0.3750
Binomial Probability of 2 or less out of 3 = 0.8750
```

Runs Test

EXAMPLE:
The figure below (Fig. 10.12) shows a data set with 14 values in a file labeled "RunsTest.tab". The Runs Test option was selected from the NonParametric submenu under the Analyses menu. The next figure (Fig. 10.13) displays the dialogue box used for specifying the variable to analyze and the results of clicking the compute button.

CASE/VAR.	VAR1	VAR2	VAR3	VAR4
CASE 1	1.00	1.00	1	1.00
CASE 2	0.00	1.00	1	1.00
CASE 3	0.00	2.00	1	1.00
CASE 4	1.00	3.00	2	0.00
CASE 5	0.00	4.00	2	0.00
CASE 6	0.00	5.00	2	0.00
CASE 7	0.00	4.00	3	0.00
CASE 8	1.00	3.00	3	0.00
CASE 9	1.00	3.00	3	1.00
CASE 10	0.00	2.00	5	1.00
CASE 11	1.00	1.00	4	1.00
CASE 12	0.00	0.00	3	1.00
CASE 13	0.00	1.00	2	0.00
CASE 14	0.00	1.00	1	0.00

Fig. 10.12 A sample file for the runs test

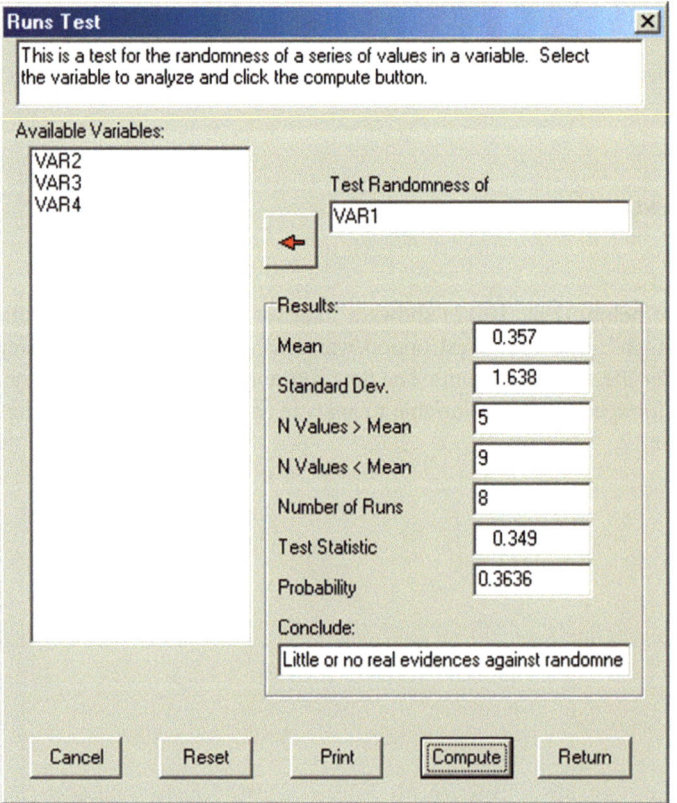

Fig. 10.13 The Runs Dialog form

Kendall's Tau and Partial Tau

```
Ranks with 12 cases.
Variables
                  X          Y          Z
      1       3.000      2.000      1.500
      2       4.000      6.000      1.500
      3       2.000      5.000      3.500
      4       1.000      1.000      3.500
      5       8.000     10.000      5.000
      6      11.000      9.000      6.000
      7      10.000      8.000      7.000
      8       6.000      3.000      8.000
      9       7.000      4.000      9.000
     10      12.000     12.000     10.500
     11       5.000      7.000     10.500
     12       9.000     11.000     12.000

Kendall Tau for File: C:\Projects\Delphi\OPENSTAT\TauData.TAB

Kendall Tau for variables X and Y
Tau = 0.6667 z = 3.017 probability > |z| = 0.001

Kendall Tau for variables X and Z
Tau = 0.3877 z = 1.755 probability > |z| = 0.040

Kendall Tau for variables Y and Z
Tau = 0.3567 z = 1.614 probability > |z| = 0.053

Partial Tau = 0.6136

NOTE: Probabilities are for large N (>10)
```

At the time this program was written, the distribution of the Partial Tau was unknown (see Siegel 1956, page 228) (Fig. 10.14).

Fig. 10.14 Kendal's Tau and Partial Tau dialog

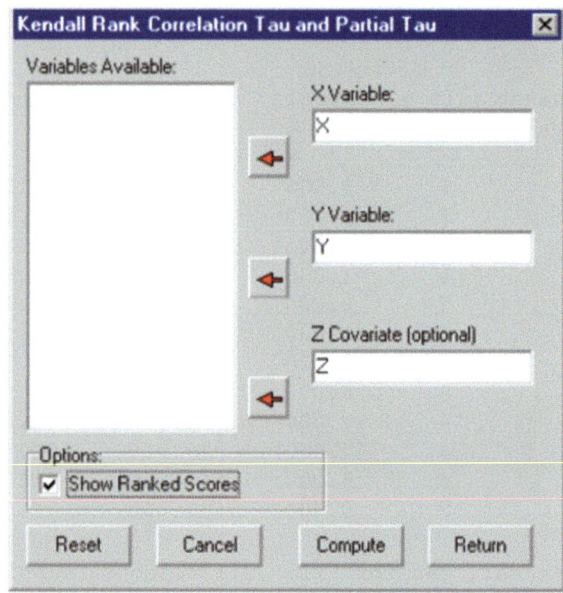

Kaplan-Meier Survival Test

CASES FOR FILE C:\OpenStat\KaplanMeier1.TEX

```
            0         Time     Event_Censored
            1            1                  2
            2            3                  2
            3            5                  2
            4            6                  1
            5            6                  1
            6            6                  1
            7            6                  1
            8            6                  1
            9            6                  1
           10            8                  1
           11            8                  1
           12            9                  2
           13           10                  1
           14           10                  1
           15           10                  2
           16           12                  1
           17           12                  1
```

18	12	1
19	12	1
20	12	1
21	12	1
22	12	2
23	12	2
24	13	2
25	15	2
26	15	2
27	16	2
28	16	2
29	18	2
30	18	2
31	20	1
32	20	2
33	22	2
34	24	1
35	24	1
36	24	2
37	27	2
38	28	2
39	28	2
40	28	2
41	30	1
42	30	2
43	32	1
44	33	2
45	34	2
46	36	2
47	36	2
48	42	1
49	44	2

We are really recording data for the "Time" variable that is sequential through the data file. We are concerned with the percent of survivors at any given time period as we progress through the observation times of the study. We record the "drop-outs" or censored subjects at each time period also. A unit cannot be censored and be one of the deaths - these are mutually exclusive.

Next we show a data file that contains both experimental and control subjects:

```
CASES FOR FILE C:\OpenStat\KaplanMeier2.TEX
```

	Time	Group	Event_Censored
1	1	1	2
2	3	2	2
3	5	1	2
4	6	1	1
5	6	1	1
6	6	2	1
7	6	2	1
8	6	2	1
9	6	2	1
10	8	2	1
11	8	2	1
12	9	1	2
13	10	1	1
14	10	1	1
15	10	1	2
16	12	1	1
17	12	1	1
18	12	1	1
19	12	1	1
20	12	2	1
21	12	2	1
22	12	1	2
23	12	2	2
24	13	1	2
25	15	1	2
26	15	2	2
27	16	1	2
28	16	2	2
29	18	2	2
30	18	2	2
31	20	2	1
32	20	1	2
33	22	2	2
34	24	1	1
35	24	2	1
36	24	1	2
37	27	1	2
38	28	2	2
39	28	2	2

Kaplan-Meier Survival Test

40	28	2	2
41	30	2	1
42	30	2	2
43	32	1	1
44	33	2	2
45	34	1	2
46	36	1	2
47	36	1	2
48	42	2	1
49	44	1	2

In this data we code the groups as 1 or 2. Censored cases are always coded 2 and Events are coded 1. This data is, in fact, the same data as shown in the previous data file. Note that in time period 6 there were 6 deaths (cases 4–9.) Again, notice that the time periods are in ascending order.

Shown below is the specification dialog for this second data file. This is followed by the output obtained when you click the compute button (Fig. 10.15).

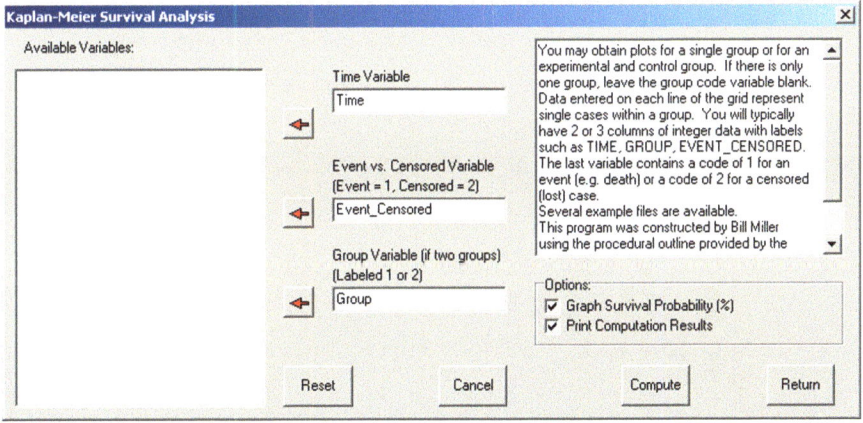

Fig. 10.15 The Kaplan-Meier dialog

Comparison of Two Groups Methd

TIME	GROUP	CENSORED	TOTAL AT RISK	EVENTS	AT RISK IN GROUP 1	EXPECTED NO. EVENTS IN 1	AT RISK IN GROUP 2	EXPECTED NO. EVENTS IN 2
0	0	0	49	0	25	0.0000	24	0.0000
1	1	1	49	0	25	0.0000	24	0.0000
3	2	1	48	0	24	0.0000	24	0.0000
5	1	1	47	0	24	0.0000	23	0.0000
6	1	0	46	6	23	3.0000	23	3.0000
6	1	0	40	0	21	0.0000	19	0.0000
6	2	0	40	0	21	0.0000	19	0.0000
6	2	0	40	0	21	0.0000	19	0.0000
6	2	0	40	0	21	0.0000	19	0.0000
6	2	0	40	0	21	0.0000	19	0.0000
8	2	0	40	2	21	1.0500	19	0.9500
8	2	0	38	0	21	0.0000	17	0.0000
9	1	1	38	0	21	0.0000	17	0.0000
10	1	0	37	2	20	1.0811	17	0.9189
10	1	0	35	0	18	0.0000	17	0.0000
10	1	1	35	0	18	0.0000	17	0.0000
12	1	0	34	6	17	3.0000	17	3.0000
12	1	0	28	0	13	0.0000	15	0.0000
12	1	0	28	0	13	0.0000	15	0.0000
12	1	0	28	0	13	0.0000	15	0.0000
12	2	0	28	0	13	0.0000	15	0.0000
12	2	0	28	0	13	0.0000	15	0.0000
12	1	1	28	0	13	0.0000	15	0.0000

Kaplan-Meier Survival Test

time	group	event	n1	d1	n2	S1	n3	S2
12	2	1	27	0	12	0.0000	15	0.0000
13	1	1	26	0	12	0.0000	14	0.0000
15	1	1	25	0	11	0.0000	14	0.0000
15	2	1	24	0	10	0.0000	14	0.0000
16	1	1	23	0	10	0.0000	13	0.0000
16	2	1	22	0	9	0.0000	13	0.0000
18	2	1	21	0	9	0.0000	12	0.0000
18	2	0	20	1	9	0.4737	11	0.5263
20	2	1	19	0	9	0.0000	10	0.0000
20	1	1	18	0	8	0.0000	10	0.0000
22	2	0	17	2	8	1.0000	9	1.0000
24	1	1	16	0	7	0.0000	9	0.0000
24	2	1	14	0	7	0.0000	8	0.0000
24	1	1	14	0	6	0.0000	7	0.0000
27	1	1	13	0	5	0.0000	7	0.0000
28	2	1	12	0	5	0.0000	7	0.0000
28	2	1	11	0	5	0.0000	7	0.0000
28	2	1	10	0	5	0.5556	6	0.4444
30	2	0	9	1	5	0.0000	5	0.0000
30	1	1	8	0	5	0.7143	4	0.2857
32	2	0	7	1	4	0.0000	3	0.0000
33	1	1	6	0	4	0.0000	2	0.0000
34	2	1	5	0	3	0.0000	2	0.0000
36	1	1	4	0	2	0.0000	1	0.0000
36	1	1	3	0	1	0.5000	1	0.5000
42	2	0	2	1	1	0.0000	1	0.0000
44	1	1	0	0	1	0.0000	0	0.0000

TIME	DEATHS	GROUP	AT RISK	PROPORTION SURVIVING	CUMULATIVE PROP.SURVIVING
1	0	1	25	0.0000	1.0000
3	0	2	24	0.0000	1.0000
5	0	1	24	0.0000	1.0000
6	6	1	23	0.9130	0.9130
6	0	1	21	0.0000	0.9130
6	0	2	19	0.0000	0.8261
6	0	2	19	0.0000	0.8261
6	0	2	19	0.0000	0.8261
6	0	2	19	0.0000	0.8261
8	2	2	19	0.8947	0.7391
8	0	2	17	0.0000	0.7391
9	0	1	21	0.0000	0.9130
10	2	1	20	0.9000	0.8217
10	0	1	18	0.0000	0.8217
10	0	1	18	0.0000	0.8217
12	6	1	17	0.7647	0.6284
12	0	1	13	0.0000	0.6284
12	0	1	13	0.0000	0.6284
12	0	1	13	0.0000	0.6284
12	0	2	15	0.0000	0.6522
12	0	2	15	0.0000	0.6522
12	0	1	13	0.0000	0.6284
12	0	2	15	0.0000	0.6522
13	0	1	12	0.0000	0.6284
15	0	1	11	0.0000	0.6284
15	0	2	14	0.0000	0.6522
16	0	1	10	0.0000	0.6284
16	0	2	13	0.0000	0.6522
18	0	2	12	0.0000	0.6522
18	0	2	11	0.0000	0.6522
20	1	2	10	0.9000	0.5870
20	0	1	9	0.0000	0.6284
22	0	2	9	0.0000	0.5870
24	2	1	8	0.8750	0.5498
24	0	2	7	0.0000	0.5136
24	0	1	7	0.0000	0.5498
27	0	1	6	0.0000	0.5498
28	0	2	7	0.0000	0.5136
28	0	2	6	0.0000	0.5136
28	0	2	5	0.0000	0.5136
30	1	2	4	0.7500	0.3852
30	0	2	3	0.0000	0.3852
32	1	1	5	0.8000	0.4399
33	0	2	2	0.0000	0.3852
34	0	1	4	0.0000	0.4399
36	0	1	3	0.0000	0.4399
36	0	1	2	0.0000	0.4399
42	1	2	1	0.0000	0.0000
44	0	1	1	0.0000	0.4399

Kaplan-Meier Survival Test

```
Total Expected Events for Experimental Group =   11.375
Observed Events for Experimental Group =   10.000
Total Expected Events for Control Group =   10.625
Observed Events for Control Group =   12.000
Chisquare = 0.344 with probability =   0.442
Risk = 0.778, Log Risk = -0.250, Std.Err. Log Risk =   0.427
95 Percent Confidence interval for Log Risk = (-1.087,0.586)
95 Percent Confidence interval for Risk = (0.337,1.796)
```

EXPERIMENTAL GROUP CUMULATIVE PROBABILITY

CASE	TIME	DEATHS	CENSORED	CUM.PROB.
1	1	0	1	1.000
3	5	0	1	1.000
4	6	6	0	0.913
5	6	0	0	0.913
12	9	0	1	0.913
13	10	2	0	0.822
14	10	0	0	0.822
15	10	0	1	0.822
16	12	6	0	0.628
17	12	0	0	0.628
18	12	0	0	0.628
19	12	0	0	0.628
22	12	0	1	0.628
24	13	0	1	0.628
25	15	0	1	0.628
27	16	0	1	0.628
32	20	0	1	0.628
34	24	2	0	0.550
36	24	0	1	0.550
37	27	0	1	0.550
43	32	1	0	0.440
45	34	0	1	0.440
46	36	0	1	0.440
47	36	0	1	0.440
49	44	0	1	0.440

CONTROL GROUP CUMULATIVE PROBABILITY

CASE	TIME	DEATHS	CENSORED	CUM.PROB.
2	3	0	1	1.000
6	6	0	0	0.826
7	6	0	0	0.826
8	6	0	0	0.826
9	6	0	0	0.826
10	8	2	0	0.739
11	8	0	0	0.739
20	12	0	0	0.652
21	12	0	0	0.652
23	12	0	1	0.652

26	15	0	1	0.652
28	16	0	1	0.652
29	18	0	1	0.652
30	18	0	1	0.652
31	20	1	0	0.587
33	22	0	1	0.587
35	24	0	0	0.514
38	28	0	1	0.514
39	28	0	1	0.514
40	28	0	1	0.514
41	30	1	0	0.385
42	30	0	1	0.385
44	33	0	1	0.385
48	42	1	0	0.000

The chi-square coefficient as well as the graph indicates no difference was found between the experimental and control group beyond what is reasonably expected through random selection from the same population (Fig. 10.16).

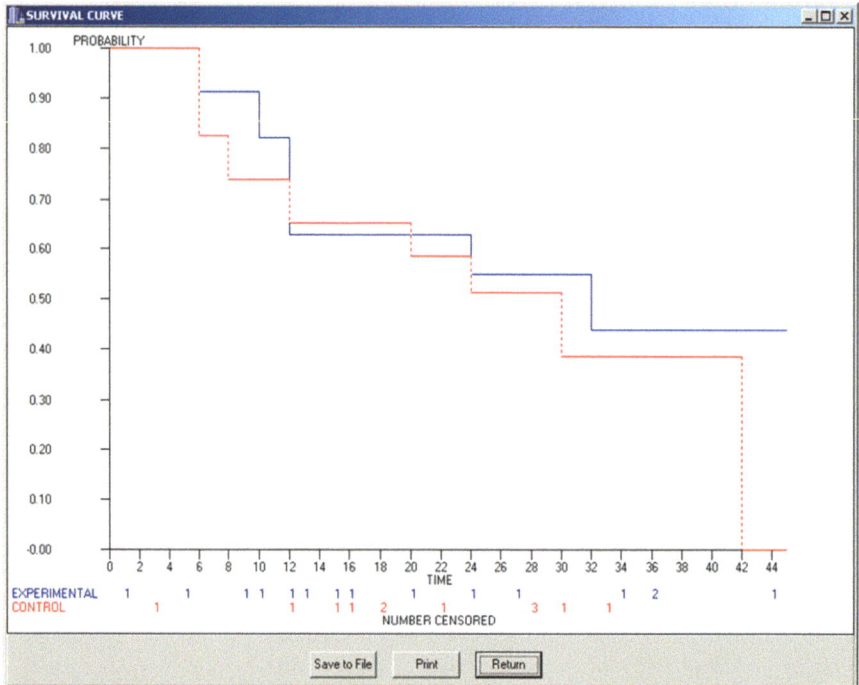

Fig. 10.16 Experimental and control curves

The Kolmogorov-Smirnov Test

The figure below (Fig. 10.17) illustrates an analysis of data collected for five values with the frequency observed for each value in a separate variable:

When you elect the Kolmogorov-Smirnov option under the Nonparametric analyses option, the following dialogue appears (Fig. 10.18):

You can see that we elected to enter values and frequencies and are comparing to a theoretically equal distribution of values. The results obtained are shown below (Fig. 10.19):

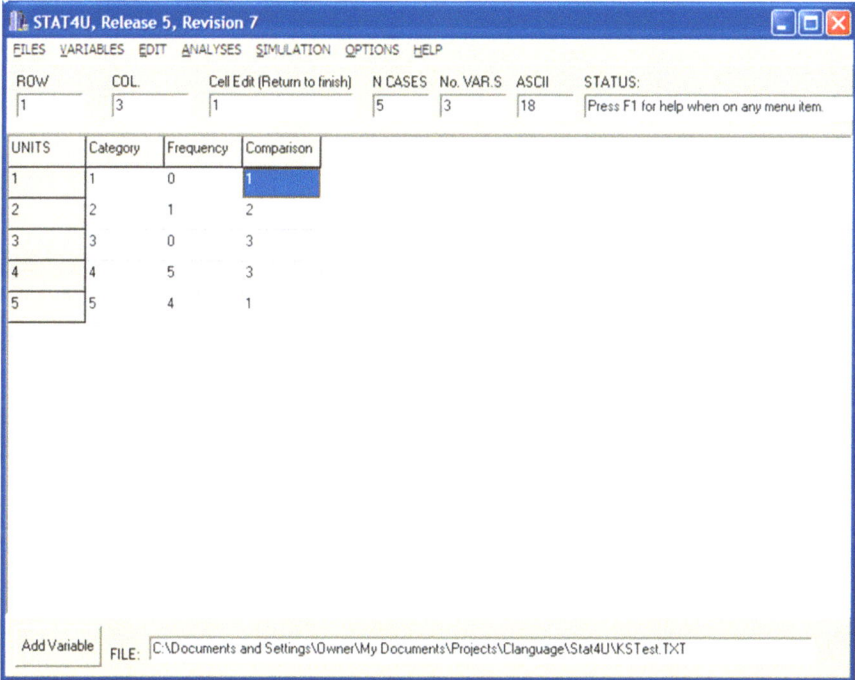

Fig. 10.17 A sample file for the Kolmogorov-Smirnov test

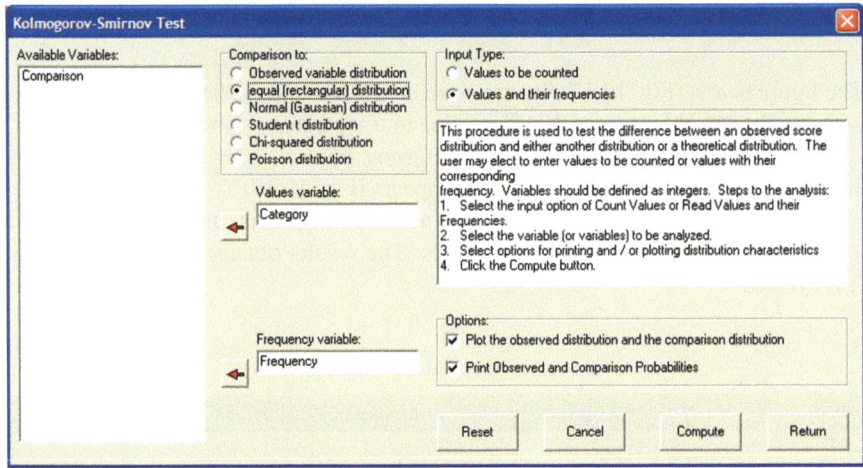

Fig. 10.18 Dialog for the Kolmogorov-Smirnov test

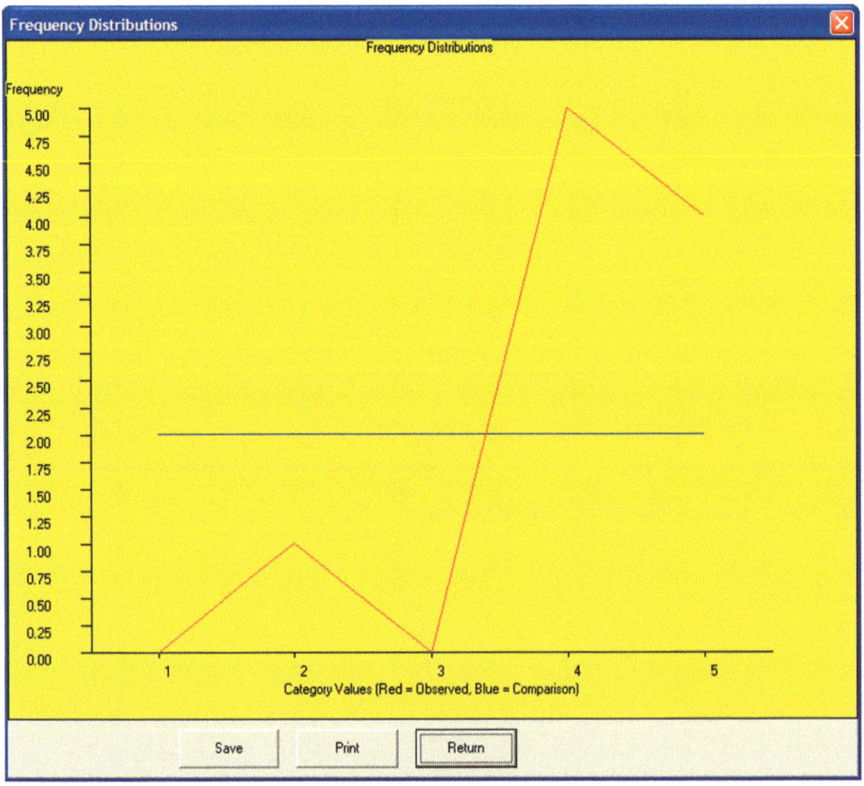

Fig. 10.19 Frequency distribution plot for the Kolmogorov-Smirnov test

The Kolmogorov-Smirnov Test

```
Kolmogorov-Smirnov Test
Analysis of variable Category
   FROM    UP TO   FREQ.    PCNT    CUM.FREQ.   CUM.PCNT.    %ILE RANK
   1.00    2.00      0      0.00       0.00        0.00         0.00
   2.00    3.00      1      0.10       1.00        0.10         0.05
   3.00    4.00      0      0.00       1.00        0.10         0.10
   4.00    5.00      5      0.50       6.00        0.60         0.35
   5.00    6.00      4      0.40      10.00        1.00         0.80

Kolmogorov-Smirnov Analysis of Category and equal (rectangular) distribution
Observed Mean = 4.200 for 10 cases in 5 categories
Standard Deviation = 0.919

Kolmogorov-Smirnov Distribution Comparison
CATEGORY         OBSERVED          COMPARISON
 VALUES        PROBABILITIES      PROBABILITIES
    1              0.000              0.200
    2              0.100              0.200
    3              0.000              0.200
    4              0.500              0.200
    5              0.400              0.200

Kolmogorov-Smirnov Distribution Comparison
CATEGORY         OBSERVED          COMPARISON
 VALUE          CUM. PROB.         CUM. PROB.
    1              0.000              0.200
    2              0.100              0.400
    3              0.100              0.600
    4              0.600              0.800
    5              1.000              1.000
    6              1.000              1.000
Kolmogorov-Smirnov Statistic D = 0.500 with probability > D = 0.013
```

The difference between the observed and theoretical comparison data would not be expected to occur by chance very often (about one in a hundred times) and one would probably reject the hypothesis that the observed distribution comes from a chance distribution (equally likely frequency in each category.)

It is constructive to compare the same observed distribution with the comparison variable and with the normal distribution variable (both are viable alternatives.)

Chapter 11
Measurement

The Item Analysis Program

Classical item analysis is used to estimate the reliability of test scores obtained from measures of subjects on some attribute such as achievement, aptitude or intelligence. In classical test theory, the obtained score for an individual on items is theorized to consist of a "true score" component and an "error score" component. Errors are typically assumed to be normally distributed with a mean of zero over all the subjects measured.

Several methods are available to estimate the reliability of the measures and vary according to the assumptions made about the scores. The Kuder-Richardson estimates are based on the product-moment correlation (or covariance) among items of the observed test scores and those of a theoretical "parallel" test form. The Cronbach and Hoyt estimates utilize a treatment by subjects analysis of variance design which yields identical results to the KR#20 method when item scores are dichotomous (0 and 1) values.

When you select the Classical Item Analysis procedure you will use the following dialogue box to specify how your test is to be analyzed. If the test consists of multiple sub-tests, you may define a scale for each sub-test by specifying those items belonging to each sub-test. The procedure will need to know how to determine the correct and incorrect responses. If your data are already 0 and 1 scores, the most simple method is to simply include, as the first record in your file, a case with 1's for each item. If your data consists of values ranging, say, between 1 and 5 corresponding to alternative choices, you will either include a first case with the correct choice values or indicate you wish to Prompt for Correct Responses (as numbers when values are numbers.) If items are to be assigned different weights, you can assign those weights by selecting the "Assign Item Weights scoring option. The scored item matrix will be printed if you elect it on the output options. Three different reliability methods are available. You can select them all if you like (Fig. 11.1).

Shown below is a sample output obtained from the Classical Item Analysis procedure followed by an item characteristic curve plot for one of the items. The file used was "itemdat.LAZ" (Figs. 11.2, 11.3).

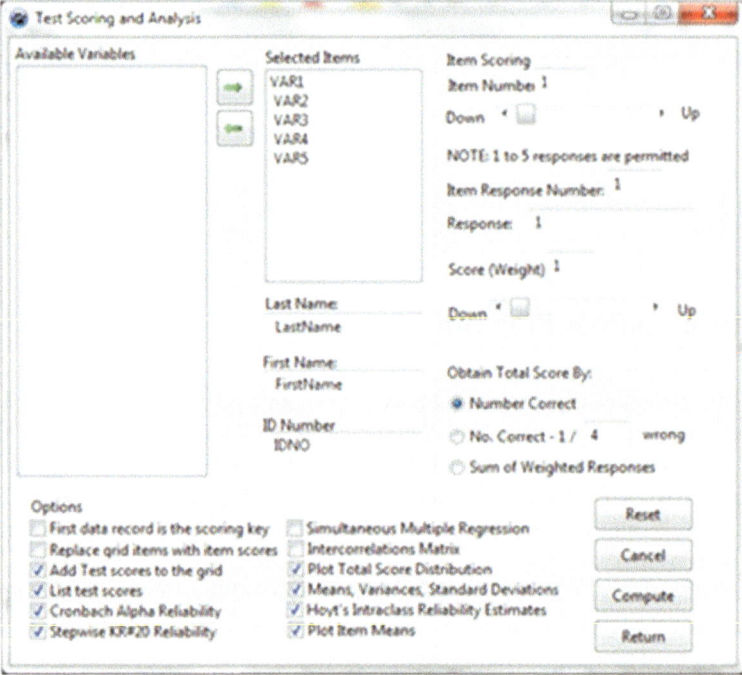

Fig. 11.1 Classical item analysis dialog

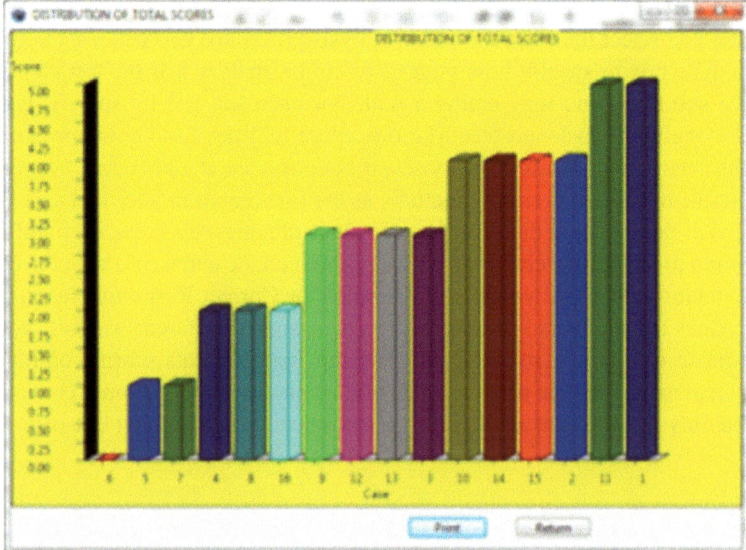

Fig. 11.2 Distribution of test scores (classical analysis)

The Item Analysis Program

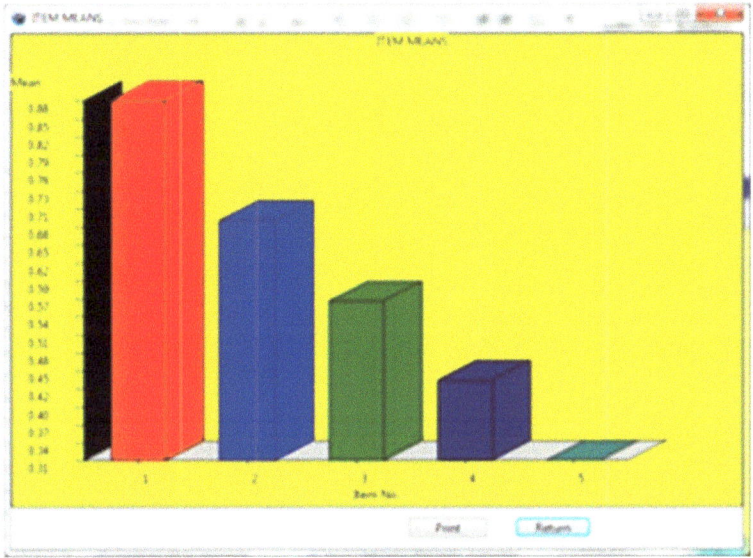

Fig. 11.3 Item means plot

```
TEST SCORING REPORT

PERSON ID NUMBER     FIRST NAME     LAST NAME      TEST SCORE
               1     Bill           Miller         5.00
               2     Barb           Benton         4.00
               3     Tom            Richards       3.00
               4     Keith          Thomas         2.00
               5     Bob            King           1.00
               6     Rob            Moreland       0.00
               7     Sandy          Landis         1.00
               8     Vernil         Moore          2.00
               9     Dick           Tyler          3.00
              10     Harry          Cook           4.00
              11     Claude         Rains          5.00
              12     Clark          Kent           3.00
              13     Bill           Clinton        3.00
              14     George         Bush           4.00
              15     Tom            Jefferson      4.00
              16     Abe            Lincoln        2.00
Alpha Reliability Estimate for Test = 0.6004 S.E. of Measurement = 0.920
Analysis of Variance for Hoyt Reliabilities
```

SOURCE	D.F.	SS	MS	F	PROB
Subjects	15	6.35	0.42	2.50	0.01
Within	64	13.20	0.21		
Items	4	3.05	0.76	4.51	0.00
Error	60	10.15	0.17		
Total	79	19.55			

Hoyt Unadjusted Test Rel. for scale TOTAL = 0.5128 S.E. of Measurement = 0.000
Hoyt Adjusted Test Rel. for scale TOTAL = 0.6004 S.E. of Measurement = 0.000
Hoyt Unadjusted Item Rel. for scale TOTAL = 0.1739 S.E. of Measurement = 0.000
Hoyt Adjusted Item Rel. for scale TOTAL = 0.2311 S.E. of Measurement = 0.000
Item and Total Score Intercorrelations with 16 cases.

Variables

	VAR1	VAR2	VAR3	VAR4	VAR5
VAR1	1.000	0.153	0.048	-0.048	0.255
VAR2	0.153	1.000	0.493	0.323	0.164
VAR3	0.048	0.493	1.000	0.270	0.323
VAR4	-0.048	0.323	0.270	1.000	0.221
VAR5	0.255	0.164	0.323	0.221	1.000
TOTAL	0.369	0.706	0.727	0.615	0.634

Variables

	TOTAL
VAR1	0.369
VAR2	0.706
VAR3	0.727
VAR4	0.615
VAR5	0.634
TOTAL	1.000

Means with 16 valid cases.

Variables	VAR1	VAR2	VAR3	VAR4	VAR5
	0.875	0.688	0.563	0.438	0.313

Variables	TOTAL
	2.875

Variances with 16 valid cases.

Variables	VAR1	VAR2	VAR3	VAR4	VAR5
	0.117	0.229	0.263	0.263	0.229

Variables	TOTAL
	2.117

Standard Deviations with 16 valid cases.

The Item Analysis Program

```
Variables          VAR1       VAR2       VAR3       VAR4       VAR5
                   0.342      0.479      0.512      0.512      0.479

Variables          TOTAL
                   1.455
KR#20 = 0.6591 for the test with mean = 1.250 and variance = 0.733
Item       Mean       Variance       Pt.Bis.r
  2        0.688        0.229         0.8538
  3        0.563        0.263         0.8737
KR#20 = 0.6270 for the test with mean = 1.688 and variance = 1.296
Item       Mean       Variance       Pt.Bis.r
  2        0.688        0.229         0.7875
  3        0.563        0.263         0.7787
  4        0.438        0.263         0.7073
KR#20 = 0.6310 for the test with mean = 2.000 and variance = 1.867
Item       Mean       Variance       Pt.Bis.r
  2        0.688        0.229         0.7135
  3        0.563        0.263         0.7619
  4        0.438        0.263         0.6667
  5        0.313        0.229         0.6116
KR#20 = 0.6004 for the test with mean = 2.875 and variance = 2.117
Item       Mean       Variance       Pt.Bis.r
  2        0.688        0.229         0.7059
  3        0.563        0.263         0.7267
  4        0.438        0.263         0.6149
  5        0.313        0.229         0.6342
  1        0.875        0.117         0.3689
Item and Total Score Intercorrelations with 16 cases.

Variables
                   VAR1       VAR2       VAR3       VAR4       VAR5
       VAR1       1.000      0.153      0.048     -0.048      0.255
       VAR2       0.153      1.000      0.493      0.323      0.164
       VAR3       0.048      0.493      1.000      0.270      0.323
       VAR4      -0.048      0.323      0.270      1.000      0.221
       VAR5       0.255      0.164      0.323      0.221      1.000
      TOTAL       0.369      0.706      0.727      0.615      0.634

Variables
                  TOTAL
       VAR1       0.369
       VAR2       0.706
       VAR3       0.727
       VAR4       0.615
       VAR5       0.634
      TOTAL       1.000

Means with 16 valid cases.
```

```
Variables          VAR1        VAR2        VAR3        VAR4        VAR5
                   0.875       0.688       0.563       0.438       0.313

Variables          TOTAL
                   2.875
```

Variances with 16 valid cases.

```
Variables          VAR1        VAR2        VAR3        VAR4        VAR5
                   0.117       0.229       0.263       0.263       0.229

Variables          TOTAL
                   2.117
```

Standard Deviations with 16 valid cases.

```
Variables          VAR1        VAR2        VAR3        VAR4        VAR5
                   0.342       0.479       0.512       0.512       0.479

Variables          TOTAL
                   1.455
```
Determinant of correlation matrix = 0.5209

Multiple Correlation Coefficients for Each Variable

```
Variable         R          R2          F        Prob.>F      DF1        DF2
  VAR1         0.327      0.107       0.330      0.852         4         11
  VAR2         0.553      0.306       1.212      0.360         4         11
  VAR3         0.561      0.315       1.262      0.342         4         11
  VAR4         0.398      0.158       0.516      0.726         4         11
  VAR5         0.436      0.190       0.646      0.641         4         11
```

Betas in Columns with 16 cases.

```
Variables
                 VAR1        VAR2        VAR3        VAR4        VAR5
   VAR1        -1.000       0.161      -0.082      -0.141       0.262
   VAR2         0.207      -1.000       0.442       0.274      -0.083
   VAR3        -0.107       0.447      -1.000       0.082       0.303
   VAR4        -0.149       0.226       0.067      -1.000       0.178
   VAR5         0.289      -0.071       0.257       0.185      -1.000
```

Standard Errors of Prediction
```
Variable      Std.Error
   VAR1         0.377
   VAR2         0.466
   VAR3         0.495
   VAR4         0.549
   VAR5         0.503
```

Raw Regression Coefficients with 16 cases.

```
Variables
                  VAR1         VAR2         VAR3         VAR4         VAR5
       VAR1      -1.000        0.225       -0.123       -0.211        0.367
       VAR2       0.147       -1.000        0.473        0.293       -0.083
       VAR3      -0.071        0.418       -1.000        0.082        0.283
       VAR4      -0.099        0.211        0.067       -1.000        0.167
       VAR5       0.206       -0.071        0.275        0.199       -1.000

Variable        Constant
    VAR1         0.793
    VAR2         0.186
    VAR3         0.230
    VAR4         0.313
    VAR5        -0.183
```

Analysis of Variance: Treatment by Subject and Hoyt Reliability

The Within Subjects Analysis of Variance involves the repeated measurement of the same unit of observation. These repeated observations are arranged as variables (columns) in the Main Form grid for the cases (grid rows.) If only two measures are administered, you will probably use the matched pairs (dependent) t-test method. When more than two measures are administered, you may use the repeated measures ANOVA method to test the equality of treatment level means in the population sampled. Since within subjects analysis is a part of the Hoyt Intraclass reliability estimation procedure, you may use this procedure to complete the analysis (see the Measurement procedures under the Analyses menu on the Main Form.) (Figs. 11.4, 11.5)

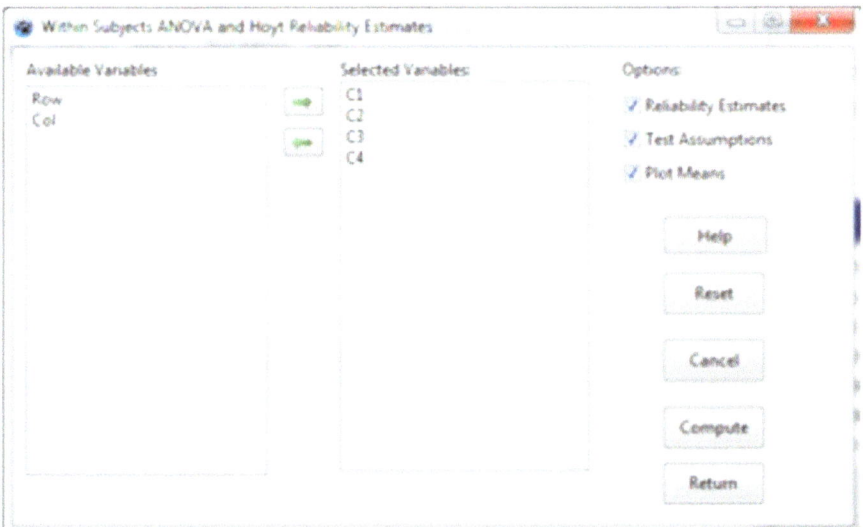

Fig. 11.4 Hoyt reliability by ANOVA

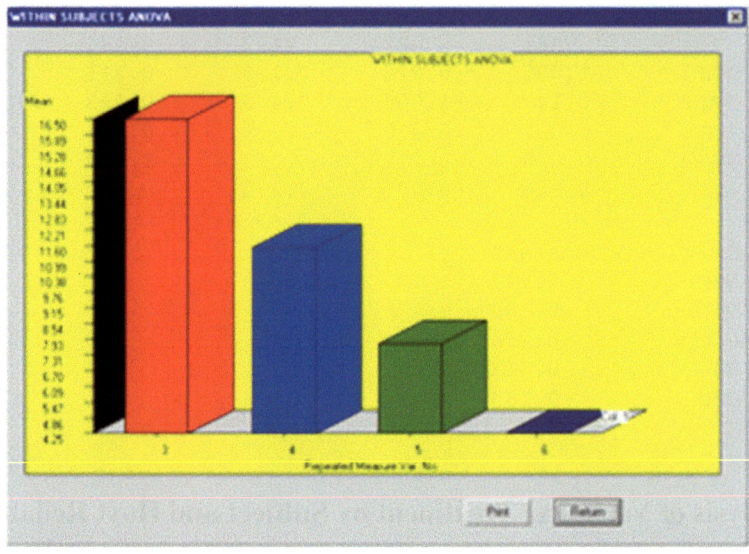

Fig. 11.5 Within subjects ANOVA plot

The output from an example analysis is shown below:

Treatments by Subjects (AxS) ANOVA Results.

Data File = C:\lazarus\Projects\LazStats\LazStatsData\ABRDATA.LAZ

```
-------------------------------------------------------------------
SOURCE                  DF          SS         MS         F    Prob. > F
-------------------------------------------------------------------
SUBJECTS                11     181.000     16.455
WITHIN SUBJECTS         36    1077.000     29.917
   TREATMENTS            3     991.500    330.500   127.561       0.000
   RESIDUAL             33      85.500      2.591
-------------------------------------------------------------------
   TOTAL                47    1258.000     26.766
-------------------------------------------------------------------
```

```
TREATMENT (COLUMN) MEANS AND STANDARD DEVIATIONS
VARIABLE     MEAN      STD.DEV.
C1          16.500       2.067
C2          11.500       2.431
C3           7.750       2.417
C4           4.250       2.864
```

Mean of all scores = 10.000 with standard deviation = 5.174

```
RELIABILITY ESTIMATES
TYPE OF ESTIMATE                    VALUE
Unadjusted total reliability       -0.818
Unadjusted item reliability        -0.127
Adjusted total (Cronbach)           0.843
Adjusted item reliability           0.572


BOX TEST FOR HOMOGENEITY OF VARIANCE-COVARIANCE MATRIX

SAMPLE COVARIANCE MATRIX with 12 cases.

Variables
              C1          C2          C3          C4
    C1      4.273       2.455       1.227       1.318
    C2      2.455       5.909       4.773       5.591
    C3      1.227       4.773       5.841       5.432
    C4      1.318       5.591       5.432       8.205

ASSUMED POP. COVARIANCE MATRIX with 12 cases.

Variables
              C1          C2          C3          C4
    C1      6.057       0.693       0.693       0.693
    C2      0.114       5.977       0.614       0.614
    C3      0.114       0.103       5.914       0.551
    C4      0.114       0.103       0.093       5.863

Determinant of variance-covariance matrix = 81.6
Determinant of homogeneity matrix = 1.26E003
ChiSquare = 108.149 with 8 degrees of freedom
Probability of larger chisquare = 9.66E-007
```

Kuder-Richardson #21 Reliability

The Kuder-Richardson formula #20 was developed from Classical Test Theory (true-score theory). A shorter form of the estimate can be made using only the mean, standard deviation and number of test items if one can assume that the inter-item covariances are equal. Below is the form which appears when this procedure is selected from the Measurement option of the Analyses menu (Fig. 11.6):

Note that we have entered the maximum score (total number of items), the test mean, and the test standard deviation. When you click the Compute button, the estimate is shown in the labeled box.

Fig. 11.6 Kuder-Richardson Formula 20 Reliability form

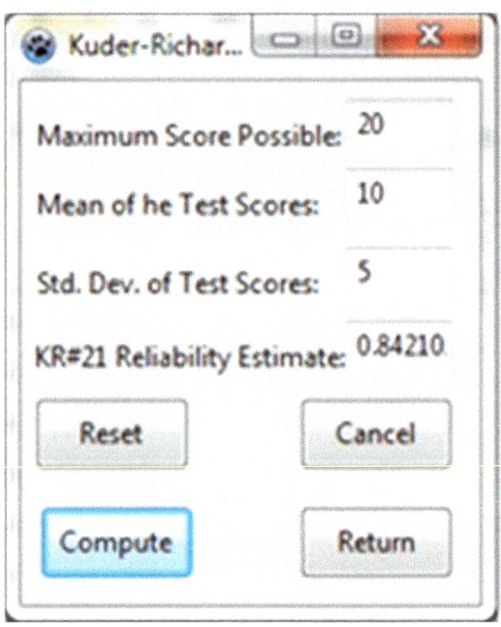

Weighted Composite Test Reliability

The reliability for a combination of tests, each of which has its own estimate of reliability and a weight assigned to it, may be computed. This composite will typically be greater than any one test by itself due to the likelihood that the subtests are correlated positively among themselves. Since teachers typically assign course grades based on a combination of individual tests administered over the time period of a course, this reliability estimate in built into the Grading System. See the description and examples in that section. A file labeled "CompRel.LAZ" is used in the example below (Fig. 11.7):

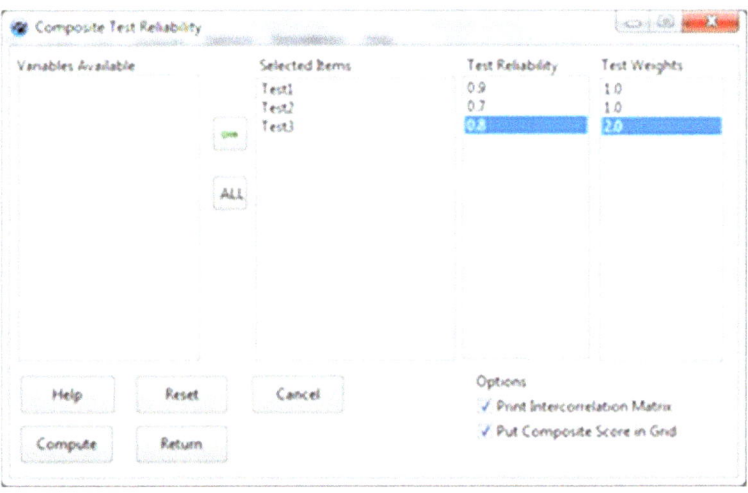

Fig. 11.7 Composite test reliability dialog

```
Composite Test Reliability
File Analyzed: C:\lazarus\Projects\LazStats\LazStatsData\CompRel.LAZ
Correlations Among Tests with 10 cases.
Variables
            Test1       Test2       Test3
    Test1   1.000       0.927       0.952
    Test2   0.927       1.000       0.855
    Test3   0.952       0.855       1.000

Means with 10 valid cases.

Variables   Test1       Test2       Test3
            5.500       5.500       7.500

Variances with 10 valid cases.

Variables   Test1       Test2       Test3
            9.167       9.167       9.167

Standard Deviations with 10 valid cases.

Variables   Test1       Test2       Test3
            3.028       3.028       3.028
```

```
Test Weights with 10 valid cases.

Variables     Test1      Test2      Test3
              1.000      1.000      2.000

Test Reliabilities with 10 valid cases.

Variables     Test1      Test2      Test3
              0.900      0.700      0.800
Composite reliability = 0.929
```

Rasch One Parameter Item Analysis

Item Response Theory (IRT) is another theoretical view of subject responses to items on a test. IRT suggests that items may possess one or more characteristics (parameters) that may be estimated. In the theory developed by George Rasch, one parameter, item difficulty, is estimated (in addition to the estimate of individual subject "ability" parameters.) Utilizing maximum-likelihood methods and log difficulty and log ability parameter estimates, the Rasch method attempts to estimate subject and item parameters that are "independent" of one another. This is unlike Classical theory in which the item difficulty (proportion of subjects passing an item) is directly a function of the ability of the subjects sampled. IRT is sometimes also considered to be a "Latent Trait Theory" due to the assumption that all of the items are measures of the same underlying "trait". Several tests of the "fit" of the item responses to this assumption are typically included in programs to estimate Rasch parameters. Other IRT procedures posit two or three parameters, the others being the "slope" and the "chance" parameters. The slope is the rate at which the probability of getting an item correct increases with equal units of increase in subject ability. The chance parameter is the probability of obtaining the item correct by chance alone. In the Rasch model, the chance probability is assumed to be zero and the slope parameter assumed to be equal for all items. The file labeled "itemdat.LAZ" is used for our example (Fig. 11.8).

Shown below is a sample of output from a test analyzed by the Rasch model. The model cannot make ability estimates for subjects that miss all items or get all items correct so they are screened out. Parameters estimated are given in log units. Also shown is one of the item information function curve plots. Each item provides the maximum discrimination (information) at that point where the log ability of the subject is approximately the same as the log difficulty of the item. In examining the output you will note that item 1 does not appear to fit the assumptions of the Rasch model as measured by the chi-square statistic (Figs. 11.9, 11.10, 11.11, 11.12).

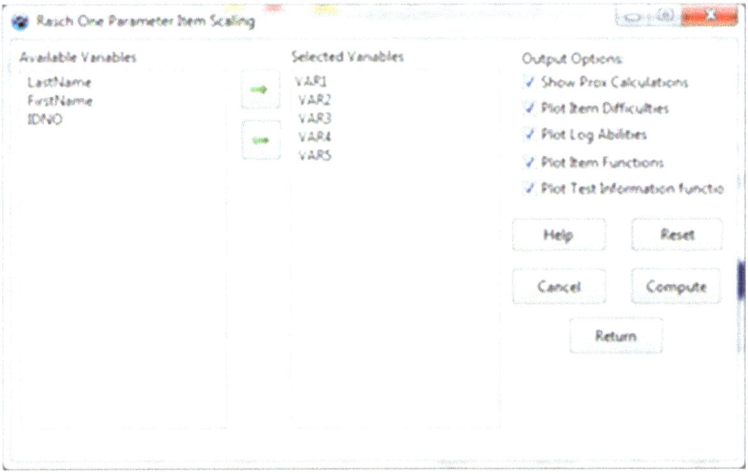

Rasch One Parameter Item Analysis

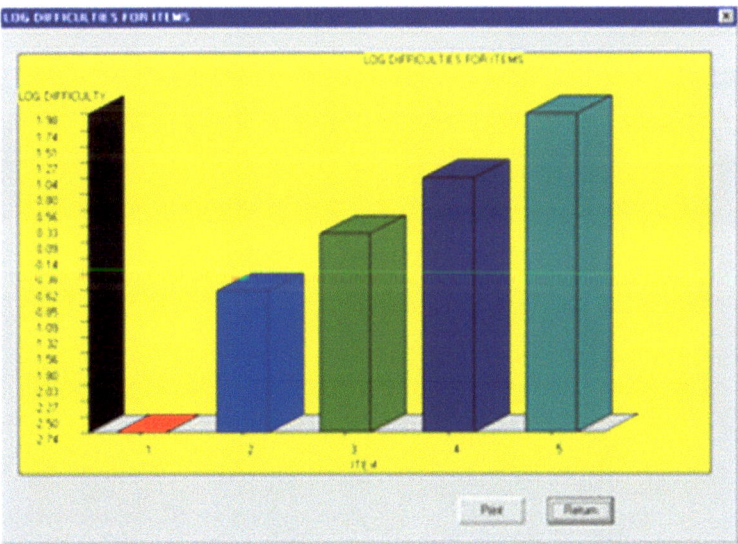

Fig. 11.8 The Rasch item analysis dialog

Fig. 11.9 Rasch item log difficulty estimate plot

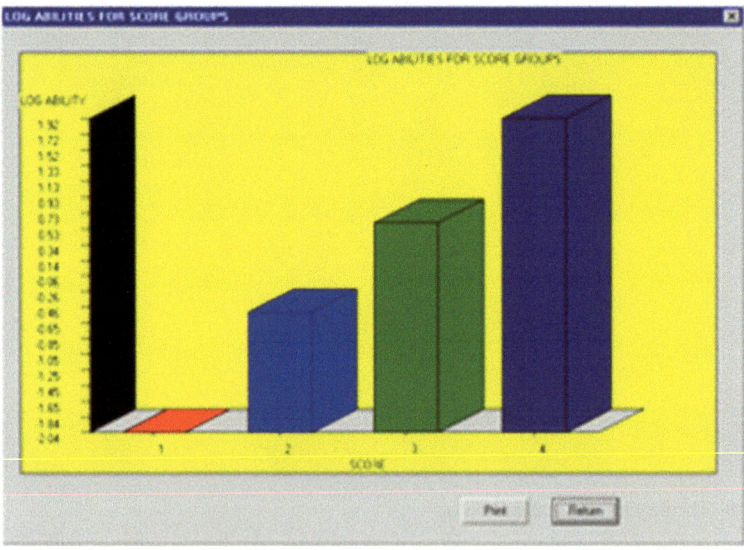

Fig. 11.10 Rasch log score estimates

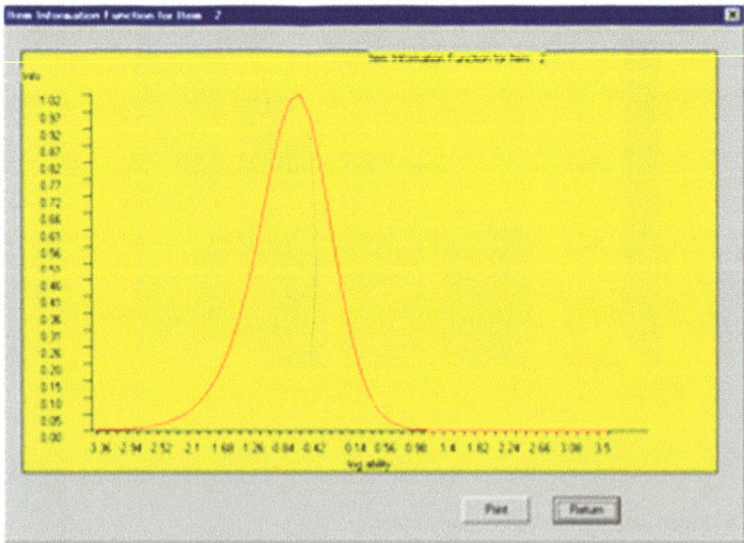

Fig. 11.11 A Rasch item characteristic curve

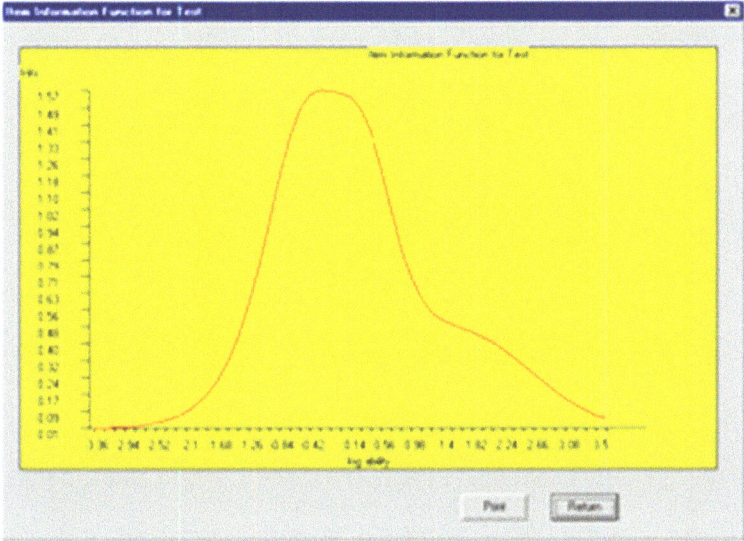

Fig. 11.12 A Rasch test information curve

```
Rasch One-Parameter Logistic Test Scaling (Item Response Theory)
Written by William G. Miller

case   1 eliminated.  Total score was   5
Case   2 Total Score :=   4 Item scores 1 1 1 1 0
Case   3 Total Score :=   3 Item scores 1 1 1 0 0
Case   4 Total Score :=   2 Item scores 1 1 0 0 0
Case   5 Total Score :=   1 Item scores 1 0 0 0 0
case   6 eliminated. Total score was 0
Case   7 Total Score :=   1 Item scores 1 0 0 0 0
Case   8 Total Score :=   2 Item scores 1 1 0 0 0
Case   9 Total Score :=   3 Item scores 1 1 1 0 0
Case  10 Total Score :=   4 Item scores 1 1 1 1 0
case  11 eliminated. Total score was 5
Case  12 Total Score :=   3 Item scores 1 0 1 0 1
Case  13 Total Score :=   3 Item scores 0 1 1 1 0
Case  14 Total Score :=   4 Item scores 1 1 1 0 1
Case  15 Total Score :=   4 Item scores 1 1 0 1 1
Case  16 Total Score :=   2 Item scores 1 0 0 1 0

Total number of score groups :=   4
```

```
Matrix of Item Failures in Score Groups
Score Group      1     2     3     4     Total
ITEM
           1     0     0     1     0       1
           2     2     1     1     0       4
           3     2     3     0     1       6
           4     2     2     3     1       8
           5     2     3     3     2      10
Total            2     3     4     4      13

Item  Log Odds  Deviation  Squared Deviation
  1    -2.48      -2.13         4.54
  2    -0.81      -0.46         0.21
  3    -0.15       0.20         0.04
  4     0.47       0.83         0.68
  5     1.20       1.56         2.43
Score Frequency Log Odds Freq.x Log Freq.x Log Odds Squared
  1      2       -1.39      -2.77            3.84
  2      3       -0.41      -1.22            0.49
  3      4        0.41       1.62            0.66
  4      4        1.39       5.55            7.69

Prox values and Standard Errors
Item   Scale Value   Standard Error
  1      -2.730         1.334
  2      -0.584         0.770
  3       0.258         0.713
  4       1.058         0.731
  5       1.999         0.844
Y expansion factor := 1.2821

Score Scale Value Standard Error
  1      -1.910         1.540
  2      -0.559         1.258
  3       0.559         1.258
  4       1.910         1.540
X expansion factor =    1.3778
Maximum Likelihood Iteration Number  0
Maximum Likelihood Iteration Number  1
Maximum Likelihood Iteration Number  2
Maximum Likelihood Iteration Number  3

Maximum Likelihood Estimates

Item    Log Difficulty
  1       -2.74
  2       -0.64
  3        0.21
  4        1.04
  5        1.98
```

```
Score   Log Ability
  1      -2.04
  2      -0.54
  3       0.60
  4       1.92
```

```
Goodness of Fit Test for Each Item
Item    Chi-Squared    Degrees of    Probability
No.     Value          Freedom       of Larger Value
 1       29.78             9           0.0005
 2        8.06             9           0.5283
 3       10.42             9           0.3177
 4       12.48             9           0.1875
 5        9.00             9           0.4371
```

```
Item Data Summary
ITEM   PT.BIS.R.   BIS.R.   SLOPE   PASSED   FAILED   RASCH DIFF
 1      -0.064    -0.117    -0.12    12.00      1       -2.739
 2       0.648     0.850     1.61     9.00      4       -0.644
 3       0.679     0.852     1.63     7.00      6        0.207
 4       0.475     0.605     0.76     5.00      8        1.038
 5       0.469     0.649     0.85     3.00     10        1.981
```

Guttman Scalogram Analysis

Guttman scales are those measurement instruments composed of items which, ideally, form a hierarchy in which the total score of a subject can indicate the actual response (correct or incorrect) of each item. Items are arranged in order of the proportion of subjects passing the item and subjects are grouped and sequenced by their total scores. If the items measure consistently, a triangular pattern should emerge. A coefficient of "reproducibility" is obtained which may be interpreted in a manner similar to test reliability.

Dichotomously scored (0 and 1) items representing the responses of subjects in your data grid rows are the variables (grid columns) analyzed. Select the items to analyze in the same manner as you would for the Classical Item Analysis or the Rasch analysis. When you click the OK button, you will immediately be presented with the results on the output form. An example is shown below (Fig. 11.13).

Fig. 11.13 Guttman scalogram analysis dialog

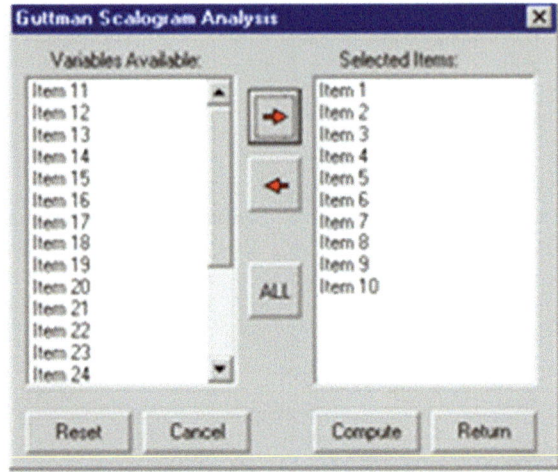

```
                    GUTTMAN SCALOGRAM ANALYSIS
                           Cornell Method
No. of Cases := 101.  No. of items :=  10
RESPONSE MATRIX
Subject Row    Item Number
Label   Sum    Item 10    Item 9    Item 1    Item 3    Item 5    Item 2
                0   1    0   1    0   1    0   1    0   1    0   1
  1      10     0   1    0   1    0   1    0   1    0   1    0   1
  6      10     0   1    0   1    0   1    0   1    0   1    0   1
 20      10     0   1    0   1    0   1    0   1    0   1    0   1
 46      10     0   1    0   1    0   1    0   1    0   1    0   1
 68      10     0   1    0   1    0   1    0   1    0   1    0   1
 77      10     0   1    0   1    0   1    0   1    0   1    0   1
 50       9     0   1    0   1    0   1    1   0    0   1    0   1
 39       9     1   0    0   1    0   1    0   1    0   1    0   1
etc.
TOTALS         53      48       52       49      51       50      51      50  50     51   48      53
ERRORS          3      22       19        9       5       20      13      10  10     10   10      13

Subject Row    Item Number
Label   Sum    Item 8    Item 6    Item 4    Item 7
                0   1    0   1    0   1    0   1
  1      10     0   1    0   1    0   1    0   1
  6      10     0   1    0   1    0   1    0   1
etc.
 65       0     1   0    1   0    1   0    1   0
 10       0     1   0    1   0    1   0    1   0
 89       0     1   0    1   0    1   0    1   0

TOTALS         46      55       44       57      44       57      41       60
ERRORS         11      11       17        3      12       11      11       15

Coefficient of Reproducibility := 0.767
```

Successive Interval Scaling

Successive Interval Scaling was developed as an approximation of Thurstone's Paired Comparisons method for estimation of scale values and dispersion of scale values for items designed to measure attitudes. Typically, five to nine categories are used by judges to indicate the degree to which an item expresses an attitude (if a subject agrees with the item) between very negative to very positive. Once scale values are estimated, the items responded to by subjects are scored by obtaining the median scale value of those items to which the subject agrees.

To obtain Successive interval scale values, select that option under the Measurement group in the Analyses menu on the main form. The specifications form below will appear. Select those items (variables) you wish to scale. The data analyzed consists of rows representing judges and columns representing the scale value chosen for an item by a judge. The file labeled "sucsintv.LAZ" is used as an example file (Fig. 11.14).

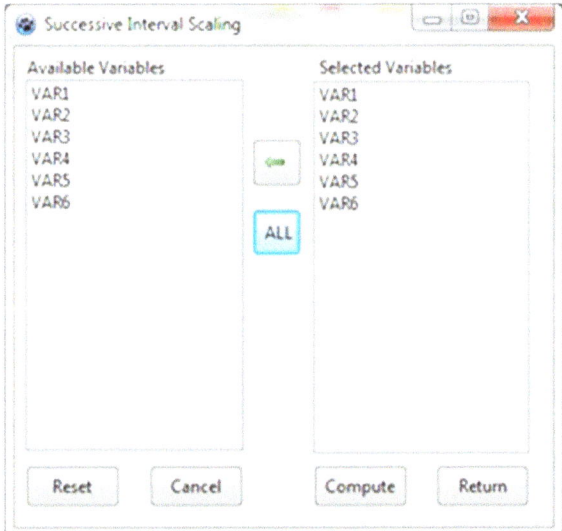

Fig. 11.14 Successive interval scaling dialog

When you click the OK button on the box above, the results will appear on the printout form. An example of results are presented below.

```
                  SUCCESSIVE INTERVAL SCALING RESULTS
                0- 1    1- 2    2- 3    3- 4    4- 5    5- 6    6- 7
         VAR1
Frequency          0       0       0       0       4       4       4
Proportion     0.000   0.000   0.000   0.000   0.333   0.333   0.333
Cum. Prop.     0.000   0.000   0.000   0.000   0.333   0.667   1.000
Normal z         -       -       -       -    -0.431   0.431     -
         VAR2
Frequency          0       0       1       3       4       4       0
Proportion     0.000   0.000   0.083   0.250   0.333   0.333   0.000
Cum. Prop.     0.000   0.000   0.083   0.333   0.667   1.000   1.000
Normal z         -       -    -1.383  -0.431   0.431     -       -
         VAR3
Frequency          0       0       4       3       4       1       0
Proportion     0.000   0.000   0.333   0.250   0.333   0.083   0.000
Cum. Prop.     0.000   0.000   0.333   0.583   0.917   1.000   1.000
Normal z         -       -    -0.431   0.210   1.383     -       -
         VAR4
Frequency          0       3       4       5       0       0       0
Proportion     0.000   0.250   0.333   0.417   0.000   0.000   0.000
Cum. Prop.     0.000   0.250   0.583   1.000   1.000   1.000   1.000
Normal z         -    -0.674   0.210     -       -       -       -
         VAR5
Frequency          5       4       3       0       0       0       0
Proportion     0.417   0.333   0.250   0.000   0.000   0.000   0.000
Cum. Prop.     0.417   0.750   1.000   1.000   1.000   1.000   1.000
Normal z     -0.210   0.674     -       -       -       -       -
         VAR6
Frequency          1       2       2       2       2       2       1
Proportion     0.083   0.167   0.167   0.167   0.167   0.167   0.083
Cum. Prop.     0.083   0.250   0.417   0.583   0.750   0.917   1.000
Normal z     -1.383  -0.674  -0.210   0.210   0.674   1.383     -

                         INTERVAL WIDTHS
               2- 1    3- 2    4- 3    5- 4    6- 5
       VAR1      -       -       -       -     0.861
       VAR2      -       -     0.952   0.861     -
       VAR3      -       -     0.641   1.173     -
       VAR4      -     0.885     -       -       -
       VAR5    0.885     -       -       -       -
       VAR6    0.709   0.464   0.421   0.464   0.709

Mean Width     0.80    0.67    0.67    0.83    0.78
No. Items         2       2       3       3       2
Std. Dev.s     0.02    0.09    0.07    0.13    0.01
Cum. Means     0.80    1.47    2.14    2.98    3.76
```

```
ESTIMATES OF SCALE VALUES AND THEIR DISPERSIONS
Item No. Ratings Scale Value Discriminal Dispersion
      VAR1       12          3.368         1.224
      VAR2       12          2.559         0.822
      VAR3       12          1.919         0.811
      VAR4       12          1.303         1.192
      VAR5       12          0.199         1.192
      VAR6       12          1.807         0.759

Z scores Estimated from Scale values
              0- 1    1- 2    2- 3    3- 4    4- 5    5- 6    6- 7
     VAR1  -3.368  -2.571  -1.897  -1.225  -0.392   0.392
     VAR2  -2.559  -1.762  -1.088  -0.416   0.416   1.201
     VAR3  -1.919  -1.122  -0.448   0.224   1.057   1.841
     VAR4  -1.303  -0.506   0.169   0.840   1.673   2.458
     VAR5  -0.199   0.598   1.272   1.943   2.776   3.000
     VAR6  -1.807  -1.010  -0.336   0.336   1.168   1.953

Cumulative Theoretical Proportions
              0- 1    1- 2    2- 3    3- 4    4- 5    5- 6    6- 7
     VAR1   0.000   0.005   0.029   0.110   0.347   0.653   1.000
     VAR2   0.005   0.039   0.138   0.339   0.661   0.885   1.000
     VAR3   0.028   0.131   0.327   0.589   0.855   0.967   1.000
     VAR4   0.096   0.306   0.567   0.800   0.953   0.993   1.000
     VAR5   0.421   0.725   0.898   0.974   0.997   0.999   1.000
     VAR6   0.035   0.156   0.369   0.631   0.879   0.975   1.000
```

Average Discrepency Between Theoretical and Observed Cumulative Proportions = 0.050

Maximum discrepency = 0.200 found in item VAR4

Differential Item Functioning

Anyone developing tests today should be sensitive to the fact that some test items may present a bias for one or more subgroups in the population to which the test is administered. For example, because of societal value systems, boys and girls may be exposed to quite different learning experiences during their youth. A word test in mathematics may unintentionally give an advantage to one gender group over another simply by the examples used in the item. To identify possible bias in an item, one can examine the differential item functioning of each item for the sub-groups to which the test is administered. The Mantel-Haenszel test statistic may be applied to test the difference on the item characteristic curve for the difference between a "focus" group and a "reference" group. We will demonstrate using a data set in which 40 items have been administered to 1,000 subjects in one group and 1,000 subjects in another group. The groups are simply coded 1 and 2 for the reference and

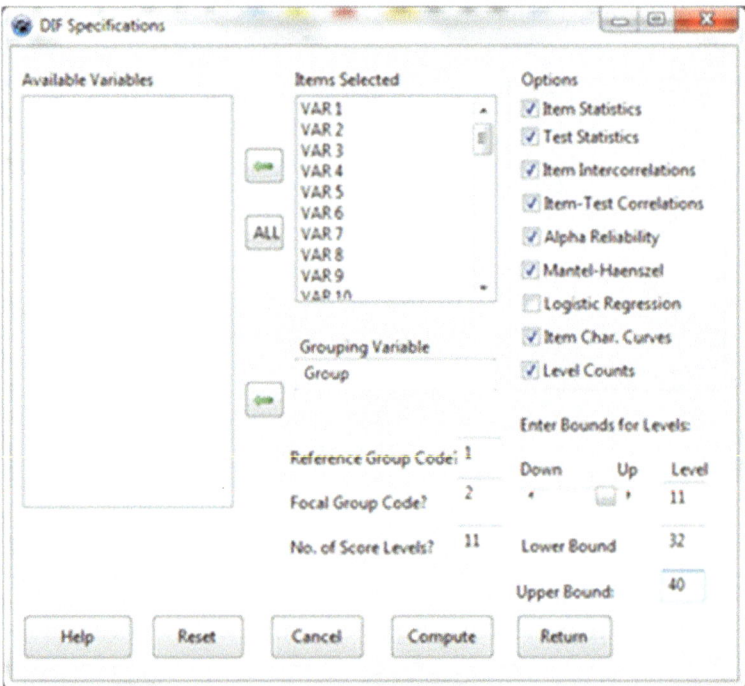

Fig. 11.15 Differential item functioning dialog

focus groups. Since there may be very few (or no) subjects that get a specific total score, we will group the total scores obtained by subjects into groups of 4 so that we are comparing subjects in the groups that have obtained total item scores of 0 to 3, 4 to 7, ..., 40 to 43. As you will see, even this grouping is too small for several score groups and we should probably change the score range for the lowest and highest scores to a larger range of scores in another run.

When you elect to do this analysis, the specification form above appears (Fig. 11.15):

On the above form you specify the items to be analyzed and also the variable defining the reference and focus group codes. You may then specify the options desired by clicking the corresponding buttons for the desired options. You also enter the number of score groups to be used in grouping the subject's total scores. When this is specified, you then enter the lowest and highest score for each of those score groups. When you have specified the low and hi score for the first group, click the right arrow on the "slider" bar to move to the next group. You will see that the lowest score has automatically been set to one higher than the previous group's highest score to save you time in entering data. You do not, of course, have to use the same size for the range of each score group. Using too large a range of scores may cut down the sensitivity of the test to differences between the groups. Fairly large samples of subjects is necessary for a reasonable analysis. Once you have completed the specifications, click the Compute button and you will see the following results are obtained (we elected to

Differential Item Functioning

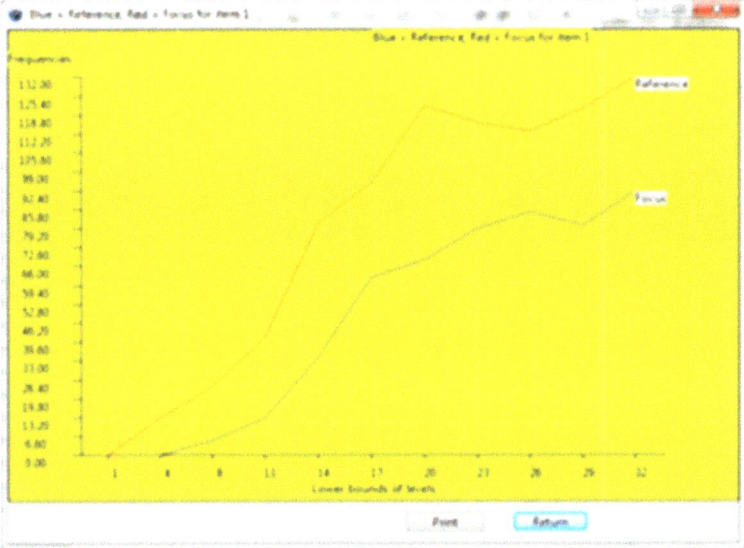

Fig. 11.16 Differential item function curves

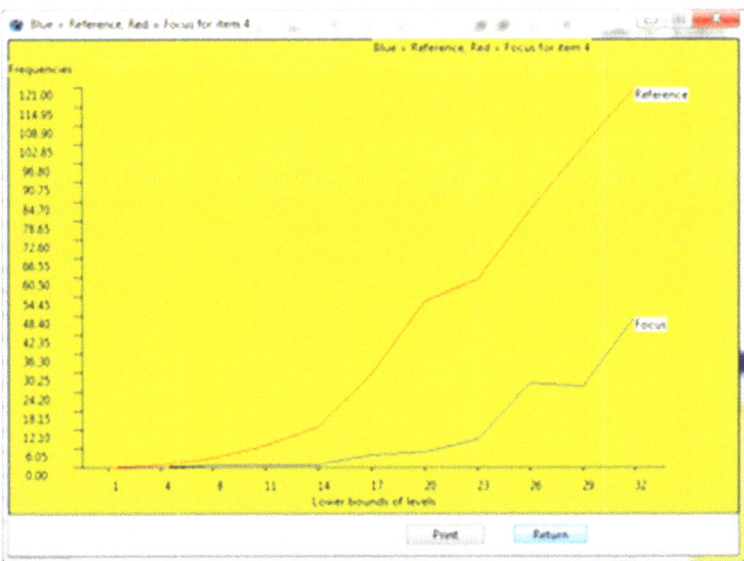

Fig. 11.17 Another item differential functioning curve

print the descriptive statistics, correlations and item plots) (Figs. 11.16, 11.17):
Mantel-Haenszel DIF Analysis adapted by Bill Miller from
EZDIF written by Niels G. Waller

Total Means with 2000 valid cases.

Variables	VAR 1	VAR 2	VAR 3	VAR 4	VAR 5
	0.688	0.064	0.585	0.297	0.451
Variables	VAR 6	VAR 7	VAR 8	VAR 9	VAR 10
	0.806	0.217	0.827	0.960	0.568
Variables	VAR 11	VAR 12	VAR 13	VAR 14	VAR 15
	0.350	0.291	0.725	0.069	0.524
Variables	VAR 16	VAR 17	VAR 18	VAR 19	VAR 20
	0.350	0.943	0.545	0.017	0.985
Variables	VAR 21	VAR 22	VAR 23	VAR 24	VAR 25
	0.778	0.820	0.315	0.203	0.982
Variables	VAR 26	VAR 27	VAR 28	VAR 29	VAR 30
	0.834	0.700	0.397	0.305	0.223
Variables	VAR 31	VAR 32	VAR 33	VAR 34	VAR 35
	0.526	0.585	0.431	0.846	0.115
Variables	VAR 36	VAR 37	VAR 38	VAR 39	VAR 40
	0.150	0.817	0.909	0.793	0.329

Total Variances with 2000 valid cases.

Variables	VAR 1	VAR 2	VAR 3	VAR 4	VAR 5
	0.215	0.059	0.243	0.209	0.248
Variables	VAR 6	VAR 7	VAR 8	VAR 9	VAR 10
	0.156	0.170	0.143	0.038	0.245
Variables	VAR 11	VAR 12	VAR 13	VAR 14	VAR 15
	0.228	0.206	0.199	0.064	0.250
Variables	VAR 16	VAR 17	**VAR 18**	**VAR 19**	**VAR 20**
	0.228	**0.054**	0.248	0.017	0.015
Variables	VAR 21	VAR 22	VAR 23	VAR 24	VAR 25
	0.173	0.148	0.216	0.162	0.018
Variables	VAR 26	VAR 27	VAR 28	VAR 29	VAR 30
	0.139	0.210	0.239	0.212	0.173
Variables	VAR 31	VAR 32	VAR 33	VAR 34	VAR 35
	0.249	0.243	0.245	0.130	0.102
Variables	VAR 36	VAR 37	VAR 38	VAR 39	VAR 40
	0.128	0.150	0.083	0.164	0.221

Total Standard Deviations with 2000 valid cases.

Variables	VAR 1	VAR 2	VAR 3	VAR 4	VAR 5
	0.463	0.244	0.493	0.457	0.498
Variables	VAR 6	VAR 7	VAR 8	VAR 9	VAR 10
	0.395	0.412	0.379	0.196	0.495
Variables	VAR 11	VAR 12	VAR 13	VAR 14	VAR 15

Differential Item Functioning

	0.477	0.454	0.447	0.253	0.500
Variables	VAR 16	VAR 17	VAR 18	VAR 19	VAR 20
	0.477	0.233	0.498	0.129	0.124
Variables	VAR 21	VAR 22	VAR 23	VAR 24	VAR 25
	0.416	0.384	0.465	0.403	0.135
Variables	VAR 26	VAR 27	VAR 28	VAR 29	VAR 30
	0.372	0.459	0.489	0.461	0.416
Variables	VAR 31	VAR 32	VAR 33	VAR 34	VAR 35
	0.499	0.493	0.495	0.361	0.319
Variables	VAR 36	VAR 37	VAR 38	VAR 39	VAR 40
	0.357	0.387	0.288	0.405	0.470

Total Score: Mean = 21.318, Variance = 66.227, Std.Dev. = 8.138

Reference group size = 1000, Focus group size = 1000
Correlations Among Items with 2000 cases.

Variables

	VAR 1	VAR 2	VAR 3	VAR 4	VAR 5
VAR 1	1.000	0.162	0.389	0.308	0.406
VAR 2	0.162	1.000	0.190	0.275	0.259
VAR 3	0.389	0.190	1.000	0.368	0.382
VAR 4	0.308	0.275	0.368	1.000	0.423
VAR 5	0.406	0.259	0.382	0.423	1.000
VAR 6	0.260	0.102	0.239	0.199	0.225
VAR 7	0.203	0.226	0.237	0.255	0.274
VAR 8	0.253	0.103	0.257	0.188	0.234
VAR 9	0.160	0.053	0.154	0.077	0.123
VAR 10	0.243	0.169	0.279	0.244	0.260
VAR 11	0.257	0.191	0.279	0.272	0.308
VAR 12	0.210	0.217	0.230	0.248	0.252
VAR 13	0.272	0.128	0.262	0.217	0.272
VAR 14	0.144	0.181	0.164	0.166	0.172
VAR 15	0.255	0.174	0.304	0.265	0.287
VAR 16	0.232	0.213	0.251	0.268	0.272
VAR 17	0.209	0.064	0.206	0.151	0.168
VAR 18	0.276	0.192	0.278	0.259	0.261
VAR 19	0.080	0.061	0.087	0.084	0.060
VAR 20	0.151	0.033	0.100	0.073	0.097
VAR 21	0.271	0.124	0.277	0.208	0.244
VAR 22	0.263	0.122	0.270	0.213	0.231
VAR 23	0.250	0.190	0.275	0.254	0.282
VAR 24	0.206	0.230	0.227	0.261	0.279
VAR 25	0.116	0.036	0.118	0.073	0.102
VAR 26	0.248	0.105	0.248	0.202	0.247
VAR 27	0.300	0.130	0.310	0.230	0.280
VAR 28	0.257	0.225	0.275	0.276	0.306
VAR 29	0.287	0.202	0.290	0.290	0.308
VAR 30	0.239	0.215	0.240	0.241	0.271
VAR 31	0.263	0.161	0.288	0.281	0.279

VAR 32	0.251	0.178	0.316	0.228	0.264
VAR 33	0.247	0.187	0.272	0.298	0.295
VAR 34	0.269	0.094	0.301	0.205	0.244
VAR 35	0.151	0.189	0.180	0.181	0.206
VAR 36	0.213	0.229	0.209	0.236	0.253
VAR 37	0.234	0.107	0.233	0.180	0.241
VAR 38	0.203	0.075	0.206	0.156	0.196
VAR 39	0.230	0.123	0.274	0.221	0.248
VAR 40	0.273	0.211	0.255	0.284	0.289

Variables

	VAR 6	VAR 7	VAR 8	VAR 9	VAR 10
VAR 1	0.260	0.203	0.253	0.160	0.243
VAR 2	0.102	0.226	0.103	0.053	0.169
VAR 3	0.239	0.237	0.257	0.154	0.279
VAR 4	0.199	0.255	0.188	0.077	0.244
VAR 5	0.225	0.274	0.234	0.123	0.260
VAR 6	1.000	0.196	0.267	0.217	0.281
VAR 7	0.196	1.000	0.193	0.095	0.253
VAR 8	0.267	0.193	1.000	0.189	0.285
VAR 9	0.217	0.095	0.189	1.000	0.198
VAR 10	0.281	0.253	0.285	0.198	1.000
VAR 11	0.235	0.302	0.237	0.129	0.300
VAR 12	0.202	0.229	0.198	0.103	0.268
VAR 13	0.308	0.202	0.256	0.177	0.299
VAR 14	0.108	0.222	0.098	0.055	0.177
VAR 15	0.268	0.278	0.264	0.163	0.335
VAR 16	0.240	0.290	0.251	0.129	0.302
VAR 17	0.238	0.114	0.261	0.224	0.201
VAR 18	0.277	0.288	0.250	0.183	0.311
VAR 19	0.055	0.118	0.060	0.027	0.076
VAR 20	0.133	0.066	0.114	0.140	0.103
VAR 21	0.308	0.202	0.299	0.167	0.306
VAR 22	0.304	0.177	0.277	0.183	0.290
VAR 23	0.253	0.322	0.217	0.111	0.326
VAR 24	0.207	0.321	0.189	0.091	0.285
VAR 25	0.224	0.063	0.192	0.086	0.135
VAR 26	0.312	0.192	0.292	0.190	0.292
VAR 27	0.284	0.247	0.299	0.156	0.320
VAR 28	0.257	0.295	0.247	0.150	0.348
VAR 29	0.248	0.320	0.206	0.108	0.293
VAR 30	0.186	0.327	0.179	0.103	0.251
VAR 31	0.273	0.281	0.261	0.169	0.323
VAR 32	0.245	0.269	0.308	0.164	0.344
VAR 33	0.284	0.291	0.234	0.147	0.336
VAR 34	0.292	0.191	0.251	0.210	0.305
VAR 35	0.157	0.232	0.149	0.074	0.204
VAR 36	0.149	0.305	0.163	0.086	0.211
VAR 37	0.338	0.183	0.271	0.167	0.240
VAR 38	0.254	0.158	0.259	0.228	0.229
VAR 39	0.282	0.197	0.278	0.236	0.278

VAR 40	0.227	0.290	0.222	0.121	0.281

Variables

	VAR 11	VAR 12	VAR 13	VAR 14	VAR 15
VAR 1	0.257	0.210	0.272	0.144	0.255
VAR 2	0.191	0.217	0.128	0.181	0.174
VAR 3	0.279	0.230	0.262	0.164	0.304
VAR 4	0.272	0.248	0.217	0.166	0.265
VAR 5	0.308	0.252	0.272	0.172	0.287
VAR 6	0.235	0.202	0.308	0.108	0.268
VAR 7	0.302	0.229	0.202	0.222	0.278
VAR 8	0.237	0.198	0.256	0.098	0.264
VAR 9	0.129	0.103	0.177	0.055	0.163
VAR 10	0.300	0.268	0.299	0.177	0.335
VAR 11	1.000	0.270	0.295	0.228	0.337
VAR 12	0.270	1.000	0.224	0.223	0.249
VAR 13	0.295	0.224	1.000	0.145	0.301
VAR 14	0.228	0.223	0.145	1.000	0.171
VAR 15	0.337	0.249	0.301	0.171	1.000
VAR 16	0.317	0.309	0.283	0.220	0.312
VAR 17	0.150	0.120	0.252	0.067	0.195
VAR 18	0.313	0.291	0.290	0.184	0.332
VAR 19	0.074	0.103	0.072	0.026	0.087
VAR 20	0.075	0.071	0.113	0.034	0.099
VAR 21	0.246	0.239	0.293	0.135	0.300
VAR 22	0.227	0.194	0.338	0.122	0.273
VAR 23	0.328	0.312	0.285	0.204	0.325
VAR 24	0.298	0.267	0.220	0.212	0.300
VAR 25	0.078	0.088	0.173	0.037	0.129
VAR 26	0.232	0.194	0.336	0.116	0.256
VAR 27	0.280	0.221	0.346	0.152	0.327
VAR 28	0.336	0.302	0.284	0.225	0.353
VAR 29	0.301	0.264	0.279	0.216	0.299
VAR 30	0.316	0.252	0.228	0.192	0.263
VAR 31	0.313	0.275	0.333	0.182	0.325
VAR 32	0.298	0.265	0.306	0.184	0.346
VAR 33	0.321	0.262	0.320	0.203	0.321
VAR 34	0.229	0.176	0.308	0.116	0.248
VAR 35	0.241	0.262	0.162	0.275	0.212
VAR 36	0.293	0.264	0.183	0.263	0.249
VAR 37	0.218	0.198	0.285	0.123	0.274
VAR 38	0.181	0.161	0.261	0.086	0.248
VAR 39	0.225	0.229	0.314	0.114	0.271
VAR 40	0.325	0.278	0.264	0.206	0.285

Variables

	VAR 16	VAR 17	VAR 18	VAR 19	VAR 20
VAR 1	0.232	0.209	0.276	0.080	0.151
VAR 2	0.213	0.064	0.192	0.061	0.033
VAR 3	0.251	0.206	0.278	0.087	0.100
VAR 4	0.268	0.151	0.259	0.084	0.073

VAR 5	0.272	0.168	0.261	0.060	0.097
VAR 6	0.240	0.238	0.277	0.055	0.133
VAR 7	0.290	0.114	0.288	0.118	0.066
VAR 8	0.251	0.261	0.250	0.060	0.114
VAR 9	0.129	0.224	0.183	0.027	0.140
VAR 10	0.302	0.201	0.311	0.076	0.103
VAR 11	0.317	0.150	0.313	0.074	0.075
VAR 12	0.309	0.120	0.291	0.103	0.071
VAR 13	0.283	0.252	0.290	0.072	0.113
VAR 14	0.220	0.067	0.184	0.026	0.034
VAR 15	0.312	0.195	0.332	0.087	0.099
VAR 16	1.000	0.154	0.315	0.138	0.084
VAR 17	0.154	1.000	0.193	0.032	0.230
VAR 18	0.315	0.193	1.000	0.089	0.089
VAR 19	0.138	0.032	0.089	1.000	0.017
VAR 20	0.084	0.230	0.089	0.017	1.000
VAR 21	0.244	0.245	0.305	0.061	0.128
VAR 22	0.235	0.270	0.268	0.041	0.120
VAR 23	0.348	0.158	0.334	0.102	0.085
VAR 24	0.331	0.114	0.244	0.116	0.053
VAR 25	0.085	0.157	0.136	0.018	0.133
VAR 26	0.218	0.288	0.284	0.048	0.129
VAR 27	0.278	0.241	0.302	0.069	0.112
VAR 28	0.321	0.183	0.340	0.099	0.077
VAR 29	0.356	0.145	0.306	0.115	0.083
VAR 30	0.296	0.122	0.267	0.106	0.048
VAR 31	0.325	0.166	0.319	0.094	0.084
VAR 32	0.300	0.197	0.343	0.095	0.091
VAR 33	0.293	0.185	0.299	0.120	0.101
VAR 34	0.232	0.269	0.292	0.056	0.148
VAR 35	0.274	0.089	0.231	0.050	0.045
VAR 36	0.267	0.104	0.251	0.075	0.053
VAR 37	0.199	0.200	0.259	0.062	0.119
VAR 38	0.178	0.221	0.214	0.042	0.171
VAR 39	0.235	0.192	0.276	0.067	0.126
VAR 40	0.303	0.127	0.296	0.139	0.079

Variables

	VAR 21	VAR 22	VAR 23	VAR 24	VAR 25
VAR 1	0.271	0.263	0.250	0.206	0.116
VAR 2	0.124	0.122	0.190	0.230	0.036
VAR 3	0.277	0.270	0.275	0.227	0.118
VAR 4	0.208	0.213	0.254	0.261	0.073
VAR 5	0.244	0.231	0.282	0.279	0.102
VAR 6	0.308	0.304	0.253	0.207	0.224
VAR 7	0.202	0.177	0.322	0.321	0.063
VAR 8	0.299	0.277	0.217	0.189	0.192
VAR 9	0.167	0.183	0.111	0.091	0.086
VAR 10	0.306	0.290	0.326	0.285	0.135

Differential Item Functioning

VAR 11	0.246	0.227	0.328	0.298	0.078
VAR 12	0.239	0.194	0.312	0.267	0.088
VAR 13	0.293	0.338	0.285	0.220	0.173
VAR 14	0.135	0.122	0.204	0.212	0.037
VAR 15	0.300	0.273	0.325	0.300	0.129
VAR 16	0.244	0.235	0.348	0.331	0.085
VAR 17	0.245	0.270	0.158	0.114	0.157
VAR 18	0.305	0.268	0.334	0.244	0.136
VAR 19	0.061	0.041	0.102	0.116	0.018
VAR 20	0.128	0.120	0.085	0.053	0.133
VAR 21	1.000	0.285	0.243	0.225	0.159
VAR 22	0.285	1.000	0.228	0.182	0.167
VAR 23	0.243	0.228	1.000	0.336	0.085
VAR 24	0.225	0.182	0.336	1.000	0.069
VAR 25	0.159	0.167	0.085	0.069	1.000
VAR 26	0.276	0.326	0.222	0.189	0.178
VAR 27	0.298	0.303	0.304	0.228	0.112
VAR 28	0.285	0.260	0.350	0.286	0.104
VAR 29	0.265	0.245	0.311	0.261	0.091
VAR 30	0.211	0.198	0.306	0.272	0.074
VAR 31	0.296	0.286	0.307	0.270	0.130
VAR 32	0.292	0.315	0.303	0.285	0.133
VAR 33	0.281	0.279	0.337	0.307	0.082
VAR 35	0.162	0.140	0.231	0.246	0.049
VAR 36	0.184	0.153	0.279	0.289	0.058
VAR 37	0.285	0.273	0.243	0.178	0.146
VAR 38	0.274	0.236	0.170	0.147	0.176
VAR 39	0.283	0.298	0.261	0.221	0.150
VAR 40	0.263	0.228	0.319	0.308	0.080

Variables

	VAR 26	VAR 27	VAR 28	VAR 29	VAR 30
VAR 1	0.248	0.300	0.257	0.287	0.239
VAR 2	0.105	0.130	0.225	0.202	0.215
VAR 3	0.248	0.310	0.275	0.290	0.240
VAR 4	0.202	0.230	0.276	0.290	0.241
VAR 5	0.247	0.280	0.306	0.308	0.271
VAR 6	0.312	0.284	0.257	0.248	0.186
VAR 7	0.192	0.247	0.295	0.320	0.327
VAR 8	0.292	0.299	0.247	0.206	0.179
VAR 9	0.190	0.156	0.150	0.108	0.103
VAR 10	0.292	0.320	0.348	0.293	0.251
VAR 11	0.232	0.280	0.336	0.301	0.316
VAR 12	0.194	0.221	0.302	0.264	0.252
VAR 13	0.336	0.346	0.284	0.279	0.228
VAR 14	0.116	0.152	0.225	0.216	0.192
VAR 15	0.256	0.327	0.353	0.299	0.263
VAR 16	0.218	0.278	0.321	0.356	0.296
VAR 17	0.288	0.241	0.183	0.145	0.122

VAR 18	0.284	0.302	0.340	0.306	0.267
VAR 19	0.048	0.069	0.099	0.115	0.106
VAR 20	0.129	0.112	0.077	0.083	0.048
VAR 21	0.276	0.298	0.285	0.265	0.211
VAR 22	0.326	0.303	0.260	0.245	0.198
VAR 23	0.222	0.304	0.350	0.311	0.306
VAR 24	0.189	0.228	0.286	0.261	0.272
VAR 25	0.178	0.112	0.104	0.091	0.074
VAR 26	1.000	0.329	0.246	0.246	0.194
VAR 27	0.329	1.000	0.311	0.306	0.244
VAR 28	0.246	0.311	1.000	0.329	0.315
VAR 29	0.246	0.306	0.329	1.000	0.269
VAR 30	0.194	0.244	0.315	0.269	1.000
VAR 31	0.269	0.305	0.298	0.322	0.289
VAR 32	0.284	0.335	0.308	0.294	0.271
VAR 33	0.283	0.302	0.328	0.333	0.297
VAR 34	0.279	0.294	0.241	0.247	0.189
VAR 35	0.123	0.188	0.236	0.272	0.236
VAR 36	0.165	0.196	0.243	0.297	0.296
VAR 37	0.307	0.271	0.251	0.241	0.163
VAR 38	0.293	0.225	0.217	0.172	0.157
VAR 39	0.287	0.310	0.285	0.247	0.202
VAR 40	0.215	0.296	0.332	0.309	0.293

Variables

	VAR 31	VAR 32	VAR 33	VAR 34	VAR 35
VAR 1	0.263	0.251	0.247	0.269	0.151
VAR 2	0.161	0.178	0.187	0.094	0.189
VAR 3	0.288	0.316	0.272	0.301	0.180
VAR 4	0.281	0.228	0.298	0.205	0.181
VAR 5	0.279	0.264	0.295	0.244	0.206
VAR 6	0.273	0.245	0.284	0.292	0.157
VAR 7	0.281	0.269	0.291	0.191	0.232
VAR 8	0.261	0.308	0.234	0.251	0.149
VAR 9	0.169	0.164	0.147	0.210	0.074
VAR 10	0.323	0.344	0.336	0.305	0.204
VAR 11	0.313	0.298	0.321	0.229	0.241
VAR 12	0.275	0.265	0.262	0.176	0.262
VAR 13	0.333	0.306	0.320	0.308	0.162
VAR 14	0.182	0.184	0.203	0.116	0.275
VAR 15	0.325	0.346	0.321	0.248	0.212
VAR 16	0.325	0.300	0.293	0.232	0.274
VAR 17	0.166	0.197	0.185	0.269	0.089
VAR 18	0.319	0.343	0.299	0.292	0.231
VAR 19	0.094	0.095	0.120	0.056	0.050
VAR 20	0.084	0.091	0.101	0.148	0.045
VAR 21	0.296	0.292	0.281	0.319	0.162
VAR 22	0.286	0.315	0.279	0.308	0.140
VAR 23	0.307	0.303	0.337	0.224	0.231

Differential Item Functioning

VAR 24	0.270	0.285	0.307	0.188	0.246
VAR 25	0.130	0.133	0.082	0.168	0.049
VAR 26	0.269	0.284	0.283	0.279	0.123
VAR 27	0.305	0.335	0.302	0.294	0.188
VAR 28	0.298	0.308	0.328	0.241	0.236
VAR 29	0.322	0.294	0.333	0.247	0.272
VAR 30	0.289	0.271	0.297	0.189	0.236
VAR 31	1.000	0.334	0.309	0.264	0.204
VAR 32	0.334	1.000	0.347	0.295	0.218
VAR 33	0.309	0.347	1.000	0.249	0.259
VAR 34	0.264	0.295	0.249	1.000	0.145
VAR 35	0.204	0.218	0.259	0.145	1.000
VAR 36	0.233	0.246	0.284	0.156	0.274
VAR 37	0.261	0.246	0.277	0.278	0.134
VAR 38	0.208	0.231	0.205	0.241	0.109
VAR 39	0.286	0.259	0.262	0.279	0.134
VAR 40	0.294	0.292	0.341	0.216	0.252

Variables

	VAR 36	VAR 37	VAR 38	VAR 39	VAR 40
VAR 1	0.213	0.234	0.203	0.230	0.273
VAR 2	0.229	0.107	0.075	0.123	0.211
VAR 3	0.209	0.233	0.206	0.274	0.255
VAR 4	0.236	0.180	0.156	0.221	0.284
VAR 5	0.253	0.241	0.196	0.248	0.289
VAR 6	0.149	0.338	0.254	0.282	0.227
VAR 7	0.305	0.183	0.158	0.197	0.290
VAR 8	0.163	0.271	0.259	0.278	0.222
VAR 9	0.086	0.167	0.228	0.236	0.121
VAR 10	0.211	0.240	0.229	0.278	0.281
VAR 11	0.293	0.218	0.181	0.225	0.325
VAR 12	0.264	0.198	0.161	0.229	0.278
VAR 13	0.183	0.285	0.261	0.314	0.264
VAR 14	0.263	0.123	0.086	0.114	0.206
VAR 15	0.249	0.274	0.248	0.271	0.285
VAR 16	0.267	0.199	0.178	0.235	0.303
VAR 17	0.104	0.200	0.221	0.192	0.127
VAR 18	0.251	0.259	0.214	0.276	0.296
VAR 19	0.075	0.062	0.042	0.067	0.139
VAR 20	0.053	0.119	0.171	0.126	0.079
VAR 21	0.184	0.285	0.274	0.283	0.263
VAR 22	0.153	0.273	0.236	0.298	0.228
VAR 23	0.279	0.243	0.170	0.261	0.319
VAR 24	0.289	0.178	0.147	0.221	0.308
VAR 25	0.058	0.146	0.176	0.150	0.080
VAR 26	0.165	0.307	0.293	0.287	0.215
VAR 27	0.196	0.271	0.225	0.310	0.296
VAR 28	0.243	0.251	0.217	0.285	0.332
VAR 29	0.297	0.241	0.172	0.247	0.309

VAR 30	0.296	0.163	0.157	0.202	0.293
VAR 31	0.233	0.261	0.208	0.286	0.294
VAR 32	0.246	0.246	0.231	0.259	0.292
VAR 33	0.284	0.277	0.205	0.262	0.341
VAR 34	0.156	0.278	0.241	0.279	0.216
VAR 35	0.274	0.134	0.109	0.134	0.252
VAR 36	1.000	0.155	0.118	0.180	0.288
VAR 37	0.155	1.000	0.250	0.276	0.204
VAR 38	0.118	0.250	1.000	0.242	0.181
VAR 39	0.180	0.276	0.242	1.000	0.262
VAR 40	0.288	0.204	0.181	0.262	1.000

Item-Total Correlations with 2000 valid cases.

Variables	VAR 1	VAR 2	VAR 3	VAR 4	VAR 5
	0.527	0.352	0.556	0.514	0.563
Variables	VAR 6	VAR 7	VAR 8	VAR 9	VAR 10
	0.507	0.509	0.488	0.302	0.579
Variables	VAR 11	VAR 12	VAR 13	VAR 14	VAR 15
	0.566	0.502	0.556	0.352	0.586
Variables	VAR 16	VAR 17	VAR 18	VAR 19	VAR 20
	0.564	0.371	0.582	0.171	0.200
Variables	VAR 21	VAR 22	VAR 23	VAR 24	VAR 25
	0.532	0.511	0.574	0.511	0.235
Variables	VAR 26	VAR 27	VAR 28	VAR 29	VAR 30
	0.507	0.570	0.591	0.569	0.507
Variables	VAR 31	VAR 32	VAR 33	VAR 34	VAR 35
	0.580	0.584	0.590	0.501	0.411
Variables	VAR 36	VAR 37	VAR 38	VAR 39	VAR 40
	0.465	0.482	0.415	0.513	0.556

Conditioning Levels

Lower	Upper
1	3
4	7
8	10
11	13
14	16
17	19
20	22
23	25
26	28

Differential Item Functioning

```
29          31
32          40
```

And so on for all items. Note the difference for the two item plots shown above! Next, the output reflects multiple passes to "fit" the data for the M-H test:

```
COMPUTING M-H CHI-SQUARE, PASS # 1

Cases in Reference Group

                    Score Level Counts by Item
Variables
              VAR 1       VAR 2       VAR 3       VAR 4       VAR 5
   1-  3        6           6           6           6           6
   4-  7       38          38          38          38          38
   8- 10       47          47          47          47          47
  11- 13       65          65          65          65          65
  14- 16      101         101         101         101         101
  17- 19      113         113         113         113         113
  20- 22      137         137         137         137         137
  23- 25      121         121         121         121         121
  26- 28      114         114         114         114         114
  29- 31      124         124         124         124         124
  32- 40      132         132         132         132         132

                    Score Level Counts by Item
Variables
              VAR 6       VAR 7       VAR 8       VAR 9       VAR 10
   1-  3        6           6           6           6           6
   4-  7       38          38          38          38          38
   8- 10       47          47          47          47          47
  11- 13       65          65          65          65          65
  14- 16      101         101         101         101         101
  17- 19      113         113         113         113         113
  20- 22      137         137         137         137         137
  23- 25      121         121         121         121         121
  26- 28      114         114         114         114         114
  29- 31      124         124         124         124         124
  32- 40      132         132         132         132         132

                    Score Level Counts by Item
Variables
              VAR 11      VAR 12      VAR 13      VAR 14      VAR 15
   1-  3        6           6           6           6           6
   4-  7       38          38          38          38          38
   8- 10       47          47          47          47          47
  11- 13       65          65          65          65          65
  14- 16      101         101         101         101         101
  17- 19      113         113         113         113         113
  20- 22      137         137         137         137         137
  23- 25      121         121         121         121         121
```

26- 28	114	114	114	114	114
29- 31	124	124	124	124	124
32- 40	132	132	132	132	132

Score Level Counts by Item

Variables

	VAR 16	VAR 17	VAR 18	VAR 19	VAR 20
1- 3	6	6	6	6	6
4- 7	38	38	38	38	38
8- 10	47	47	47	47	47
11- 13	65	65	65	65	65
14- 16	101	101	101	101	101
17- 19	113	113	113	113	113
20- 22	137	137	137	137	137
23- 25	121	121	121	121	121
26- 28	114	114	114	114	114
29- 31	124	124	124	124	124
32- 40	132	132	132	132	132

Score Level Counts by Item

Variables

	VAR 21	VAR 22	VAR 23	VAR 24	VAR 25
1- 3	6	6	6	6	6
4- 7	38	38	38	38	38
8- 10	47	47	47	47	47
11- 13	65	65	65	65	65
14- 16	101	101	101	101	101
17- 19	113	113	113	113	113
20- 22	137	137	137	137	137
23- 25	121	121	121	121	121
26- 28	114	114	114	114	114
29- 31	124	124	124	124	124
32- 40	132	132	132	132	132

Score Level Counts by Item

Variables

	VAR 26	VAR 27	VAR 28	VAR 29	VAR 30
1- 3	6	6	6	6	6
4- 7	38	38	38	38	38
8- 10	47	47	47	47	47
11- 13	65	65	65	65	65
14- 16	101	101	101	101	101
17- 19	113	113	113	113	113
20- 22	137	137	137	137	137
23- 25	121	121	121	121	121
26- 28	114	114	114	114	114

Differential Item Functioning

```
   29- 31         124         124         124         124         124
   32- 40         132         132         132         132         132

                      Score Level Counts by Item
Variables
                   VAR 31      VAR 32      VAR 33      VAR 34      VAR 35
    1- 3             6           6           6           6           6
    4- 7            38          38          38          38          38
    8- 10           47          47          47          47          47
   11- 13           65          65          65          65          65
   14- 16          101         101         101         101         101
   17- 19          113         113         113         113         113
   20- 22          137         137         137         137         137
   23- 25          121         121         121         121         121
   26- 28          114         114         114         114         114
   29- 31          124         124         124         124         124
   32- 40          132         132         132         132         132

                      Score Level Counts by Item
Variables
                   VAR 36      VAR 37      VAR 38      VAR 39      VAR 40
    1- 3             6           6           6           6           6
    4- 7            38          38          38          38          38
    8- 10           47          47          47          47          47
   11- 13           65          65          65          65          65
   14- 16          101         101         101         101         101
   17- 19          113         113         113         113         113
   20- 22          137         137         137         137         137
   23- 25          121         121         121         121         121
   26- 28          114         114         114         114         114
   29- 31          124         124         124         124         124
   32- 40          132         132         132         132         132

Cases in Focus Group

                      Score Level Counts by Item
Variables
                   VAR 1       VAR 2       VAR 3       VAR 4       VAR 5
    1- 3             7           7           7           7           7
    4- 7            47          47          47          47          47
    8- 10           64          64          64          64          64
   11- 13           85          85          85          85          85
   14- 16          123         123         123         123         123
   17- 19          138         138         138         138         138
   20- 22          127         127         127         127         127
   23- 25          115         115         115         115         115
```

	26- 28	108	108	108	108	108
	29- 31	91	91	91	91	91
	32- 40	95	95	95	95	95

Score Level Counts by Item

Variables	VAR 6	VAR 7	VAR 8	VAR 9	VAR 10
1- 3	7	7	7	7	7
4- 7	47	47	47	47	47
8- 10	64	64	64	64	64
11- 13	85	85	85	85	85
14- 16	123	123	123	123	123
17- 19	138	138	138	138	138
20- 22	127	127	127	127	127
23- 25	115	115	115	115	115
26- 28	108	108	108	108	108
29- 31	91	91	91	91	91
32- 40	95	95	95	95	95

Score Level Counts by Item

Variables	VAR 11	VAR 12	VAR 13	VAR 14	VAR 15
1- 3	7	7	7	7	7
4- 7	47	47	47	47	47
8- 10	64	64	64	64	64
11- 13	85	85	85	85	85
14- 16	123	123	123	123	123
17- 19	138	138	138	138	138
20- 22	127	127	127	127	127
23- 25	115	115	115	115	115
26- 28	108	108	108	108	108
29- 31	91	91	91	91	91
32- 40	95	95	95	95	95

Score Level Counts by Item

Variables	VAR 16	VAR 17	VAR 18	VAR 19	VAR 20
1- 3	7	7	7	7	7
4- 7	47	47	47	47	47
8- 10	64	64	64	64	64
11- 13	85	85	85	85	85
14- 16	123	123	123	123	123
17- 19	138	138	138	138	138
20- 22	127	127	127	127	127
23- 25	115	115	115	115	115
26- 28	108	108	108	108	108

Differential Item Functioning

29- 31	91	91	91	91	91
32- 40	95	95	95	95	95

Score Level Counts by Item

Variables	VAR 21	VAR 22	VAR 23	VAR 24	VAR 25
1- 3	7	7	7	7	7
4- 7	47	47	47	47	47
8- 10	64	64	64	64	64
11- 13	85	85	85	85	85
14- 16	123	123	123	123	123
17- 19	138	138	138	138	138
20- 22	127	127	127	127	127
23- 25	115	115	115	115	115
26- 28	108	108	108	108	108
29- 31	91	91	91	91	91
32- 40	95	95	95	95	95

Score Level Counts by Item

Variables	VAR 26	VAR 27	VAR 28	VAR 29	VAR 30
1- 3	7	7	7	7	7
4- 7	47	47	47	47	47
8- 10	64	64	64	64	64
11- 13	85	85	85	85	85
14- 16	123	123	123	123	123
17- 19	138	138	138	138	138
20- 22	127	127	127	127	127
23- 25	115	115	115	115	115
26- 28	108	108	108	108	108
29- 31	91	91	91	91	91
32- 40	95	95	95	95	95

Score Level Counts by Item

Variables	VAR 31	VAR 32	VAR 33	VAR 34	VAR 35
1- 3	7	7	7	7	7
4- 7	47	47	47	47	47
8- 10	64	64	64	64	64
11- 13	85	85	85	85	85
14- 16	123	123	123	123	123
17- 19	138	138	138	138	138
20- 22	127	127	127	127	127
23- 25	115	115	115	115	115

26- 28	108	108	108	108	108
29- 31	91	91	91	91	91
32- 40	95	95	95	95	95

Score Level Counts by Item

Variables

	VAR 36	VAR 37	VAR 38	VAR 39	VAR 40
1- 3	7	7	7	7	7
4- 7	47	47	47	47	47
8- 10	64	64	64	64	64
11- 13	85	85	85	85	85
14- 16	123	123	123	123	123
17- 19	138	138	138	138	138
20- 22	127	127	127	127	127
23- 25	115	115	115	115	115
26- 28	108	108	108	108	108
29- 31	91	91	91	91	91
32- 40	95	95	95	95	95

Insufficient data found in level: 1 - 3

CODES	ITEM	SIG.	ALPHA	CHI2	P-VALUE	MH D-DIF	S.E. MH D-DIF
C R	1	***	9.367	283.535	0.000	-5.257	0.343
C R	2	***	8.741	65.854	0.000	-5.095	0.704
C R	3	***	7.923	287.705	0.000	-4.864	0.310
C R	4	***	10.888	305.319	0.000	-5.611	0.358
C R	5	***	13.001	399.009	0.000	-6.028	0.340
B	6	***	0.587	13.927	0.000	1.251	0.331
A	7	*	0.725	5.598	0.018	0.756	0.311
A	8	*	0.724	4.851	0.028	0.760	0.335
B	9	*	0.506	6.230	0.013	1.599	0.620
B	10	***	0.638	15.345	0.000	1.056	0.267
A	11		0.798	3.516	0.061	0.529	0.274
A	12	***	0.700	8.907	0.003	0.838	0.276
A	13	***	0.663	10.414	0.001	0.964	0.294
B	14	*	0.595	6.413	0.011	1.219	0.466
B	15	***	0.616	17.707	0.000	1.139	0.268
B	16	***	0.617	16.524	0.000	1.133	0.276
A	17		0.850	0.355	0.551	0.382	0.537
A	18	**	0.729	7.642	0.006	0.742	0.263
A	19		0.595	1.721	0.190	1.222	0.831
A	20		2.004	1.805	0.179	-1.633	1.073
A	21	*	0.746	4.790	0.029	0.688	0.307
A	22		0.773	2.996	0.083	0.606	0.336
B	23	***	0.573	20.155	0.000	1.307	0.289
A	24	*	0.736	4.796	0.029	0.722	0.320
A	25		0.570	1.595	0.207	1.320	0.914
B	26	***	0.554	14.953	0.000	1.388	0.354
A	27	**	0.707	7.819	0.005	0.816	0.287
A	28	*	0.750	5.862	0.015	0.675	0.272
A	29	***	0.704	7.980	0.005	0.825	0.286
A	30	*	0.769	3.845	0.050	0.618	0.305
A	31	**	0.743	6.730	0.009	0.698	0.263

A	32	*	0.762	5.551	0.018	0.640	0.266
A	33	*	0.749	6.193	0.013	0.681	0.268
A	34		0.976	0.007	1.000	0.058	0.360
A	35		0.790	1.975	0.160	0.555	0.375
A	36		0.832	1.310	0.252	0.432	0.354
A	37	*	0.721	5.148	0.023	0.770	0.329
A	38	*	0.678	4.062	0.044	0.914	0.433
A	39		0.804	2.490	0.115	0.512	0.312
A	40	***	0.664	11.542	0.001	0.963	0.279

No. of items purged in pass 1 = 5
Item Numbers:
1
2
3
4
5

One should probably combine the first two score groups (0–3 and 4–7) into one group and the last three groups into one group so that sufficient sample size is available for the comparisons of the two groups. This would, of course, reduce the number of groups from 11 in our original specifications to 8 score groups. The chi-square statistic identifies items you will want to give specific attention. Examine the data plots for those items. Differences found may suggest bias in those items. Only examination of the actual content can help in this decision. Even though two groups may differ in their item response patterns does not provide sufficient grounds to establish bias - perhaps it simply identifies a true difference in educational achievement due to other factors.

Adjustment of Reliability For Variance Change

Researchers will sometimes use a test that has been standardized on a large, heterogenous population of subjects. Such tests typically report rather high internal-consistency reliability estimates (e.g. Cronbach's estimate.) But what is the reliability if one administers the test to a much more homogeneous population? For example, assume a high school counselor administers a "College Aptitude Test" that reports a reliability of 0.95 with a standard deviation of 15 (variance of 225) and a mean of 20.0 for the national norm. What reliability would the counselor expect to obtain for her sample of students that obtain a mean of 22.8 and a standard deviation of 10.2 (variance of 104.04)? This procedure will help provide the estimate. Shown below is the specification form and our sample values entered. When the compute button is clicked, the results shown are obtained (Fig. 11.18).

Fig. 11.18 Reliability adjustment for variability dialog

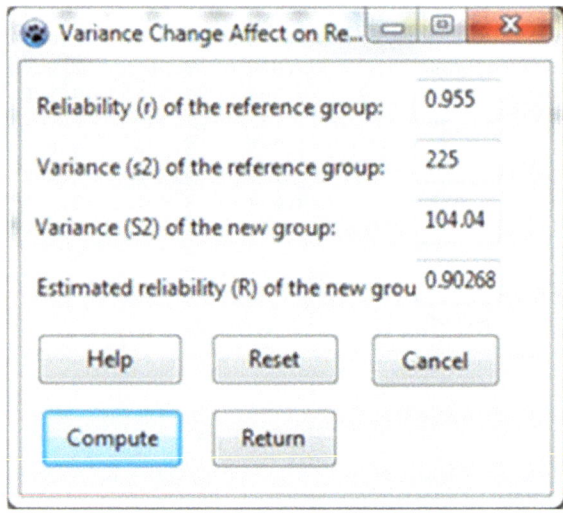

Polytomous DIF Analysis

The purpose of the differential item functioning program is to identify test or attitude items that "perform" differently for two groups - a target group and a reference group. Two procedures are provided and selected on the basis of whether the items are dichotomous (0 and 1 scoring) or consist of multiple categories (e.g. Likert responses ranging from 1 to 5.) The latter case is where the Polytomous DIF Analysis is selected. When you initiate this procedure you will see the dialogue box shown below (Fig. 11.19):

The results from an analysis of three items with five categories that have been collapsed into three category levels is shown below. A sample of 500 subject's attitude scores were observed (Fig. 11.20).

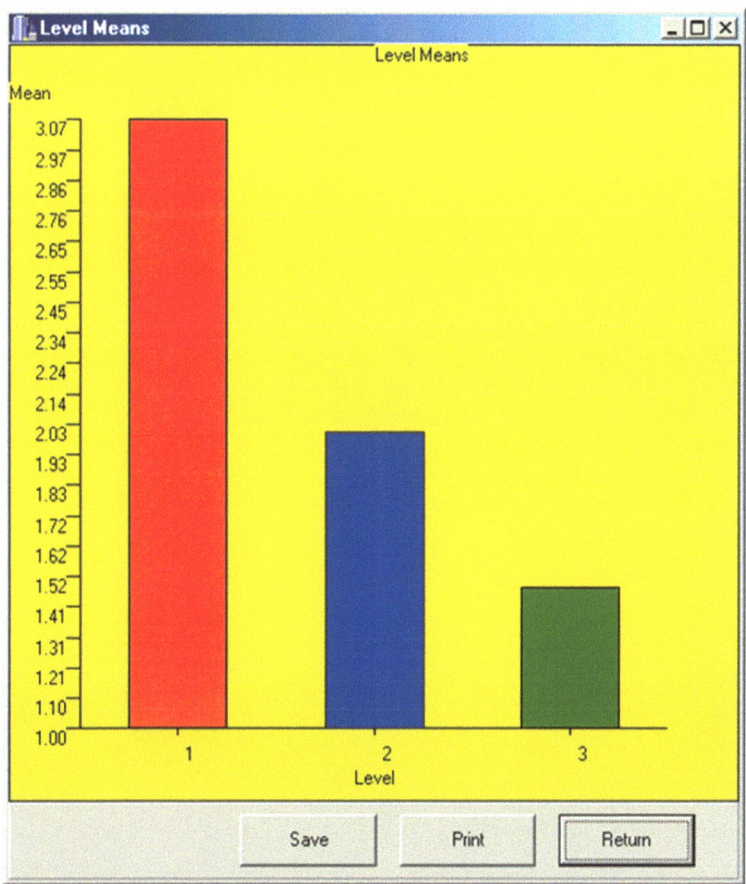

Fig. 11.19 Polytomous item differential functioning dialog

Fig. 11.20 Level means for polytomous item

Polytomous Item DIF Analysis adapted by Bill Miller from
Procedures for extending item bias detection techniques
by Catherine Welch and H.D. Hoover, 1993
Applied Measurement in Education 6(1), pages 1-19.

```
Conditioning Levels
Lower    Upper
  0        1
  2        3
  4        5
```

For Item 1:

Observed Category Frequencies

Item	Group	Level	Category Number				
			1	2	3	4	5
1	Ref.	1	46	51	39	64	48
1	Focal	1	40	41	38	46	42
1	Total	1	86	92	77	110	90
1	Ref.	2	2	0	0	0	0
1	Focal	2	1	0	0	0	0
1	Total	2	3	0	0	0	0
1	Ref.	3	12	8	1	0	0
1	Focal	3	15	6	0	0	0
1	Total	3	27	14	1	0	0

```
t-test values for Reference and Focus Means for each level
Mean Reference =      3.069 SD =      24.396 N =     248
Mean Focal     =      3.043 SD =      21.740 N =     207
Level 1 t =     -0.011 with deg. freedom =      453
Mean Reference =      2.000 SD = 2.000 N =     2
Mean Focal     =      1.000 SD = 1.000 N =     1
Level 2 t = 0.000 with deg. freedom = 0
Mean Reference =      1.476 SD = 4.262 N =     21
Mean Focal     =      1.286 SD = 4.088 N =     21
Level   3 t =    -0.144 with deg. freedom =     40
Composite z statistic = -0.076. Prob. > |z| = 0.530
Weighted Composite z statistic = -0.248. Prob. > |z| = 0.598
Generalized Mantel-Haenszel = 0.102 with D.F. = 1 and Prob. >
Chi-Sqr. = 0.749
```

```
For Item 2:

Observed Category Frequencies
Item Group Level Category Number
                           1       2       3       4       5
   2    Ref.    1         56      46      47      48      51
   2    Focal   1         37      38      49      35      48
   2    Total   1         93      84      96      83      99
   2    Ref.    2          2       0       0       0       0
   2    Focal   2          1       0       0       0       0
   2    Total   2          3       0       0       0       0
   2    Ref.    3         12       8       1       0       0
   2    Focal   3          9      11       1       0       0
   2    Total   3         21      19       2       0       0

t-test values for Reference and Focus Means for each level
Mean Reference =       2.968 SD = 23.046 N =    248
Mean Focal     =       3.092 SD = 22.466 N =    207
Level   1 t =   0.058 with deg. freedom   = 453
Mean Reference =       2.000 SD =  2.000 N    = 2
Mean Focal     =       1.000 SD =  1.000 N    = 1
Level   2 t =   0.000 with deg. freedom   = 0
Mean Reference =       1.476 SD =  4.262 N    = 21
Mean Focal     =       1.619 SD =  5.094 N    = 21
Level   3 t =   0.096 with deg. freedom   = 40
Composite z statistic = 0.075. Prob. > |z| = 0.470
Weighted Composite z statistic = 0.673. Prob. > |z| = 0.250
Generalized Mantel-Haenszel = 1.017 with D.F. = 1 and Prob. >
Chi-Sqr. = 0.313

Observed Category Frequencies
Item Group Level Category Number
                           1       2       3       4       5
   3    Ref.    1         35      38      52      68      55
   3    Focal   1         42      41      37      42      45
   3    Total   1         77      79      89     110     100
   3    Ref.    2          2       0       0       0       0
   3    Focal   2          1       0       0       0       0
   3    Total   2          3       0       0       0       0
   3    Ref.    3          8      10       3       0       0
   3    Focal   3          7      10       4       0       0
   3    Total   3         15      20       7       0       0
```

```
t-test values for Reference and Focus Means for each level
Mean Reference =       3.282 SD = 26.866 N =    248
Mean Focal      =      3.034 SD = 21.784 N =    207
Level    1 t =    -0.107 with deg. freedom =    453
Mean Reference =       2.000 SD = 2.000 N =    2
Mean Focal      =      1.000 SD = 1.000 N =    1
Level    2 t =     0.000 with deg. freedom =    0
Mean Reference =       1.762 SD = 4.898 N =    21
Mean Focal      =      1.857 SD = 5.102 N =    21
Level    3 t =     0.060 with deg. freedom =    40
Composite z statistic = -0.023. Prob. > |z| = 0.509
Weighted Composite z statistic = -1.026. Prob. > |z| = 0.848
Generalized Mantel-Haenszel =      3.248 with D.F. = 1 and Prob.
> ChiSqr. =   0.071
```

Generate Test Data

To help you become familiar with some of the measurement procedures, you can experiment by creating "artificial" item responses to a test. When you select the option to generate simulated test data, you complete the information in the following specification form. An example is shown. Before you begin, be sure you have closed any open file already in the data grid since the data that is generated will be placed in that grid (Fig. 11.21).

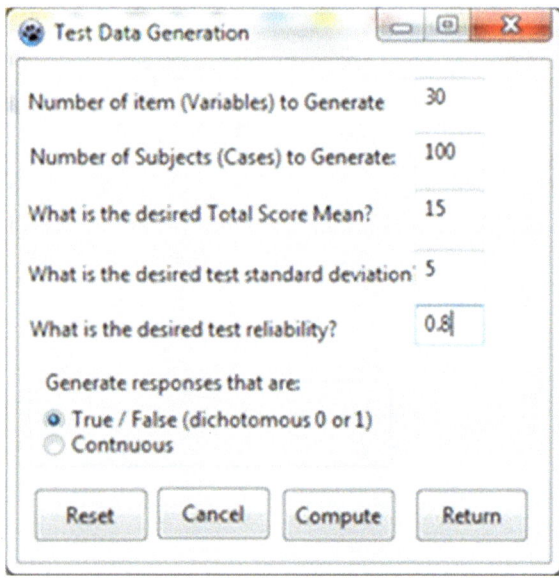

Fig. 11.21 The item generation dialog

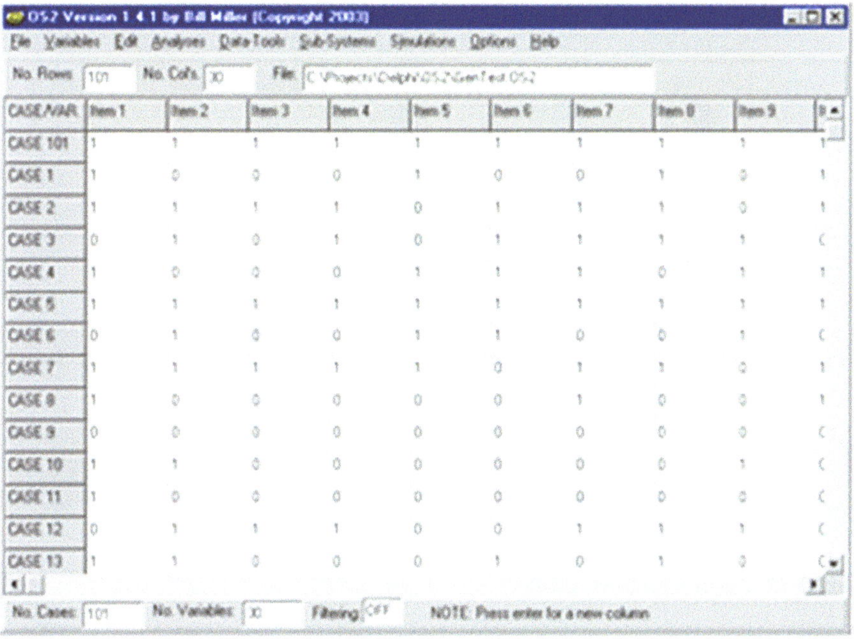

Fig. 11.22 Generated item data in the main grid

Shown above is a "snap-shot" of the generated test item responses. An additional row has been inserted for the first case which consists of all 1's. It will serve as the "correct" response for scoring each of the item responses of the subsequent cases. You can save your generated file for future analyses or other work (Fig. 11.22).

Notice that in our example we specified the creation of test data that would have a reliability of 0.8 for 30 items administered to 100 students. If we analyze this data with our Classical Test Analysis procedure, we obtain the following output:

Alpha Reliability Estimate for Test = 0.8997 S.E. of Measurement = 2.406

Clearly, the test generated from our population specifications yielded a somewhat higher reliability than the 0.8 specified for the reliability. Have we learned something about sampling variability? If you request that the total be placed in the data grid when you use analyze the test, you can also use the descriptive statistics procedure to obtain the sample mean, etc. as shown below:

DISTRIBUTION PARAMETER ESTIMATES

TOTAL (N=100) Sum = 1560.000
Mean = 15.600 Variance = 55.838 Std.Dev. = 7.473
Std.Error of Mean = 0.747
Range = 29.000 Minimum = 1.000 Maximum = 30.000
Skewness = −0.134 Std. Error of Skew = 0.241
Kurtosis = −0.935 Std. Error Kurtosis = 0.478

Fig. 11.23 Plot of generated test data

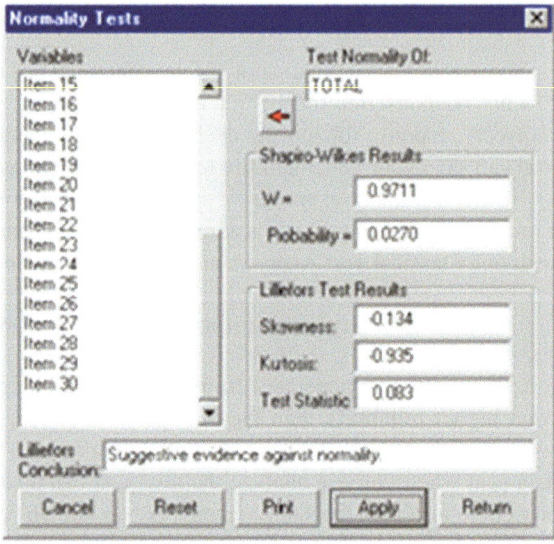

Fig. 11.24 Test of normality for generated data

The frequencies procedure can plot the total score distribution of our sample with the normal curve as a reference to obtain (Fig. 11.23):

A test of normality of the total scores suggests a possibility that the obtained scores are not normally distributed as shown in the normality test form above (Fig. 11.24):

Spearman-Brown Reliability Prophecy

The Spearman-Brown "Prophecy" formula has been a corner-stone of many instructional text books in measurement theory. Based on "Classical True-Score" theory, it provides an estimate of what the reliability of a test of a different length would be based on the initial test's reliability estimate. It assumes the average difficulty and inter-item covariances of the extended (or trimmed) test are the same as the original test. If these assumptions are valid, it is a reasonable estimator. Shown below is the specification form which appears when you elect this Measurement option from the Analyses menu (Fig. 11.25):

You can see that in an example, that when a test with an initial reliability of 0.8 is doubled (the multiplier k=2) that the new test is expected to have a reliability of 0.89 approximately. The program may be useful for reducing a test (perhaps by randomly selecting items to delete) that requires too long to administer and has an initially high internal consistency reliability estimate. For example, assume a test of 200 items has a reliability of .95. What is the estimate if the test is reduced by one-half? If the new reliability of 0.9 is satisfactory, considerable time and money may be saved!

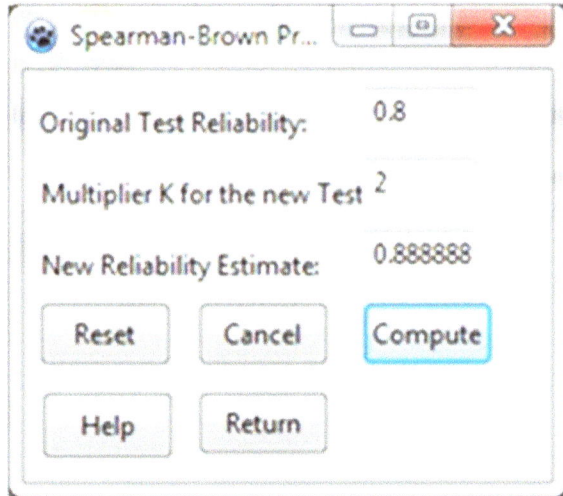

Fig. 11.25 Spearman-Brown Prophecy dialog

Chapter 12
Statistical Process Control

XBAR Chart

An Example

We will use the file labeled boltsize.txt to demonstrate the XBAR Chart procedure. Load the file and select the option Statistics/Statistical Process Control/Control Charts/XBAR Chart from the menu. The file contains two variables, lot number and bolt length. These values have been entered in the specification form which is shown below. Notice that the form also provides the option to enter and use a specific "target" value for the process as well as specification levels which may have been provided as guidelines for determining whether or not the process was in control for a given sample (Fig. 12.1).

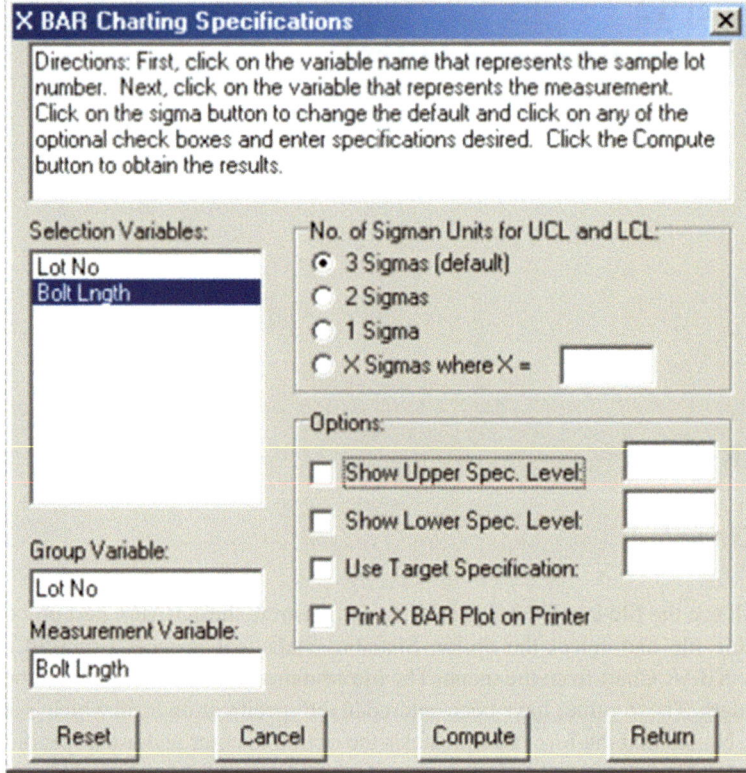

Fig. 12.1 XBAR chart dialog

Pressing the Compute button results in the following (Fig. 12.2):

XBAR Chart

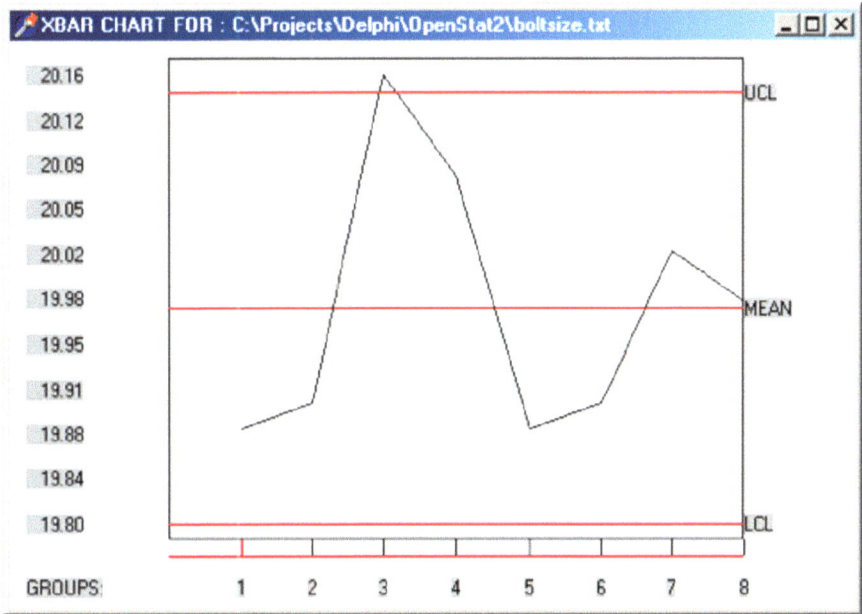

Fig. 12.2 XBAR chart for boltsize

```
X Bar Chart Results

Group  Size   Mean    Std.Dev.
  1     5     19.88    0.37
  2     5     19.90    0.29
  3     5     20.16    0.27
  4     5     20.08    0.29
  5     5     19.88    0.49
  6     5     19.90    0.39
  7     5     20.02    0.47
  8     5     19.98    0.43
Grand Mean = 19.97, Std.Dev. = 0.359, Standard Error of Mean = 0.06
Lower Control Limit = 19.805, Upper Control Limit = 20.145
```

If, in addition, we specify a target value of 20 for our bolt and upper and lower specification levels (tolerance) of 20.1 and 19.9, we would obtain the chart shown below (Fig. 12.3):

In this chart we can see that the mean of the samples falls slightly below the specified target value and that samples 3 and 5 appear to have bolts outside the tolerance specifications.

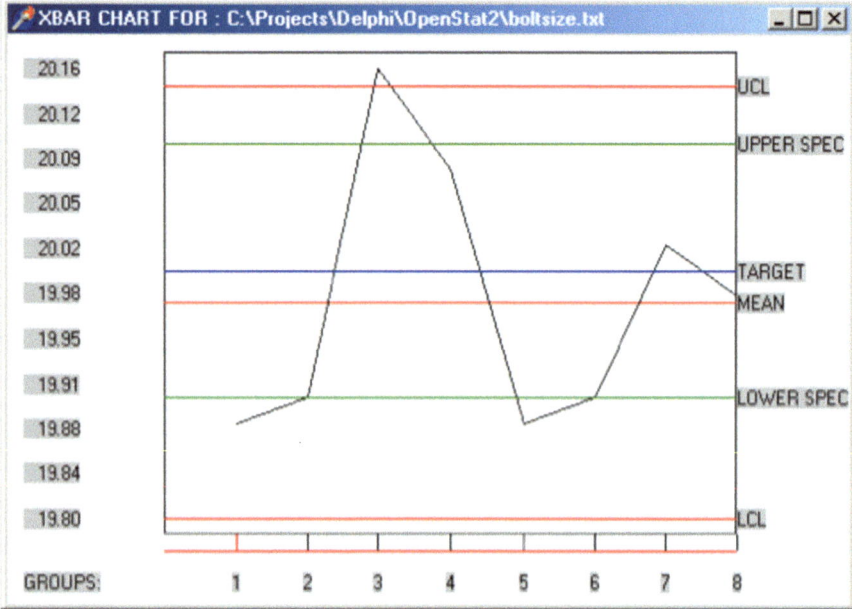

Fig. 12.3 XBAR chart plot with target specifications

Range Chart

As tools wear the products produced may begin to vary more and more widely around the values specified for them. The mean of a sample may still be close to the specified value but the range of values observed may increase. The result is that more and more parts produced may be under or over the specified value. Therefore quality assurance personnel examine not only the mean (XBAR chart) but also the range of values in their sample lots. Again, examine the boltsize.txt file with the option Statistics/Statistical Process Control/Control Charts/Range Chart. Shown below is the specification form and the results (Figs. 12.4, 12.5):

Range Chart

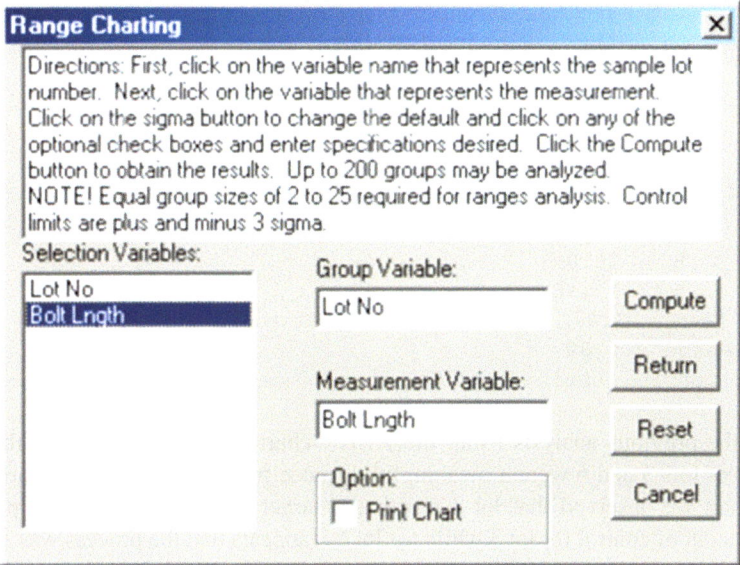

Fig. 12.4 Range chart dialog

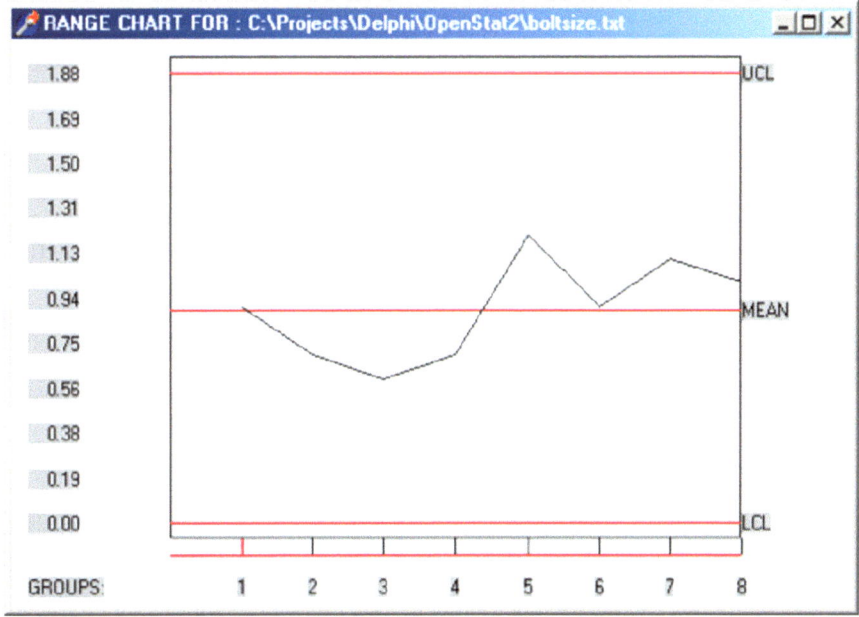

Fig. 12.5 Range chart plot

```
X Bar Chart Results
Group  Size   Mean   Range  Std.Dev.
  1      5    19.88  0.90   0.37
  2      5    19.90  0.70   0.29
  3      5    20.16  0.60   0.27
  4      5    20.08  0.70   0.29
  5      5    19.88  1.20   0.49
  6      5    19.90  0.90   0.39
  7      5    20.02  1.10   0.47
  8      5    19.98  1.00   0.43
Grand Mean = 19.97, Std.Dev. = 0.359, Standard Error of Mean = 0.06
Mean Range = 0.89
Lower Control Limit = 0.000, Upper Control Limit = 1.876
```

In the previous analysis using the XBAR chart procedure we found that the means of lots 3 and 6 were a meaningful distance from the target specification. In this chart we observed that lot 3 also had a larger range of values. The process appears out of control for lot 3 while for lot 6 it appears that the process was simply requiring adjustment toward the target value. In practice we would more likely see a pattern of increasing ranges as a machine becomes "loose" due to wear even though the averages may still be "on target".

S Control Chart

The sample standard deviation, like the range, is also an indicator of how much values vary in a sample. While the range reflects the difference between largest and smallest values in a sample, the standard deviation reflects the square root of the average squared distance around the mean of the values. We desire to reduce this variability in our processes so as to produce products as similar to one another as is possible. The S control chart plot the standard deviations of our sample lots and allows us to see the impact of adjustments and improvements in our manufacturing processes.

Examine the boltsize.txt data with the S Control Chart. Shown below is the specification form for the analysis and the results obtained (Figs. 12.6, 12.7):

S Control Chart

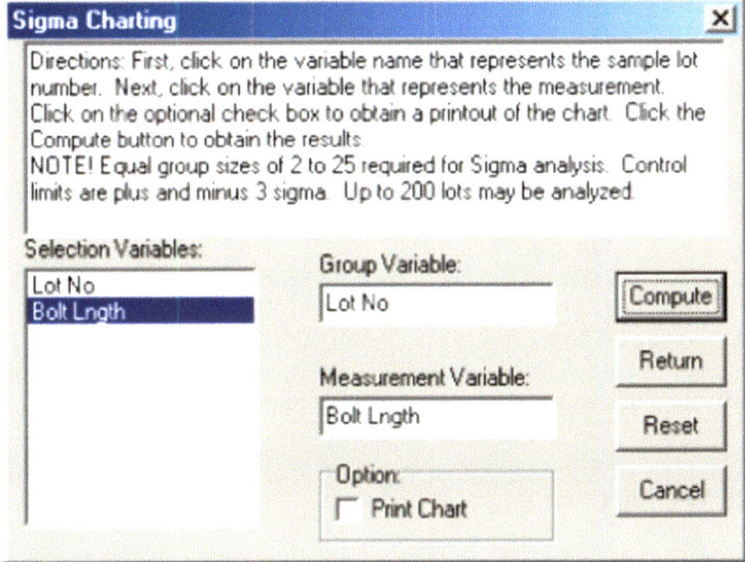

Fig. 12.6 Sigma chart dialog

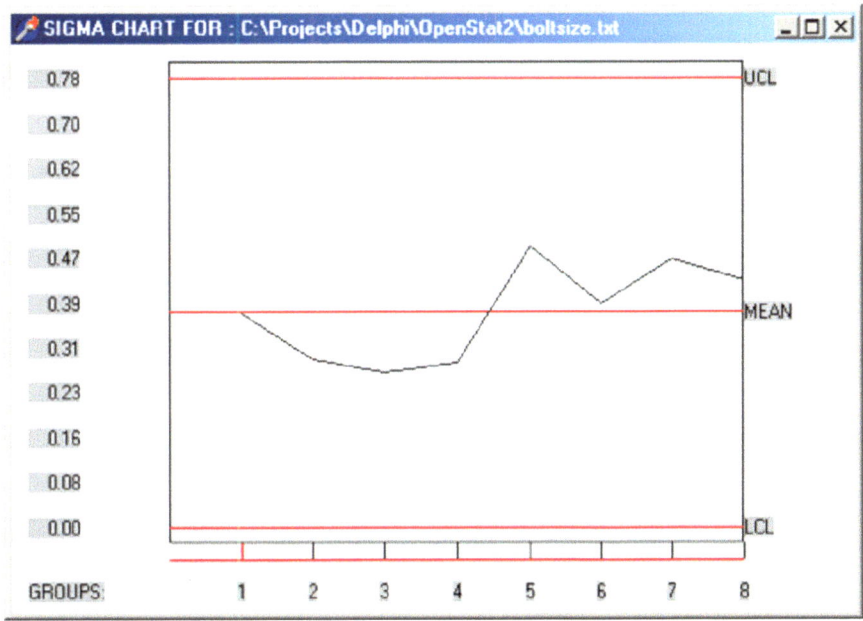

Fig. 12.7 Sigma chart plot

```
Group    Size    Mean     Std.Dev.
  1       5      19.88    0.37
  2       5      19.90    0.29
  3       5      20.16    0.27
  4       5      20.08    0.29
  5       5      19.88    0.49
  6       5      19.90    0.39
  7       5      20.02    0.47
  8       5      19.98    0.43
Grand Mean = 19.97, Std.Dev. = 0.359, Standard Error of Mean = 0.06
Mean Sigma = 0.37
Lower Control Limit = 0.000, Upper Control Limit = 0.779
```

The pattern of standard deviations is similar to that of the Range Chart.

CUSUM Chart

The specification form for the CUSUM chart is shown below for the data file labeled boltsize.txt. We have specified our desire to detect shifts of 0.02 in the process and are using the 0.05 and 0.20 probabilities for the two types of errors (Figs. 12.8, 12.9).

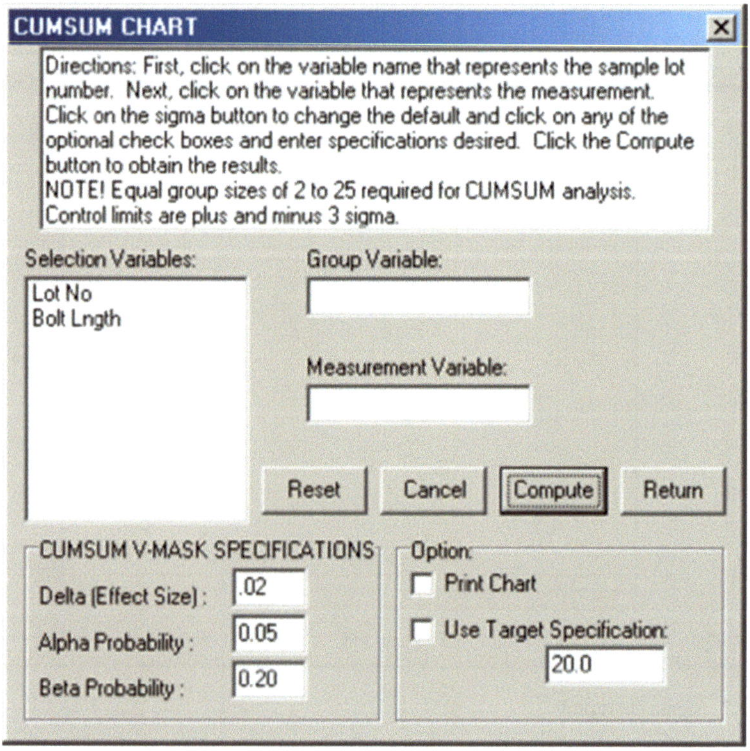

Fig. 12.8 CUMSUM chart dialog

CUSUM Chart

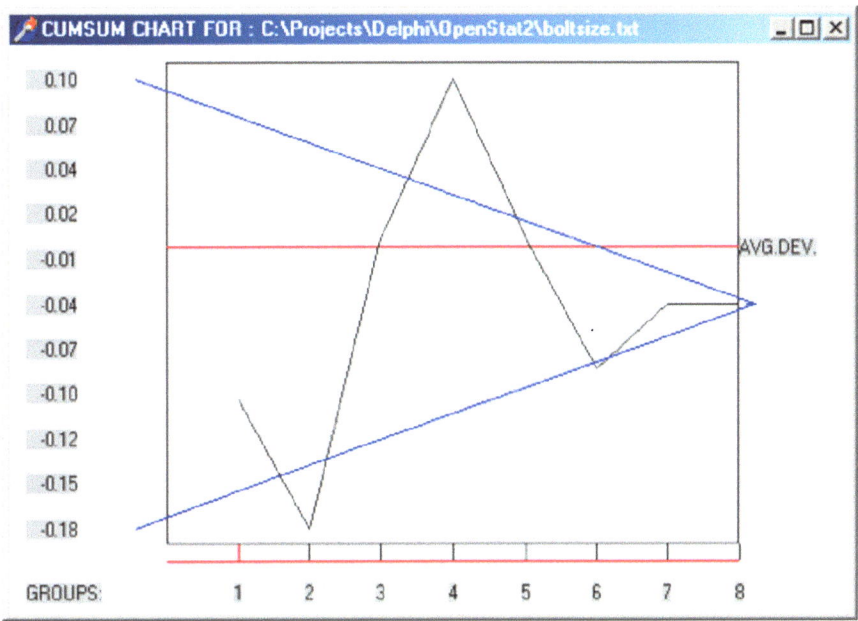

Fig. 12.9 CUMSUM chart plot

```
CUMSUM Chart Results

Group   Size    Mean    Std.Dev.    Cum.Dev. of mean from Target

1       5       19.88   0.37        -0.10
2       5       19.90   0.29        -0.18
3       5       20.16   0.27         0.00
4       5       20.08   0.29         0.10
5       5       19.88   0.49         0.00
6       5       19.90   0.39        -0.08
7       5       20.02   0.47        -0.04
8       5       19.98   0.43        -0.04
Mean of group deviations = -0.005
Mean of all observations = 19.975
Std. Dev. of Observations = 0.359
Standard Error of Mean = 0.057
Target Specification = 19.980
Lower Control Limit = 19.805, Upper Control Limit = 20.145
```

The results are NOT typical in that it appears that we have a process that is moving into control instead of out of control. Movement from lot 1 to 2 and from lot 3 to 4 indicate movement to out-of-control while the remaining values appear to be closer to in-control. If one checks the "Use the target value:" (of 20.0) the mask would indicate that lot 3 to 4 had moved to an out-of-control situation.

p Chart

To demonstrate the p Chart we will utilize a file labeled pchart.txt. Load the file and select the Analyses/Statistical Process Control/p Chart option. The specification form is shown below along with the results obtained after clicking the Compute Button (Figs. 12.10, 12.11):

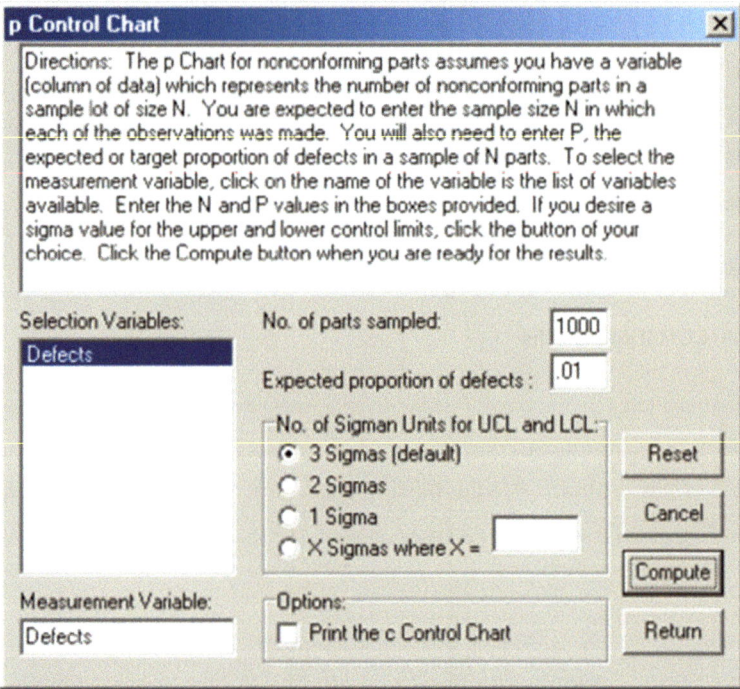

Fig. 12.10 p control chart dialog

p Chart

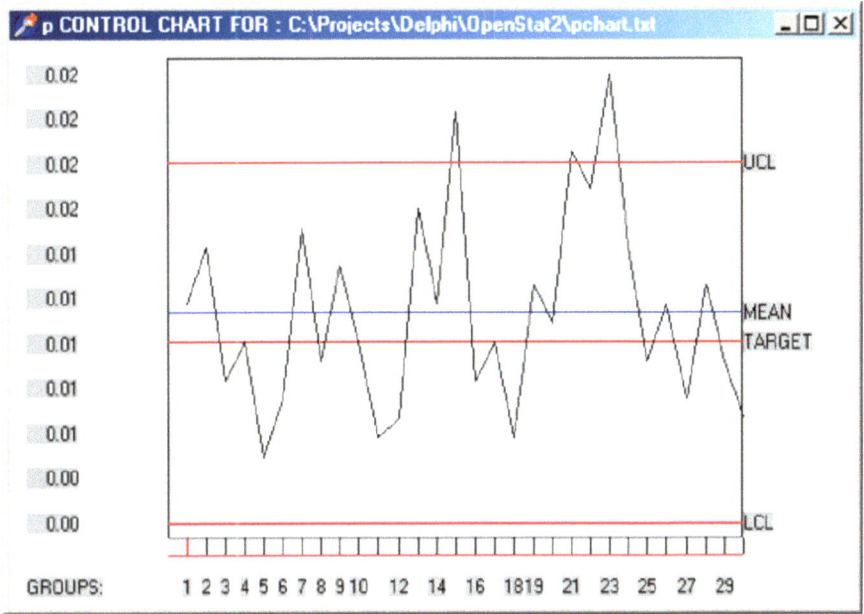

Fig. 12.11 p control chart plot

```
Target proportion = 0.0100
Sample size for each observation = 1000
Average proportion observed = 0.0116
Defects p Control Chart Results
```

Sample No.	Proportion
1	0.012
2	0.015
3	0.008
4	0.010
5	0.004
6	0.007
7	0.016
8	0.009
9	0.014
10	0.010
11	0.005
12	0.006
13	0.017
14	0.012
15	0.022
16	0.008
17	0.010
18	0.005
19	0.013

```
            20                  0.011
            21                  0.020
            22                  0.018
            23                  0.024
            24                  0.015
            25                  0.009
            26                  0.012
            27                  0.007
            28                  0.013
            29                  0.009
            30                  0.006
Target proportion = 0.0100
Sample size for each observation = 1000
Average proportion observed = 0.0116
```

Several of the sample lots (N=1,000) had disproportionately high defect rates and would bear further examination of what may have been occurring in the process at those points.

Defect (Non-conformity) c Chart

The previous section discusses the proportion of defects in samples (p Chart.) This section examines another defect process in which there is a count of defects in a sample lot. In this chart it is assumed that the occurrence of defects are independent, that is, the occurrence of a defect in one lot is unrelated to the occurrence in another lot. It is expected that the count of defects is quite small compared to the total number of parts potentially defective. For example, in the production of light bulbs, it is expected that in a sample of 1,000 bulbs, only a few would be defective. The underlying assumed distribution model for the count chart is the Poisson distribution where the mean and variance of the counts are equal. Illustrated below is an example of processing a file labeled cChart.txt (Figs. 12.12, 12.13).

Defect (Non-conformity) c Chart

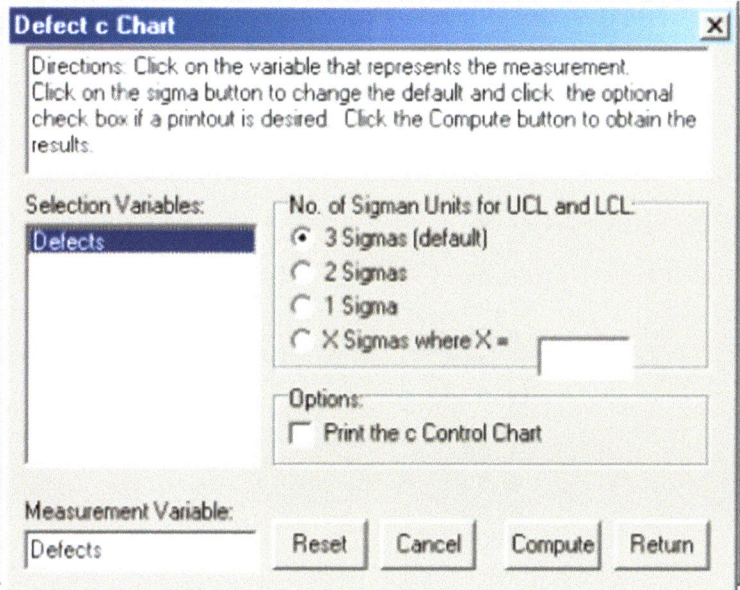

Fig. 12.12 Defect c chart dialog

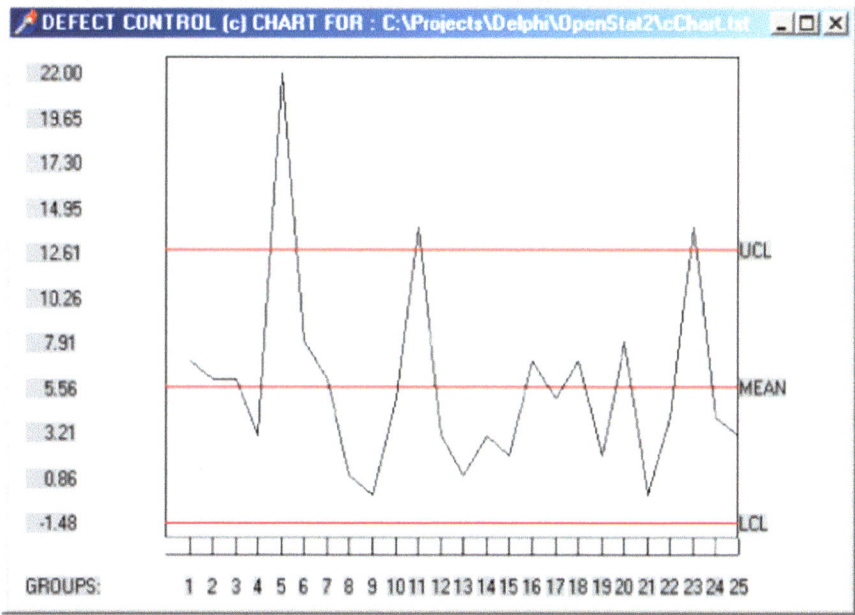

Fig. 12.13 Defect control chart plot

```
Defects c Control Chart Results
Sample     Number of Noncomformities
  1                  7.00
  2                  6.00
  3                  6.00
  4                  3.00
  5                 22.00
  6                  8.00
  7                  6.00
  8                  1.00
  9                  0.00
 10                  5.00
 11                 14.00
 12                  3.00
 13                  1.00
 14                  3.00
 15                  2.00
 16                  7.00
 17                  5.00
 18                  7.00
 19                  2.00
 20                  8.00
 21                  0.00
 22                  4.00
 23                 14.00
 24                  4.00
 25                  3.00
Total Nonconformities = 141.00
No. of samples = 25
Poisson mean and variance = 5.640
Lower Control Limit = -1.485, Upper Control Limit = 12.765
```

The count of defects for three of the 25 objects is greater than the upper control limit of three standard deviations.

Defects Per Unit U Chart

The specification form and results for the computation following the click of the Compute button are shown below (Figs. 12.14, 12.15):

Defects Per Unit U Chart

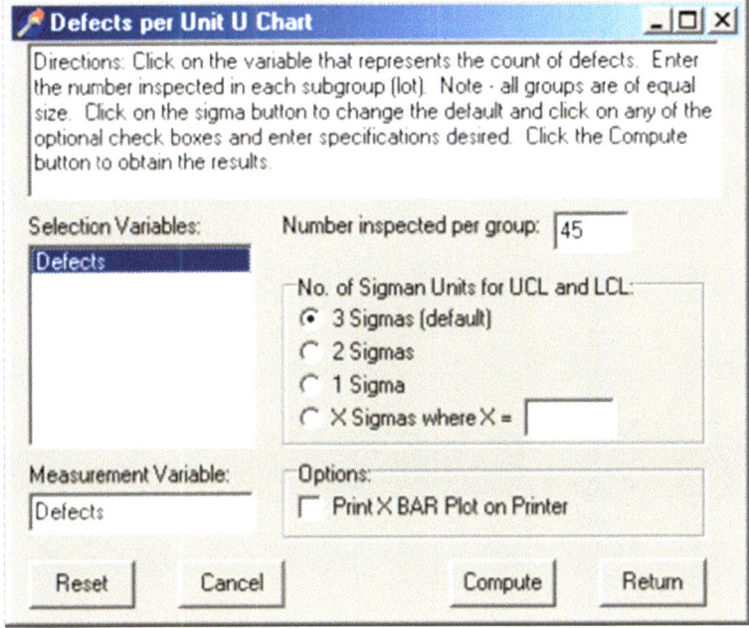

Fig. 12.14 Defects U chart dialog

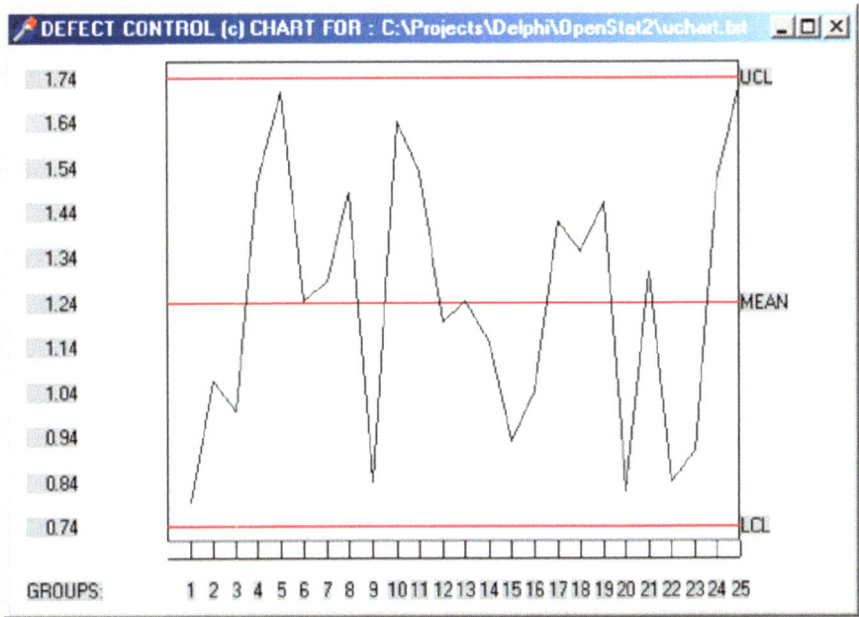

Fig. 12.15 Defect control chart plot

```
Sample    No Defects    Defects Per Unit
   1        36.00            0.80
   2        48.00            1.07
   3        45.00            1.00
   4        68.00            1.51
   5        77.00            1.71
   6        56.00            1.24
   7        58.00            1.29
   8        67.00            1.49
   9        38.00            0.84
  10        74.00            1.64
  11        69.00            1.53
  12        54.00            1.20
  13        56.00            1.24
  14        52.00            1.16
  15        42.00            0.93
  16        47.00            1.04
  17        64.00            1.42
  18        61.00            1.36
  19        66.00            1.47
  20        37.00            0.82
  21        59.00            1.31
  22        38.00            0.84
  23        41.00            0.91
  24        68.00            1.51
  25        78.00            1.73
Total Nonconformities = 1399.00
No. of samples = 25
Def. / unit mean = 1.244 and variance = 0.166
Lower Control Limit = 0.745, Upper Control Limit = 1.742
```

In this example, the number of defects per unit are all within the upper and lower control limits.

Chapter 13
Linear Programming

The Linear Programming Procedure

To start the Linear Programming procedure, click on the Sub-Systems menu item and select the Linear Programming procedure. The following screen will appear (Fig. 13.1):

We have loaded a file named Metals.LPR by pressing the Load File button and selecting a file which we had already constructed to do the first problem given above. When you start a problem, you will typically enter the number of variables (X's) first. When you press the tab key to go to the next field or click on another area of the form, the grids which appear on the form will automatically reflect the correct number of columns for data entry. In the Metals problem we have 1 constraint of the 'Maximum' type, 1 constraint of the 'Minimum' type and 3 Equal constraints. When you have entered the number of each type of constraint the grids will automatically provide the correct number of rows for entry of the coefficients for those constraints. Next, we enter the 'Objective' or cost values. Notice that you do NOT enter a dollar sign, just the values for the variables - five in our example. Now we are ready to enter our constraints and the corresponding coefficients. Our first (maximum) constraint is set to 1000 to set an upper limit for the amount of metal to produce. This constraint applies to each of the variables and a value of 1.00 has been entered for the coefficients of this constraint. The one minimum constraint is entered next. In this case we have entered a value of 100 as the minimum amount to produce. Notice that the coefficients entered are ALL negative values of 1.0! You will be entering negative values for the Minimum and Equal constraints coefficients.

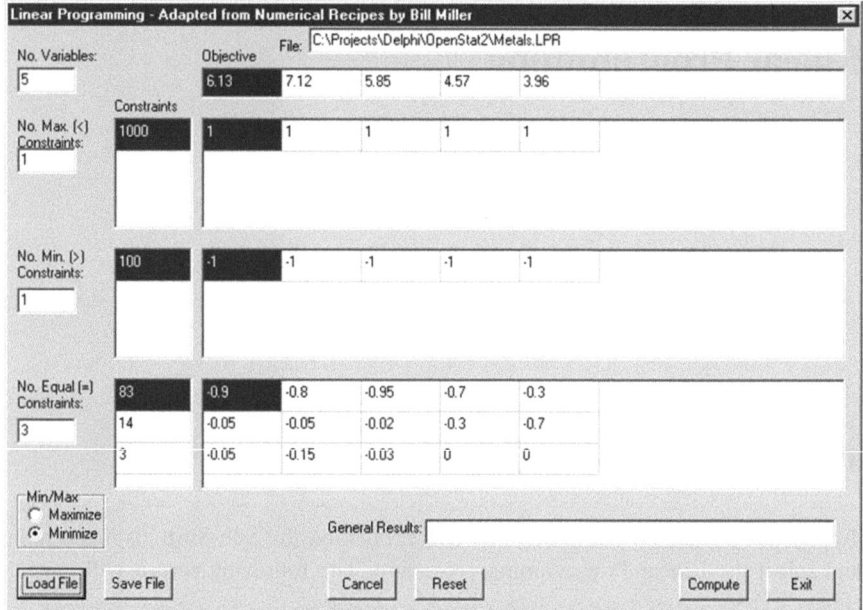

Fig. 13.1 Linear programming dialog

The constraint values themselves must all be zero or greater. We now enter the Equal constraint values and their coefficients from the second through the fourth equations. Again note that negative values are entered. Finally, we click on the Minimize button to indicate that we are minimizing the objective. We then press the Compute button to obtain the following results:

```
Linear Programming Results

                X1         X5
  z   544.8261  -0.1520   -0.7291
 Y1  1100.0000   0.0000    0.0000
 X3    47.8261  -0.7246    1.7391
 Y2     0.0000   0.0000    0.0000
 X4    41.7391  -0.0870   -2.3913
 X2    10.4348  -0.1884   -0.3478
```

The first column provides the answers we sought. The cost of our new alloy will be minimal if we combine the alloys 2, 3 and 4 with the respective percentages of 10.4, 47.8 and 41.7. Alloys 1 and 5 are not used. The z value in the first column is our objective function value (544.8).

The Linear Programming Procedure

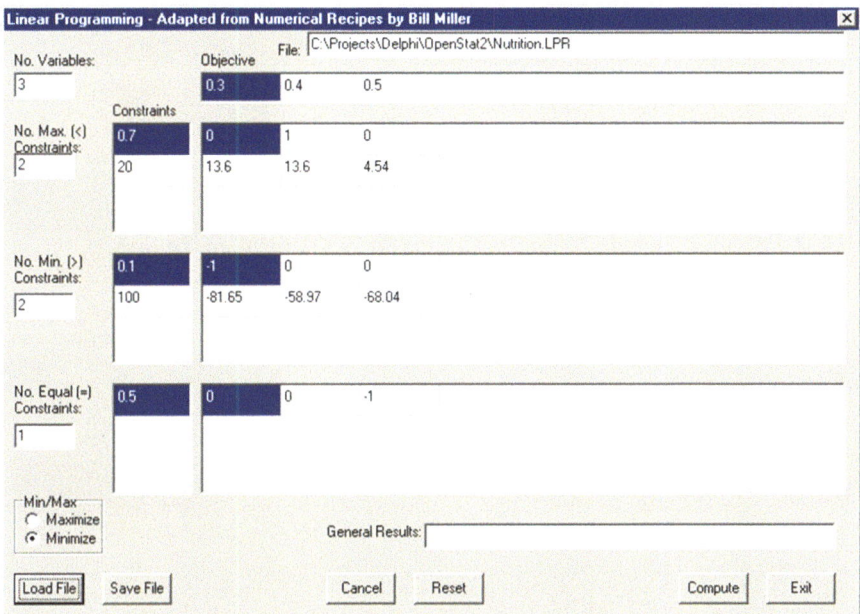

Fig. 13.2 Example specifications for a linear programming problem

Next, we will examine the second problem in which the nutritionist desires to minimize costs for the optimal food mix. We will click the Reset button on the form to clear our previous problem and load a previously saved file labeled 'Nutrition. LPR'. The form appears above (Fig. 13.2):

Again note that the minimum and equal constraint coefficients entered are negative values. When the compute button is pressed we obtain the following results:

```
Linear Programming Results

                    Y4         X2

   z    0.4924   -0.0037    -0.1833
   Y1   0.7000    0.0000     1.0000
   Y2  33.2599    0.1666     3.7777
   X1   0.8081    0.0122    -0.7222
   Y3   0.7081    0.0122    -0.7222
   X3   0.5000    0.0000     0.0000
```

In this solution we will be using .81 parts of Food A and .5 parts of Food C. Food B is not used.

The Linear Programming procedure of this program is one adapted from the Simplex program in the Numerical Recipes book listed in the bibliography (#56). The form design is one adapted from the Linear Programming program by Ane Visser of the AgriVisser consulting firm.

Chapter 14
Using MatMan

Purpose of MatMan

MatMan was written to provide a platform for performing common matrix and vector operations. It is designed to be helpful for the student learning matrix algebra and statistics as well as the researcher needing a tool for matrix manipulation. If you are already a user of the OpenStat program, you can import files that you have saved with OpenStat into a grid of MatMan.

Using MatMan

When you first start the MatMan program, you will see the main program form below. This form displays four "grids" in which matrices, row or column vectors or scalars (single values) may be entered and saved. If a grid of data has already been saved, it can be retrieved into any one of the four grids. Once you have entered data into a grid, a number of operations can be performed depending on the type of data entered (matrix, vector or scalar.) Before performing an operation, you select the grid of data to analyze by clicking on the grid with the left mouse button. If the data in the selected grid is a matrix (file extension of .MAT) you can select any one of the matrix operations by clicking on the Matrix Operations "drop-down" menu at the top of the form. If the data is a row or column vector, select an operation option from the Vector Operations menu. If the data is a single value, select an operation from the Scalar Operations menu (Fig. 14.1).

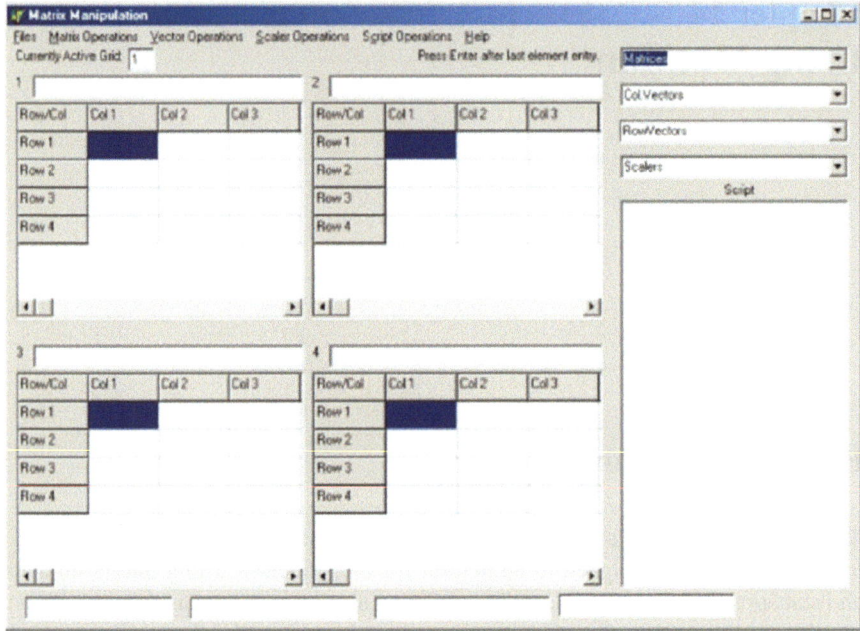

Fig. 14.1 The MatMan dialog

Using the Combination Boxes

In the upper right portion of the MatMan main form, there are four "Combo Boxes". These boxes each contain a drop-down list of file names. The top box labeled "Matrix" contains the list of files containing matrices that have been created in the current disk directory and end with an extension of .MAT. The next two combo boxes contain similar lists of column or row vectors that have been created and are in the current disk directory. The last contains name of scalar files that have been saved in the current directory. These combo boxes provide documentation as to the names of current files already in use. In addition, they provide a "short-cut" method of opening a file and loading it into a selected grid.

Files Loaded at the Start of MatMan

Five types of files are loaded when you first start the execution of the MatMan program. The program will search for files in the current directory that have file extensions of .MAT, .CVE, .RVE, .SCA and .OPT. The first four types of files are simply identified and their names placed into the corresponding combination boxes of matrices, column vectors, row vectors and scalars. The last, options, is a file which contains only two integers: a 1 if the script should NOT contain File Open operations when it is generated or a 0 and a 1 if the script should NOT contain File Save operations when a script is generated or a 0. Since File Open and File Save operations are not actually executed when a script or script line is executed, they are in a script only for documentation purposes and may be left out.

Clicking the Matrix List Items

A list of Matrix files in the current directory will exist in the Matrix "Drop-Down" combination box when the MatMan program is first started. By clicking on one of these file names, you can directly load the referenced file into a grid of your selection.

Clicking the Vector List Items

A list of column and row vector files in the current directory will exist in the corresponding column vector or row vector "Drop-Down" combination boxes when the MatMan program is first started. By clicking a file name in one of these boxes, you can directly load the referenced file into a grid of your selection.

Clicking the Scalar List Items

When you click on the down arrow of the Scalar "drop-down" combination box, a list of file names appear which have been previously loaded by identifying all scalar files in the current directory. Also listed are any new scalar files that you have created during a session with MatMan. If you move your mouse cursor down to a file name and click on it, the file by that name will be loaded into the currently selected grid or a grid of your choice.

The Grids

The heart of all operations you perform involve values entered into the cells of a grid. These values may represent values in a matrix, a column vector, a row vector or a scalar. Each grid is like a spreadsheet. Typically, you select the first row and column cell by clicking on that cell with the left mouse key when the mouse cursor is positioned over that cell. To select a particular grid, click the left mouse button when the mouse cursor is positioned over any cell of that grid. You will then see that the grid number currently selected is displayed in a small text box in the upper left side of the form (directly below the menus.)

Operations and Operands

At the bottom of the form (under the grids) are four "text" boxes labeled Operation, Operand1, Operand2 and Operand3. Each time you perform an operation by use of one of the menu options, you will see an abbreviation of that operation in the Operation box. Typically there will be at least one or two operands related to that operation. The first operand is typically the name of the data file occupying the current grid and the second operand the name of the file containing the results of the operation. Some operations involve two grids, for example, adding two matrices. In these cases, the name of the two grid files involved will be in operands1 and operands2 boxes while the third operand box will contain the file for the results.

You will also notice that each operation or operand is prefixed by a number followed by a dash. In the case of the operation, this indicates the grid number from which the operation was begun. The numbers which prefix the operand labels indicate the grid in which the corresponding files were loaded or saved. The operation and operands are separated by a colon (:). When you execute a script line by double clicking an operation in the script list, the files are typically loaded into corresponding grid numbers and the operation performed.

Menus

The operations which may be performed on or with matrices, vectors and scalars are all listed as options under major menu headings shown across the top of the main form. For example, the File menu, when selected, provides a number of options for loading a grid with file data, saving a file of data from a grid, etc. Click on a menu heading and explore the options available before you begin to use MatMan. In nearly all cases, when you select a menu option you will be prompted to enter additional information. If you select an option by mistake you can normally cancel the operation.

Combo Boxes

Your main MatMan form contains what are known as "Drop-Down" combination boxes located on the right side of the form. There are four such boxes: The "Matrix" box, the "Column Vectors" box, the "Row Vectors" box and the "Scalars" box. At the right of each box is an arrow which, when clicked, results in a list of items "dropped-down" into view. Each item in a box represents the name of a matrix, vector or scalar file in the current directory or which has been created by one of the possible menu operations. By clicking on one of these items, you initiate the loading of the file containing the data for that matrix, vector or scalar. You will find this is a convenient alternative to use of the File menu for opening files which you have been working with. Incidentally, should you wish to delete an existing file, you may do so by selecting the "edit" option under the Script menu. The script editor lists all files in a directory and lets you delete a file by simply double-clicking the file name!

The Operations Script

Located on the right side of the main form is a rectangle which may contain operations and operands performed in using MatMan. This list of operations and their corresponding operands is known collectively as a "Script". If you were to perform a group of operations, for example, to complete a multiple regression analysis, you may want to save the script for reference or repeated analysis of another set of data. You can also edit the scripts that are created to remove operations you did not intend, change the file names referenced, etc. Scripts may also be printed.

Getting Help on a Topic

You obtain help on a topic by first selecting a menu item, grid or other area of the main form by placing the mouse over the item for which you want information. Once the area of interest is selected, press the F1 key on your keyboard. If a topic exists in the help file, it will be displayed. You can press the F1 key at any point to bring up the help file. A book is displayed which may be opened by double clicking it. You may also search for a topic using the help file index of keywords.

Scripts

Each time an operation is performed on grid data, an entry is made in a "Script" list shown in the right-hand portion of the form. The operation may have one to three "operands" listed with it. For example, the operation of finding the eigenvalues and eigenvectors of a matrix will have an operation of SVDInverse followed by the name of the matrix being inverted, the name of the eigenvalues matrix and the name of the eigenvectors matrix. Each part of the script entry is preceded by a grid number followed by a hyphen (–). A colon separates the parts of the entry (:). Once a series of operations have been performed the script that is produced can be saved.

Saved scripts can be loaded at a later time and re-executed as a group or each entry executed one at a time. Scripts can also be edited and re-saved. Shown below is an example script for obtaining multiple regression coefficients.

```
CURRENT SCRIPT LISTING:
FileOpen:1-newcansas
1-ColAugment:newcansas:1-X
1-FileSave:1-X.MAT
1-MatTranspose:1-X:2-XT
2-FileSave:2-XT.MAT
2-PreMatxPostMat:2-XT:1-X:3-XTX
3-FileSave:3-XTX.MAT
3-SVDInverse:3-XTX.MAT:1-XTXINV
1-FileSave:1-XTXINV.MAT
FileOpen:1-XT.MAT
FileOpen:2-Y.CVE
1-PreMatxPostVec:1-XT.MAT:2-Y.CVE:3-XTY
3-FileSave:3-XTY.CVE
FileOpen:1-XTXINV.MAT
1-PreMatxPostVec:1-XTXINV.MAT:3-XTY:4-BETAS
4-FileSave:4-Bweights.CVE
```

Print

To print a script which appears in the Script List, move your mouse to the Script menu and click on the Print option. The list will be printed on the Output Form. At the bottom of the form is a print button that you can click with the mouse to get a hard-copy output.

Clear Script List

To clear an existing script from the script list, move the mouse to the Script menu and click the Clear option. Note: you may want to save the script before clearing it if it is a script you want to reference at a later time.

Edit the Script

Occasionally you may want to edit a script you have created or loaded. For example, you may see a number of Load File or Save File operations in a script. Since these are entered only for documentation and cannot actually be executed by clicking on them, they can be removed from the script. The result is a more compact and succinct script of operations performed. You may also want to change the name of files accessed for some operations or the name of files saved following an operation so that the same operations may be performed on a new set of data. To begin editing a script, move the mouse cursor to the Script menu and click on the Edit option. A new form appears which provides options for the editing. The list of operations appears on the left side of the form and an Options box appears in the upper right portion of the form. To edit a given script operation, click on the item to be edited and then click one of the option buttons. One option is to simply delete the item. Another is to edit (modify) the item. When that option is selected, the item is copied into an "Edit Box" which behaves like a miniature word processor. You can click on the text of an operation at any point in the edit box, delete characters following the cursor with the delete key, use the backspace key to remove characters in front of the cursor, and enter characters at the cursor. When editing is completed, press the return key to place the edited operation back into the script list from which it came.

Also on the Edit Form is a "Directory Box" and a "Files Box". Shown in the directory box is the current directory you are in. The files list box shows the current files in that directory. You can delete a file from any directory by simply double-clicking the name of the file in the file list. A box will pop up to verify that you want to delete the selected file. Click OK to delete the file or click Cancel if you do not want to delete the file. CAUTION! Be careful NOT to delete an important file like MATMAN.EXE, MATMAN.HLP or other system files (files with extensions of .exe, .dll, .hlp, .inf, etc.! Files which ARE safe to delete are those you have created with MatMan. These all end with an extension of .MAT, .CVE, .RVE ,.SCA or .SCR .

Load a Script

If you have saved a script of matrix operations, you can re-load the script for execution of the entire script of operations or execution of individual script items. To load a previously saved script, move the mouse to the Script menu and click on the Load option. Alternatively, you can go to the File menu and click on the Load Script option. Operation scripts are saved in a file as text which can also be read and edited with any word processing program capable of reading ASCII text files. For examples of scripts that perform statistical operations in matrix notation, see the help book entitled Script Examples.

Save a Script

Nearly every operation selected from one of the menus creates an entry into the script list. This script provides documentation of the steps performed in carrying out a sequence of matrix, vector or scalar operations. If you save the script in a file with a meaningful name related to the operations performed, that script may be "re-used" at a later time.

Executing a Script

You may quickly repeat the execution of a single operation previously performed and captured in the script. Simply click on the script item with the left mouse button when the cursor is positioned over the item to execute. Notice that you will be prompted for the name of the file or files to be opened and loaded for that operation. You can, of course, choose a different file name than the one or ones previously used in the script item. If you wish, you can also re-execute the entire script of operations. Move your mouse cursor to the Script menu and click on the Execute option. Each operation will be executed in sequence with prompts for file names appearing before execution each operation. Note: you will want to manually save the resulting file or files with appropriate names.

Script Options

File Open and File Save operations may or may not appear in a script list depending on options you have selected and saved. Since these two operations are *not* executed when a script is re-executed, it is not necessary that they be saved in a script (other than for documentation of the steps performed.) You can choose whether or not to have these operations appear in the script as you perform matrix, vector or scalar operations. Move your mouse cursor to the Script menu and click on the Options option. A pop-up form will appear on which you can elect to save or not save the File Open and File Save operations. The default (unchecked) option is to save these operations in a script. Clicking on an option tells the program to NOT write the operation to the script. Return to the MatMan main form by clicking the Return or Cancel button.

Files

When MatMan is first started it searches the current directory of your disk for any matrices, column vectors, row vectors or scalars which have previously been saved. The file names of each matrix, vector or scalar are entered into a drop-down list box corresponding to the type of data. These list boxes are located in the upper right portion of the main form. By first selecting one of the four grids with a click of the left mouse button and then clicking on one of the file names in a drop-down list, you can automatically load the file in the selected grid. Each time you save a grid of data with a new name, that file name is also added to the appropriate file list (Matrix, Column Vector, Row Vector or Scalar.)

At the top of the main form is a menu item labeled "Files". By clicking on the Files menu you will see a list of file options as shown in the picture below. In addition to saving or opening a file for a grid, you can also import an OpenStat .txt file, import a file with tab-separated values, import a file with comma separated values

Files 345

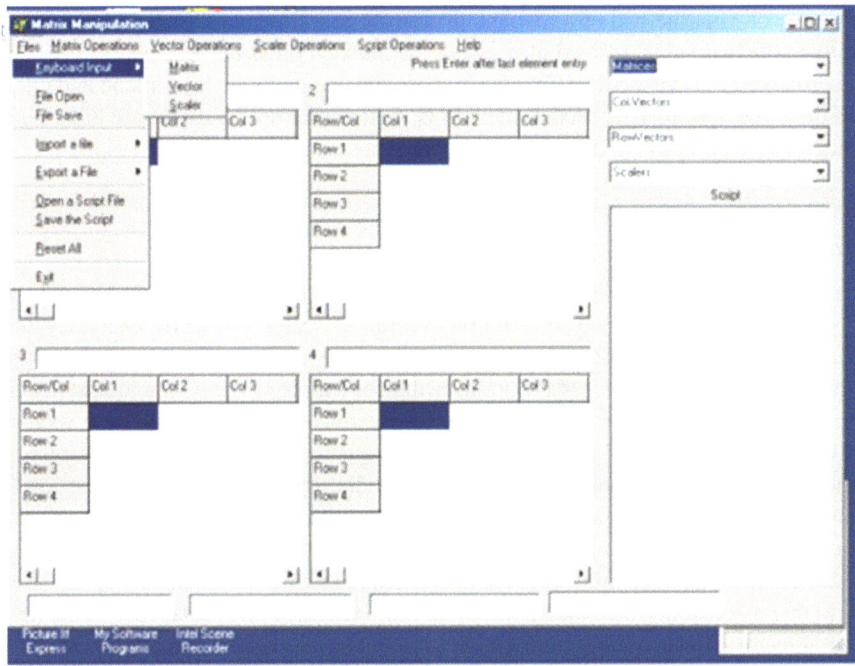

Fig. 14.2 Using the MatMan files menu

or import a file with spaces separating the values. All files saved with MatMan are ASCII text files and can be read (and edited if necessary) with any word processor program capable of reading ASCII files (for example the Windows Notepad program) (Fig. 14.2).

Keyboard Input

You can input data into a grid directly from the keyboard to create a file. The file may be a matrix, row vector, column vector or a scalar. Simply click on one of the four grids to receive your keystrokes. Note that the selected grid number will be displayed in a small box above and to the left of the grids. Next, click on the Files menu and move your cursor down to the Keyboard entry option. You will see that this option is expanded for you to indicate the type of data to be entered. Click on the type of data to be entered from the keyboard. If you selected a matrix, you will be prompted for the number of rows and columns of the matrix. For a vector, you will be prompted for the type (column or row) and the number of elements. Once the type of data to be entered and the number of elements are known, the program will "move" to the pre-selected grid and be waiting for your data entry. Click on the first cell (Row 1 and Column 1) and type your (first) value. Press the tab key to move to the next element in a row or, if at the end of a row, the first element in the next row. When you have entered the last value, instead of pressing the tab key, press the return key. You will be prompted to save the data. Of course, you can also go to the Files menu and click on the Save option. This second method is particularly useful if you are entering a very large data matrix and wish to complete it in several sessions.

File Open

If you have previously saved a matrix, vector or scalar file while executing the MatMan program, it will have been saved in the current directory (where the MatMan program resides.) MatMan saves data of a matrix type with a file extension of .MAT. Column vectors are saved with an extension of .CVE and row vectors saved with an extension of .RVE. Scalars have an extension of .SCA. When you click the File Open option in the File menu, a dialogue box appears. In the lower part of the box is an indication of the type of file. Click on this drop-down box to see the various extensions and click on the one appropriate to the type of file to be loaded. Once you have done that, the files listed in the files box will be only the files with that extension. Since the names of all matrix, vector and scalar files in the current directory are also loaded into the drop-down boxes in the upper right portion of the MatMan main form, you can also load a file by clicking on the name of the file in one of these boxes. Typically, you will be prompted for the grid number of the grid in which to load the file. The grid number is usually the one you have previously selected by clicking on a cell in one of the four grids.

File Save

Once you have entered data into a grid or have completed an operation producing a new output grid, you may save it by clicking on the save option of the File menu. Files are automatically saved with an extension which describes the type of file being saved, that is, with a .MAT, .CVE, .RVE or .SCA extension. Files are saved in the current directory unless you change to a different directory from the save dialogue box which appears when you are saving a file. It is recommended that you save files in the same directory (current directory) in which the MatMan program resides. The reason for doing this is that MatMan automatically loads the names of your files in the drop-down boxes for matrices, column vectors, row vectors and scalars.

Import a File

In addition to opening an existing MatMan file that has an extension of .MAT, .CVE, .RVE or .SCA, you may also *import* a file created by other programs.

Many word processing and spread-sheet programs allow you to save a file with the data separated by tabs, commas or spaces. You can import any one of these types of files. Since the first row of data items may be the names of variables, you will be asked whether or not the first line of data contains variable labels.

You may also import files that you have saved with the OpenStat program. These files have an extension of .TXT or .txt when saved by the OpenStat program. While they are ASCII type text files, they contain a lot of information such as variable labels, long labels, format of data, etc. MatMan simply loads the variable labels, replacing the column labels currently in a grid and then loads numeric values into the grid cells of the grid you have selected to receive the data.

Export a File

You may wish to save your data in a form which can be imported into another program such as OpenStat, Excel, MicroSoft Word, WordPerfect, etc. Many programs permit you to import data where the data elements have been separated by a tab, comma or space character. The tab character format is particularly attractive because it creates an ASCII (American Standard Code for Information Interchange) file with clearly delineated spacing among values and which may be viewed by most word processing programs.

Open a Script File

Once you have performed a number of operations on your data you will notice that each operation has been "summarized" in a list of script items located in the script list on the right side of the MatMan form. This list of operations may be saved for later reference or re-execution in a file labeled appropriate to the series of operations. To re-open a script file, go to the File Menu and select the Open a Script File option. A dialogue box will appear. Select the type of file with an extension of .SCR and you will see the previously saved script files listed. Click on the one to load and press the OK button on the dialogue form. Note that if a script is already in the script list box, the new file will be added to the existing one. You may want to clear the script list box before loading a previously saved script. Clear the script list box by selecting the Clear option under the Script Operations menu.

Save the Script

Once a series of operations have been performed on your data, the operations performed will be listed in the Script box located to the right of the MatMan form.

The series of operations may represent the completion of a data analysis such as multiple regression, factor analysis, etc. You may save this list of operations for future reference or re-execution. To save a script, select the Save Script option from the File Menu. A dialogue box will appear in which you enter the name of the file. Be sure that the type of file is selected as a .SCR file (types are selected in the drop-down box of the dialogue form.) A file extension of .SCR is automatically appended to the name you have entered. Click on the OK button to complete the saving of the script file.

Reset All

Occasionally you may want to clear all grids of data and clear all drop-down boxes of currently listed matrix, vector and scalar files. To do so, click the Clear All option under the Files Menu. Note that the script list box is NOT cleared by this operation. To clear a script, select the Clear operation under the Script Operations menu.

Entering Grid Data

Grids are used to enter matrices, vectors or scalars. Select a grid for data by moving the mouse cursor to the one of the grids and click the left mouse button. Move your mouse to the Files menu at the top of the form and click it with the left mouse button. Bring your mouse down to the Keyboard Input option. For entry of a matrix of values, click on the Matrix option. You will then be asked to verify the grid for entry. Press return if the grid number shown is correct or enter a new grid number and press return. You will then be asked to enter the name of your matrix (or vector or scalar.) Enter a descriptive name but keep it fairly short. A default extension of .MAT will automatically be appended to matrix files, a .CVE will be appended to column vectors, a .RVE appended to row vectors and a .SCA appended to a scalar. You will then be prompted for the number of rows and the number of columns for your data. Next, click on the first available cell labeled Col.1 and Row 1. Type the numeric value for the first number of your data. Press the tab key to move to the next column in a row (if you have more than one column) and enter the next value. Each time you press the tab key you will be ready to enter a value in the next cell of the grid. You can, of course, click on a particular cell to edit the value already entered or enter a new value. When you have entered the last data value, press the Enter key. A "Save" dialog box will appear with the name you previously chose. You can keep this name or enter a new name and click the OK button. If you later wish to edit values, load the saved file, make the changes desired and click on the Save option of the Files menu.

When a file is saved, an entry is made in the Script list indicating the action taken. If the file name is not already listed in one of the drop-down boxes (e.g. the matrix drop-down box), it will be added to that list.

Clearing a Grid

Individual grids are quickly reset to a blank grid with four rows and four columns by simply moving the mouse cursor over a cell of the grid and clicking the RIGHT mouse button. CAUTION! Be sure the data already in the grid has been saved if you do not want to lose it!

Inserting a Column

There may be occasions where you need to add another variable or column of data to an existing matrix of data. You may insert a new blank column in a grid by selecting the Insert Column operation under the Matrix Operations menu. First, click on an existing column in the matrix prior to or following the cell where you want the new column inserted. Click on the Insert Column option. You will be prompted to indicate whether the new column is to precede or follow the currently selected column. Indicate your choice and click the Return button.

Inserting a Row

There may be occasions where you need to add another subject or row of data to an existing matrix of data. You may insert a new blank row in a grid by selecting the Insert Row operation under the Matrix Operations menu. First, click on an existing row in the matrix prior to or following the cell where you want the new row inserted. Click on the Insert Row option. You will be prompted to indicate whether the new row is to precede or follow the number of the selected row. Indicate your choice and click the Return button.

Deleting a Column

To delete a column of data in an existing data matrix, click on the grid column to be deleted and click on the Delete Column option under the Matrix Operations menu. You will be prompted for the name of the new matrix to save. Enter the new matrix name (or use the current one if the previous one does not need to be saved) and click the OK button.

Deleting a Row

To delete a row of data in an existing data matrix, click on the grid row to be deleted and click on the Delete Row option under the Matrix Operations menu. You will be prompted for the name of the new matrix to save. Enter the new matrix name (or use the current one if the previous one does not need to be saved) and click the OK button.

Using the Tab Key

You can navigate through the cells of a grid by simply pressing the tab key. Of course, you may also click the mouse button on any cell to select that cell for data entry or editing. If you are at the end of a row of data and you press the tab key, you are moved to the first cell of the next row (if it exists.) To save a file press the Return key when located in the last row and column cell.

Using the Enter Key

If you press the Return key after entering the last data element in a matrix, vector or scalar, you will automatically be prompted to save the file. A "save" dialogue box will appear in which you enter the name of the file to save your data. Be sure the type of file to be saved is selected before you click the OK button.

Editing a Cell Value

Errors in data entry DO occur (after all, we are human aren't we?) You can edit a data element by simply clicking on the cell to be edited. If you double click the cell, it will be highlighted in blue at which time you can press the delete key to remove the cell value or enter a new value. If you simply wish to edit an existing value, click the cell so that it is NOT highlighted and move the mouse cursor to the position in the value at which you want to start editing. You can enter additional characters, press the backspace key to remove a character in front of the cursor or press the delete key to remove a character following the cursor. Press the tab key to move to the next cell or press the Return key to obtain the save dialogue box for saving your corrections.

Loading a File

Previously saved matrices, vectors or scalars are easily loaded into any one of the four grids. First select a grid to receive the data by clicking on one of the cells of the target grid. Next, click on the Open File option under the Files Menu. An "open" dialogue will appear which lists the files in your directory. The dialogue has a drop-down list of possible file types. Select the type for the file to be loaded. Only files of the selected type will then be listed. Click on the name of the file to load and click the OK button to load the file data.

Matrix Operations

Once a matrix of data has been entered into a grid you can elect to perform a number of matrix operations. The figure below illustrates the options under the Matrix Operations menu. Operations include:

Row Augment
Column Augment
Delete a Row
Delete a Column
Extract Col. Vector from Matrix
SVD Inverse
Tridiagonalize
Upper-Lower Decomposition
Diagonal to Vector
Determinant
Normalize Rows
Normalize Columns
Premultiply by : Row Vector; Matrix; Scalar
Postmultiply by : Column Vector; Matrix
Eigenvalues and Vectors
Transpose
Trace
Matrix A+Matrix B
Matrix A−Matrix B
Print

Printing

You may elect to print a matrix, vector, scalar or file. When you do, the output is placed on an "Output" form. At the bottom of this form is a button labeled 'Print' which, if clicked, will send the contents of the output form to the printer. Before printing this form, you may type in additional information, edit lines, cut and paste lines and in general edit the output to your liking. Edit operations are provided as icons at the top of the form. Note that you can also save the output to a disk file, load another output file and, in general, use the output form as a word processor.

Row Augment

You may add a row of 1's to a matrix with this operation. When the transpose of such an augmented matrix is multiplied times this matrix, a cell will be created in the resulting matrix, which contains the number of columns in the augmented matrix.

Column Augmentation

You may add a column of 1's to a matrix with this operation. When the transpose of such an augmented matrix is multiplied times this matrix, a cell will be created in the resulting matrix, which contains the number of rows in the augmented matrix. The procedure for completing a multiple regression analysis often involves column augmentation of a data matrix containing a row for each object (e.g. person) and column cells containing independent variable values. The column of 1's created from the Column Augmentation process ends up providing the intercept (regression constant) for the analysis.

Extract Col. Vector from Matrix

In many statistics programs the data matrix you begin with contains columns of data representing independent variables and one or more columns representing dependent variables. For example, in multiple regression analysis, one column of data represents the dependent variable (variable to be predicted) while one or more columns represent independent variables (predictor variables.) To analyze this data with the MatMan program, one would extract the dependent variable and save it as a column vector for subsequent operations (see the sample multiple regression script.) To extract a column vector from a matrix you first load the matrix into one of the four grids, click on a cell in the column to be extracted and then click on the Extract Col. Vector option under the Matrix Operations menu.

SVDInverse

A commonly used matrix operation is the process of finding the inverse (reciprocal) of a symmetric matrix. A variety of methods exist for obtaining the inverse (if one exists.) A common problem with some inverse methods is that they will not provide a solution if one of the variables is dependent (or some combination of) on other variables (rows or columns) of the matrix. One advantage of the "Singular Value Decomposition" method is that it typically provides a solution even when one or more dependent variables exist in the matrix. The offending variable(s) are essentially replaced by zeroes in the row and column of the dependent variable. The resulting inverse will NOT be the desired inverse.

To obtain the SVD inverse of a matrix, load the matrix into a grid and click on the SVDInverse option from the Matrix Operations menu. The results will be displayed in grid 1 of the main form. In addition, grids 2 through 4 will contain additional information which may be helpful in the analysis. Figures 1 and 2 below illustrate the results of inverting a 4 by 4 matrix, the last column of which contains values that are the sum of the first three column cells in each row (a dependent variable.)

When you obtain the inverse of a matrix, you may want to verify that the resulting inverse is, in fact, the reciprocal of the original matrix. You can do this by multiplying the original matrix times the inverse. The result should be a matrix with 1's in the diagonal and 0's elsewhere (the identity matrix.) Figure 3 demonstrates that the inverse was NOT correct, that is, did not produce an identity matrix when multiplied times the original matrix.

Figure 1. DepMat.MAT From Grid Number 1

	Columns			
	Col.1	Col.2	Col.3	Col.4
Rows				
1	5.000	11.000	2.000	18.000
2	11.000	2.000	4.000	17.000
3	2.000	4.000	1.000	7.000
4	18.000	17.000	7.000	1.000

Figure 2. DepMatInv.MAT From Grid Number 1

	Columns			
	Col.1	Col.2	Col.3	Col.4
Rows				
1	0.584	0.106	-1.764	0.024
2	0.106	-0.068	-0.111	0.024
3	-1.764	-0.111	4.802	0.024
4	0.024	0.024	0.024	-0.024

Figure 3. DepMatxDepMatInv.MAT From Grid Number 3

	Columns			
	Col.1	Col.2	Col.3	Col.4
Rows				
1	1.000	0.000	0.000	0.000
2	0.000	1.000	0.000	0.000
3	0.000	0.000	1.000	0.000
4	1.000	1.000	1.000	0.000

NOTE! This is NOT an Identity matrix.

Tridiagonalize

In obtaining the roots and vectors of a matrix, one step in the process is frequently to reduce a symmetric matrix to a tri-diagonal form. The resulting matrix is then solved more readily for the eigenvalues and eigenvectors of the original matrix. To reduce a matrix to its tridiagonal form, load the original matrix in one of the grids and click on the Tridiagonalize option under the Matrix Operations menu.

Upper-Lower Decomposition

A matrix may be decomposed into two matrices: a lower matrix (one with zeroes above the diagonal) and an upper matrix (one with zeroes below the diagonal matrix.) This process is sometimes used in obtaining the inverse of a matrix. The matrix is first decomposed into lower and upper parts and the columns of the inverse solved one at a time using a routine that solves the linear equation $AX=B$ where A is the upper/lower decomposition matrix, B are known result values of the equation and X is solved by the routine. To obtain the LU decomposition, enter or load a matrix into a grid and select the Upper-Lower Decomposition option from the Matrix Operations menu.

Diagonal to Vector

In some matrix algebra problems it is necessary to perform operations on a vector extracted from the diagonal of a matrix. The Diagonal to Vector operation extracts the diagonal elements of a matrix and creates a new column vector with those values. Enter or load a matrix into a grid and click on the Diagonal to Vector option under the Matrix Operations menu to perform this operation.

Determinant

The determinant of a matrix is a single value characterizing the matrix values. A singular matrix (one for which the inverse does not exist) will have a determinant of zero. Some ill-conditioned matrices will have a determinant close to zero. To obtain the determinant of a matrix, load or enter a matrix into a grid and select the Determinant option from among the Matrix Operations options. Shown below is the determinant of a singular matrix (row/column 4 dependent on columns 1 through 3.)

```
            Columns
            Col.1           Col.2           Col.3           Col.4
Rows
  1          5.000          11.000          2.000          18.000
  2         11.000           2.000          4.000          17.000
  3          2.000           4.000          1.000           7.000
  4         18.000          17.000          7.000          42.000

            Columns
            Col 1
Rows
  1          0.000
```

Normalize Rows or Columns

In matrix algebra the columns or rows of a matrix often represent vectors in a multi-dimension space. To make the results more interpretable, the vectors are frequently scaled so that the vector length is 1.0 in this "hyper-space" of k-dimensions. This scaling is common for statistical procedures such as Factor Analysis, Principal Component Analysis, Discriminant Analysis, Multivariate Analysis of Variance, etc. To normalize the row (or column) vectors of a matrix such as eigenvalues, load the matrix into a grid and select the Normalize Rows (or Normalize Columns) option from the Matrix Operations menu.

Pre-multiply By

A matrix may be multiplied by a row vector, another matrix or a single value (scalar.) When a row vector with N columns is multiplied times a matrix with N rows, the result is a row vector of N elements. When a matrix of N rows and M columns is multiplied times a matrix with M rows and Q columns, the result is a matrix of N rows and Q columns. Multiplying a matrix by a scalar results in each element of the matrix being multiplied by the value of the scalar.

To perform the pre-multiplication operation, first load two grids with the values of a matrix and a vector, matrix or scalar. Click on a cell of the grid containing the matrix to insure that the matrix grid is selected. Next, select the Pre-Multiply by: option and then the type of value for the pre-multiplier in the sub-options of the Matrix Operations menu. A dialog box will open asking you to enter the grid number of the matrix to be multiplied. The default value is the selected matrix grid. When you press the OK button another dialog box will prompt you for the grid number containing the row vector, matrix or scalar to be multiplied times the matrix. Enter the grid number for the pre-multiplier and press return. Finally, you will be prompted to enter the grid number where the results are to be displayed. Enter a number different than the first two grid numbers entered. You will then be prompted for the name of the file for saving the results.

Post-multiply By

A matrix may be multiplied times a column vector or another matrix. When a matrix with N rows and Q columns is multiplied times a column vector with Q rows, the result is a column vector of N elements. When a matrix of N rows and M columns is multiplied times a matrix with M rows and Q columns, the result is a matrix of N rows and Q columns.

To perform the post-multiplication operation, first load two grids with the values of a matrix and a vector or matrix. Click on a cell of the grid containing the matrix to insure that the matrix grid is selected. Next, select the Post-Multiply by: option and then the type of value for the post-multiplier in the sub-options of the Matrix Operations menu. A dialog box will open asking you to enter the grid number of the matrix multiplier. The default value is the selected matrix grid. When you press the OK button another dialog box will prompt you for the grid number containing the column vector or matrix. Enter the grid number for the post-multiplier and press return. Finally, you will be prompted to enter the grid number where the results are to be displayed. Enter a number different than the first two grid numbers entered. You will then be prompted for the name of the file for saving the results.

Eigenvalues and Vectors

Eigenvalues represent the k roots of a polynomial constructed from k equations. The equations are represented by values in the rows of a matrix. A typical equation written in matrix notation might be:

$$Y = B X$$

where X is a matrix of known "independent" values, Y is a column vector of "dependent" values and B is a column vector of coefficients which satisfies specified properties for the solution. An example is given when we solve for "least-squares" regression coefficients in a multiple regression analysis. In this case, the X matrix contains cross-products of k independent variable values for N cases, Y contains known values obtained as the product of the transpose of the X matrix times the N values for subjects and B are the resulting regression coefficients.

In other cases we might wish to transform our matrix X into another matrix V which has the property that each column vector is "orthogonal" to (un-correlated) with the other column vectors. For example, in Principal Components analysis, we seek coefficients of vectors that represent new variables that are uncorrelated but which retain the variance represented by variables in the original matrix. In this case we are solving the equation

$$VXV^T = \lambda$$

X is a symmetric matrix and λ are roots of the matrix stored as diagonal values of a matrix. If the columns of V are normalized then $V V^T = I$, the identity matrix.

Transpose

The transpose of a matrix or vector is simply the creation of a new matrix or vector where the number of rows is equal to the number of columns and the number of columns equals the number of rows of the original matrix or vector. For example, the transpose of the row vector [1 2 3 4] is the column vector:

$$\begin{matrix} 1 \\ 2 \\ 3 \\ 4 \end{matrix}$$

Similarly, given the matrix of values:

$$\begin{matrix} 1 & 2 & 3 \\ 4 & 5 & 6 \end{matrix}$$

the transpose is:

$$\begin{matrix} 1 & 4 \\ 2 & 5 \\ 3 & 6 \end{matrix}$$

You can transpose a matrix by selecting the grid in which your matrix is stored and clicking on the Transpose option under the Matrix Operations menu. A similar option is available under the Vector Operations menu for vectors.

Trace

The trace of a matrix is the sum of the diagonal values.

Matrix A + Matrix B

When two matrices of the same size are added, the elements (cell values) of the first are added to corresponding cells of the second matrix and the result stored in a corresponding cell of the results matrix. To add two matrices, first be sure both are stored in grids on the main form. Select one of the grid containing a matrix and click on the Matrix A+Matrix B option in the Matrix Operations menu. You will be prompted for the grid numbers of each matrix to be added as well as the grid number of the results. Finally, you will be asked the name of the file in which to save the results.

Matrix A−Matrix B

When two matrices of the same size are subtracted, the elements (cell values) of the second are subtracted from corresponding cells of the first matrix and the result stored in a corresponding cell of the results matrix. To subtract two matrices, first be sure both are stored in grids on the main form. Select one of the grids containing the matrix from which another will be subtracted and click on the Matrix A−Matrix B option in the Matrix Operations menu. You will be prompted for the grid numbers of each matrix as well as the grid number of the results. Finally, you will be asked the name of the file in which to save the results.

Print

To print a matrix be sure the matrix is loaded in a grid, the grid selected and then click on the print option in the Matrix Operations menu. The data of the matrix will be shown on the output form. To print the output form on your printer, click the Print button located at the bottom of the output form.

Vector Operations

A number of vector operations may be performed on both row and column vectors. Shown below is the main form with the Vector Operations menu selected. The operations you may perform are:

Transpose
Multiply by Scalar
Square Root of Elements
Reciprocal of Elements
Print
Row Vec. × Col. Vec.
Col. Vec × Row Vec.

Vector Transpose

The transpose of a matrix or vector is simply the interchange of rows with columns. Transposing a matrix results in a matrix with the first row being the previous first column, the second row being the previous second column, etc. A column vector becomes a row vector and a row vector becomes a column vector. To transpose a vector, click on the grid where the vector resides that is to be transposed. Select the Transpose Option from the Vector Operations menu and click it. Save the transposed vector in a file when the save dialogue box appears.

Multiply a Vector by a Scalar

When you multiply a vector by a scalar, each element of the vector is multiplied by the value of that scalar. The scalar should be loaded into one of the grids and the vector in another grid. Click on the Multiply by a Scalar option under the Vector Operations menu. You will be prompted for the grid numbers containing the scalar and vector. Enter those values as prompted and click the return button following each. You will then be presented a save dialogue in which you enter the name of the new vector.

Square Root of Vector Elements

You can obtain the square root of each element of a vector. Simply select the grid with the vector and click the Square Root option under the Vector Operations menu. A save dialogue will appear after the execution of the square root operations in which you indicate the name of your new vector. Note - you cannot take the square root of a vector that contains a negative value - an error will occur if you try.

Reciprocal of Vector Elements

Several statistical analysis procedures involve obtaining the reciprocal of the elements in a vector (often the diagonal of a matrix.) To obtain reciprocals, click on the grid containing the vector then click on the Reciprocal option of the Vector Operations menu. Of course, if one of the elements is zero, an error will occur! If valid values exist for all elements, you will then be presented a save dialogue box in which you enter the name of your new vector.

Print a Vector

Printing a vector is the same as printing a matrix, scalar or script. Simply select the grid to be printed and click on the Print option under the Vector Operations menu. The printed output is displayed on an output form. The output form may be printed by clicking the print button located at the bottom of the form.

Row Vector Times a Column Vector

Multiplication of a column vector by a row vector will result in a single value (scalar.) Each element of the row vector is multiplied times the corresponding element of the column vector and the products are added. The number of elements in the row vector must be equal to the number of elements in the column vector. This operation is sometimes called the "dot product" of two vectors. Following execution of this vector operation, you will be shown the save dialogue for saving the resulting scalar in a file.

Column Vector Times Row Vector

When you multiply a column vector of k elements times a row vector of k elements, the result is a k by k matrix. In the resulting matrix each row by column cell is the product of the corresponding column element of the row vector and the corresponding row element of the column vector. The result is equivalent to multiplying a k by 1 matrix times a 1 by k matrix.

Scalar Operations

The operations available in the Scalar Operations menu are:
- Square Root
- Reciprocal
- Scalar x Scalar
- Print

Square Root of a Scalar

Selecting this option under the Scalar Operations menu results in a new scalar that is the square root of the original scalar. The new value should probably be saved in a different file than the original scalar. Note that you will get an error message if you attempt to take the square root of a negative value.

Reciprocal of a Scalar

You obtain the reciprocal of a scalar by selecting the Reciprocal option under the Scalar Operations menu. You will obtain an error if you attempt to obtain the reciprocal of a value zero. Save the new scalar in a file with an appropriate label.

Scalar Times a Scalar

Sometimes you need to multiply a scalar by another scalar value. If you select this option from the Scalar Operations menu, you will be prompted for the value of the multiplier. Once the operation has been completed you should save the new scalar product in a file appropriately labeled.

Print a Scalar

Select this option to print a scalar residing in one of the four grids that you have selected. Notice that the output form contains all objects that have been printed. Should you need to print only one grid's data (matrix, vector or scalar) use the Clear All option under the Files menu.

Chapter 15
The GradeBook Program

The GradeBook Main Form

The image below will first appear when you begin the GradeBook program (Fig. 15.1):

At the bottom of the form is the "main menu". Move your mouse to one of the topics such as "OPENFILE", click on it with the left mouse button. Your typical first step is to click the box in the area marked "For Grade Book" and click the box for "Enter a Title for This Grade Book". You can then enter student information in the top "grid" of the form as shown by the example above. Once you have entered student information, you can add a new test column. One test has been added in the above example. Enter the "raw" scores for each student. Once those have been placed in the grid test area, you should enter a grading system for the test. Once that has been completed you can do a variety of analyses for the test or the class by selecting an option in the respective box of the first two blocks of options. Note that you must click the "DO ABOVE" button to implement your choice.

The Student Page Tab

The majority of the form consists of a "tabbed" series of grids. The program will begin with the "Students" grid. By clicking any one of the tabs located along the top, you can change to a different grid. The Student grid is where you will first enter the last name, first name and middle initial for each student in your class. Don't worry about the order in which you enter them - you can sort them later with a click of the mouse button! Be sure an assign an Identification Number for each student. A sequential integer will work if you don't have a school ID or social security number.

To enter the first student's last name, click on the Student 1 and Last Name row and column cell. Enter the last name. Press the tab key on your keyboard to move to the next cell for the First Name. Continue to enter information requested using the

Fig. 15.1 The GradeBook dialog

tab key to move from cell to cell. Be sure and press the Enter key following the entry of the student ID number.

You can use the four navigation keys (arrow keys) on your keyboard to move from cell to cell or click on the cell where you wish to make an entry or change. Pressing the "enter" key on the keyboard "toggles" the cell between what is known as "edit mode" or selection mode. When in selection mode the cell will be colored blue. If you make an entry when in selected mode, the previous entry is replaced by the new key strokes. When in edit mode, you can move back and forth in your entry and make deletions using the delete key or backspace key and type new characters following the cursor in the cell.

Once you have entered your students names and identification numbers, click on the File menu and select the "Save As" option by clicking on it with the left mouse button. A "dialogue box" will open up in which you enter the name of the file you have selected for your grade book. Enter a name and click on the save button.

Test Result Page Tabs

If you have entered one or more tests and the corresponding raw scores for each student, there are a variety of operations that you can perform. Once you have saved your file and re-opened it, the names of your students are automatically copied to all

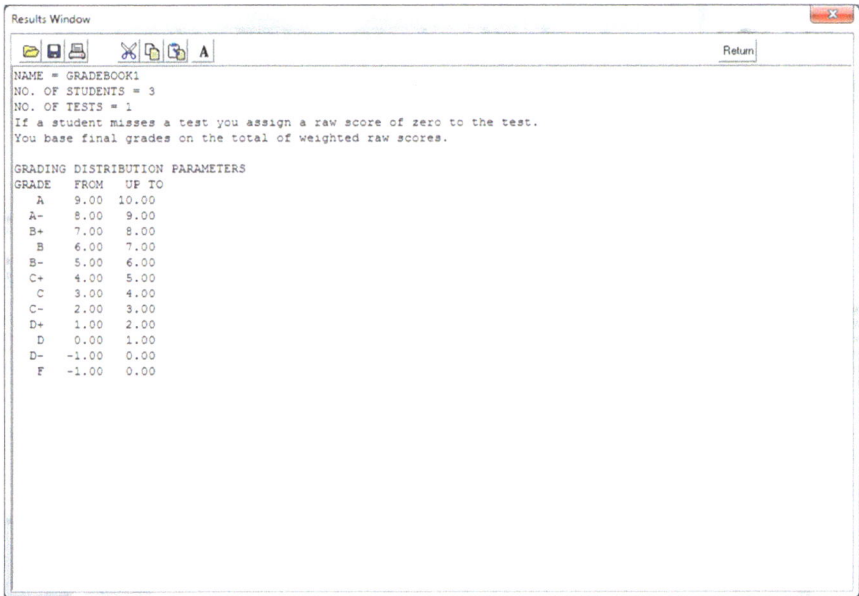

Fig. 15.2 The GradeBook summary

of the tab pages. The Test areas are used to record the scores obtained by each student on one of the tests you have administered. Once a score has been entered for each student, you can elect to calculate one or more (or all) transformations available from the main menu's "Compute" options. The previous image illustrates the selection of the possible score transformations. As an illustration of one of the options, we have elected to print a grade book summary (Fig. 15.2):

Once raw scores are entered into one of the Test pages, the user should complete the specification of the measurements and the grading procedure for each test. Ideally, the teacher knows at the beginning of a course how many tests will be administered, the possible number of points for each measure, the type of transformation to be used for grading, and the "cut-points" for each grade assignment. Shown below is the form used to specify the measurements utilized in the course. This form is obtained by clicking the Enter Grading Specifications box under the For Grade Book list of options (Fig. 15.3).

Notice that for each test, the user is expected to enter the minimum and maximum points which can be awarded for the test, quiz, essay or measurement. In addition, an estimate of reliability should be entered if a composite reliability estimate is to be obtained. Note - you can get an estimate of reliability for a test as an option under the For Selected Test options. The weight that the measure is to receive in obtaining the composite score for the course is also entered. We recommend integer values such as 1 for a quiz, 2 for major tests and perhaps 3 or 4 for tests like a midterm or final examination. Finally, there is an area for a brief note describing the purpose or nature of the measurement.

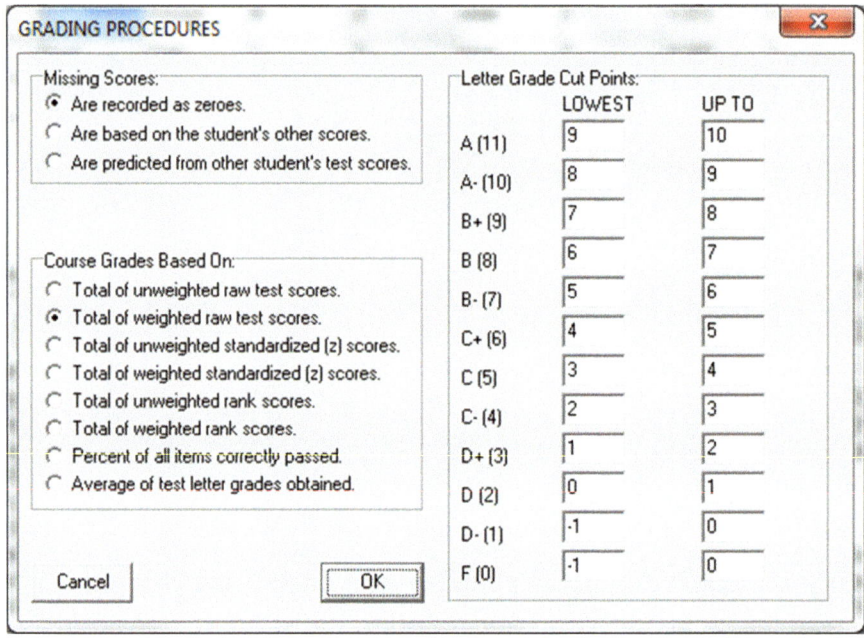

Fig. 15.3 The GradeBook Measurement Specifications form

Chapter 16
The Item Banking Program

Introduction

Teachers are confronted with large classes that often make it difficult to evaluate students on the basis of evaluations based on essay examinations, problems or creative work which permits the students to demonstrate their mastery of concepts and skills in a particular area of learning. As a consequence, a variety of test questions have been devised to sample student knowledge and skills from the larger domain of knowledge contained in a given content area. Multiple choice items, true or false items, sentence completion items, matching items and short essay items have been developed to reduce the time required to evaluate students. The test theory that has evolved around these various types of items indicates that they are quite adequate in reliably assessing differences that exist among students in the domain sampled. Many states, for example, have gone to the use of computerized testing for individuals applying for driving licenses. The individual taking these examinations are presented multiple-choice types of items drawn from a computerized item bank. If the applicant performs at a given level of competence they are then permitted to demonstrate their actual driving skills in a second evaluation stage. Many Area Educational Agencies have also developed banks of items appropriate to various instructional subjects across the school grades such as in English, mathematics, science and history. Teachers may draw items from these banks to create tests over the subject area they teach.

Many teacher-constructed items utilize a picture or photograph (for example, maps, machines, paintings, etc.) as part of one or more items in a test. These pictures may be saved in the computer as "bitmap" files and tied to specific items in the bank. When the test is printed, if a picture is used it is printed prior to the printing of the item.

Item Coding

A variety of coding schemes may be developed to categorize test items. For example, one might use the Taxonomy of Educational Objectives to classify items. If one is teaching from a text book utilized across different schools in a given district, the items might be classified by the chapter, section, page and paragraph of the content to which an item refers. One may also construct a classification structure based on a breakdown of subject matter into sub-categories of the content. For example, the broad field of statistics might be initially broken down into parametric and non-parametric statistics. These domains may be further broken into categories such as univariate, multivariate, Neyman-Pearson, Bayesian, etc. which in turn may be further broken down into topics such as theory, terminology, symbols, equations, etc.

Most classification schemes result in a classification "tree" with sub-categories representing branches from the previous category level. This item banking program lets you determine your own coding system and enter codes that classify each item. You may utilize as many levels as is practical (typically three or four.) A style of code entry is required that is consistent across all items in the bank. For example, a code of 05.13.06.01 would represent a coding structure with four levels, each level having a maximum of 99 categories at each level.

In addition to classifying items by their content, one will also need to classify items by their type, that is, whether the item is a multiple-choice item, a true-false item, a matching item within a set of matching items, etc. This program requires the user to specify one of five item types for each item.

Items may also have other characteristics. In particular, one may have experience with the use of specific items in past tests and have a reasonable approximation of the difficulty of the item. Typically, the difficulty of the item is measured (in the Classical Test Theory) by the proportion of students that pass the item. For example an item with a difficulty index of .3 is more difficult than an item with an index of .8. If one is utilizing one, two or three parameter logistic scaling (Item Response Theory) he or she may have a difficulty parameter, a discrimination parameter and a chance correct parameter to describe the item. In the area often called "Tailored Testing", items are selected to administer the student in such a manner that the estimate of student ability is obtained with relatively few items. This is done by selecting items based on their difficulty parameter and the response the student gives to each item in the sequence. This program lets you enter parameter estimates (Classical or Item Response Theory estimates) for each item.

Items stored in the item bank may be retrieved on the basis of one or more criteria. One may, for example, select items within specific code areas, item difficulty and item type. By this means one can create a test of items that cover a certain topical area, have a specific range of difficulty and are of a given type or types.

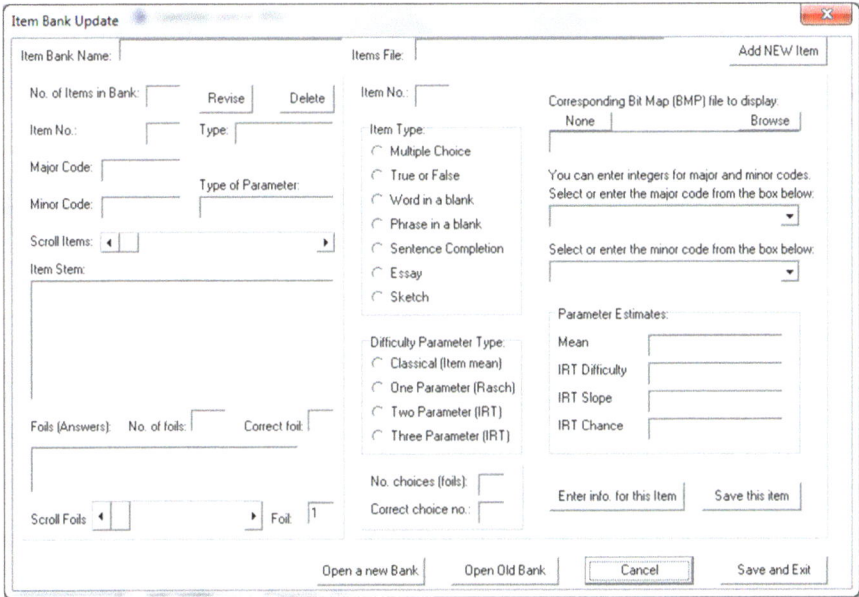

Fig. 16.1 The Item Bank form

Using the Item Bank Program

You reach the Item Banking program by clicking on the Analyses->Measurement->Item Banking menu on the main form of OpenStat. There you can click one of three choices: Enter/Edit items, Specify a Test to Administer or Generate a Test. If you click on the first submenu, you will see the above form (Fig. 16.1):

In the above form you can open a new item bank or load an existing item bank. If you create a new item bank you can enter a variety of item types into the item bank along with an estimate of the items difficulty level. Some items may have a corresponding bit map figure that you have created for the item. You can also enter a major and minor code for an item so that different tests you may want to generate have different items based on the codes selected.

Specifying a Test

If you have already created an item bank, you can then select the next option from the main menu to specify the nature of a test to generate. When you do, the following form is shown (Fig. 16.2):

Within this form you can specify a test using characteristics of the items in the item bank such as the item difficulty or item codes. A test may be printed or administered on a computer screen.

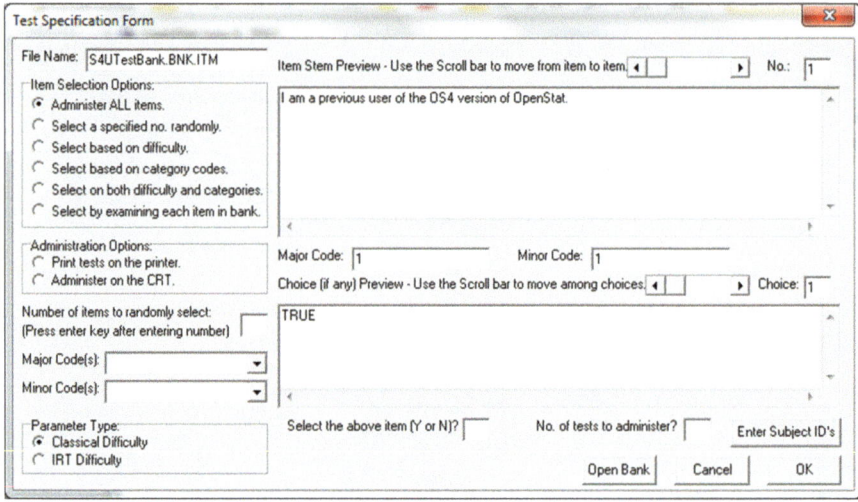

Fig. 16.2 The item banking Test Specification form

Generate a Test

This is the third option in the Item Banking system. If you have specified a test the following form is displayed (Fig. 16.3):

Notice that the form first requests the name of the previously created item bank file and it then automatically loads the test specification form previously created. The sample item bank we created only contains two items which we specified to be administered on the computer screen to a student with the ID = Student 1. If we now click the "Proceed with the test button we obtain the following prompt form (Fig. 16.4):

When the "OK" button is pressed, the test is administered or printed. Our example would display a screen as shown below (Fig. 16.5):

Following administration of the test, the total correct score is displayed.

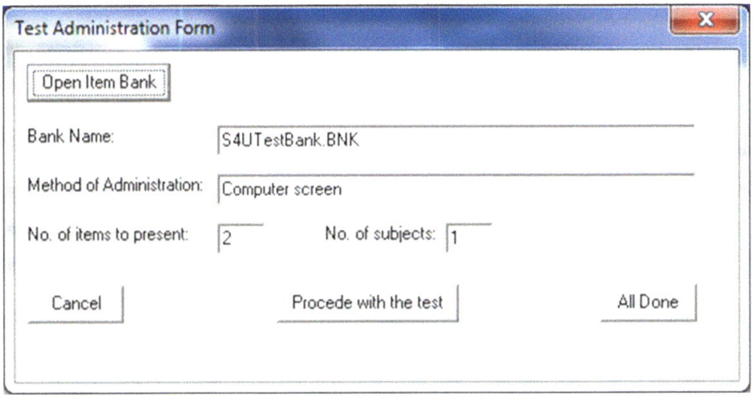

Generate a Test

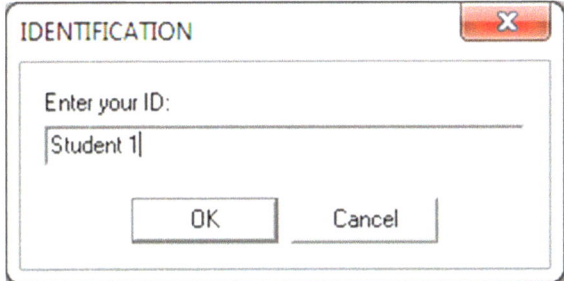

Fig. 16.3 The form to generate a test

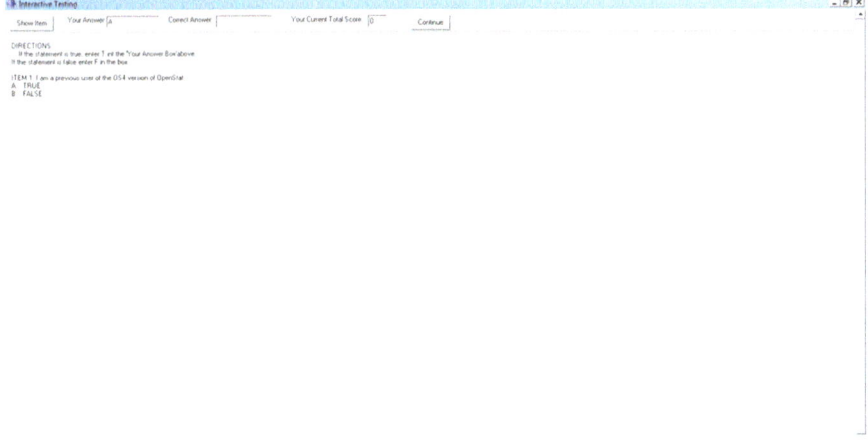

Fig. 16.4 Student verification form for a test administration

Fig. 16.5 A test displayed on the computer

Chapter 17
Neural Networks

Using the Program

The Neural Form

In the figure below (Fig. 17.1) you see a menu consisting of drop-down boxes for Files, Generation, etc. You also see a grid and a list of commands used to create "control files." The Neural program completes its work by reading a file of control commands. Each command consists of one or two parts, the parts separated by a colon (:) in the command list box. In some cases, the user provides the second part, often the name of a file. To aid the user to complete some "traditional" types of analyses, the program can automatically generate a control file in the data grid. To do this, one first clicks on the "File" in the menu and then move the mouse to the "New" option and from there to the "Control File" option. Clicking the "Control File" option modifies the grid to contain two columns with sufficient width to hold control commands. The figure below shows the File menu options (Fig. 17.2):

Once the user has indicated he or she intends to generate a new control file, the menu item labeled "Generate" is clicked and the mouse moved to the type of control file to generate. Figure 17.3 illustrates the selection of the option to generate a control file for prediction:

When the "Controls for Prediction" option is clicked, the program opens a dialog form for entering the parameters of the prediction problem. Figure 17.4 below illustrates this form:

The user supplies the name of a "Training File" and a data file containing validation data for analysis. In standard multiple regression methods, the multiple correlation coefficient represents the correlation between the predicted scores and the actual dependent variable scores. In using the neural network program, one can analyze the same data as the training data and correlate the obtained predicted scores with the original scores to obtain a similar index of prediction accuracy. In the figure below, a control file is shown that was used to predict the variable "jumps" using five independent variables (height, weight, etc.) from a file labeled "canszscaled.

374   17 Neural Networks

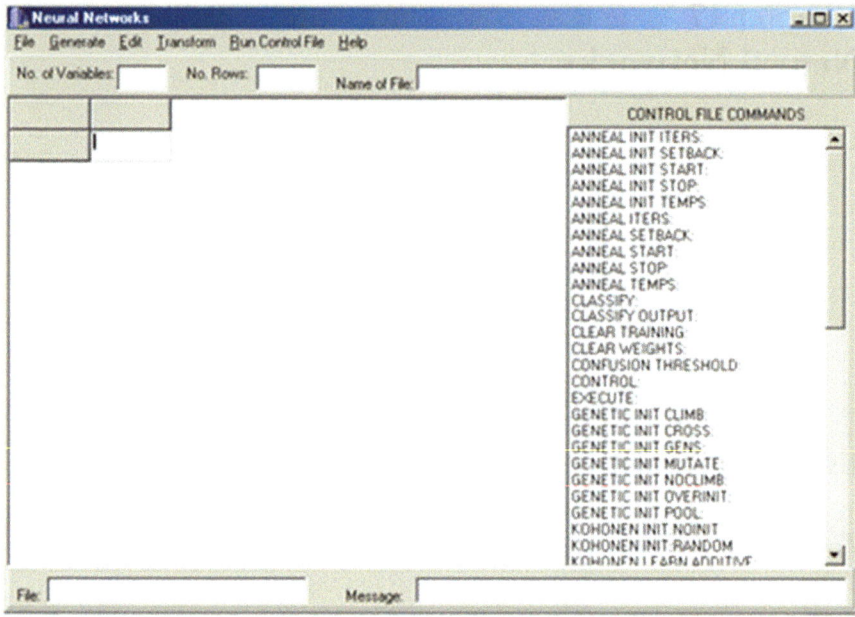

Fig. 17.1 The Neural form

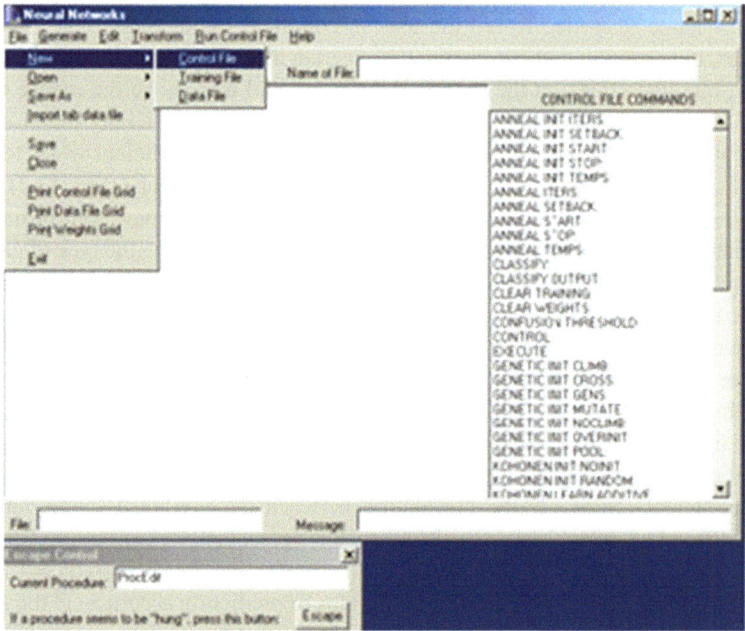

Fig. 17.2 The neural file menu

Using the Program

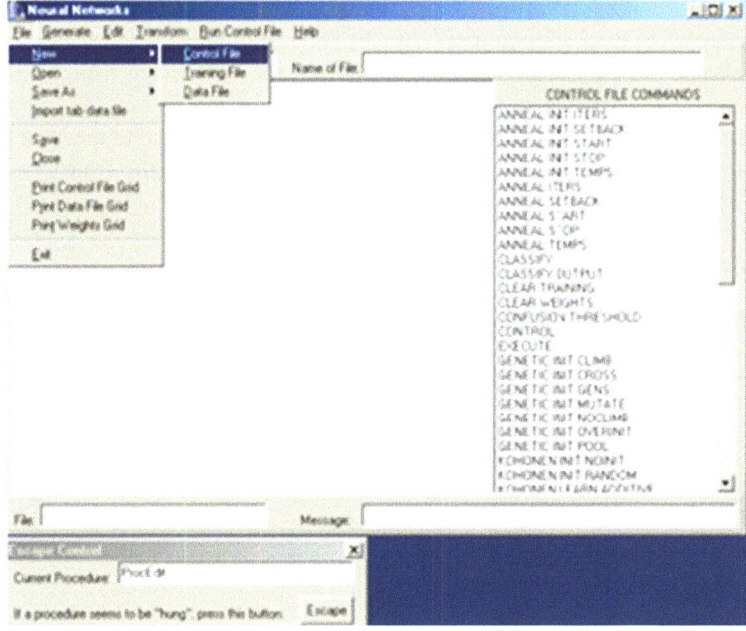

Fig. 17.3 The neural control file generation options

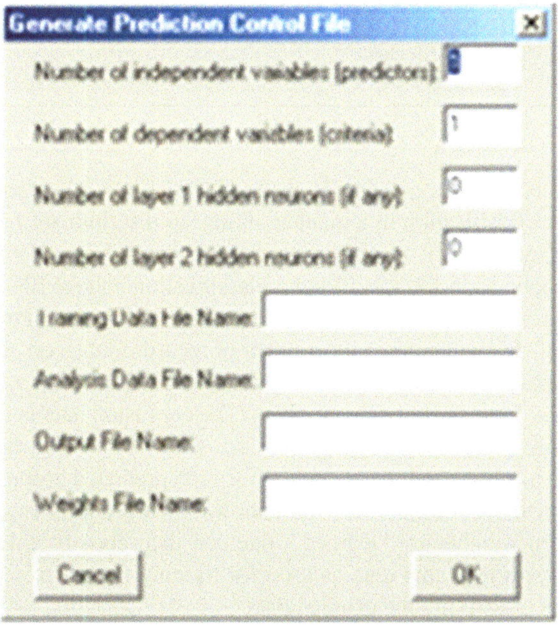

Fig. 17.4 The control file generation form for prediction problems

dat." The file consists of raw measures that have been transformed to z-scores and then re-scaled to have a range from .1 to .9. The resulting predicted scores are in a similar range but may be re-converted to z-scores for comparison with the original z-scores of the dependent variable.

Note - for users of Openstat , the file cansas.tab was imported to the Neural program and the transformation option applied using the options in the Transformations menu item.

Example Control File for Prediction

```
QUIT ERROR:.1
QUIT RETRIES:3
CONFUSION THRESHOLD:50
NETWORK MODEL:LAYER
LAYER INIT:ANNEAL
OUTPUT MODEL:GENERAL
N INPUTS:5
N OUTPUTS:1
N HIDDEN1:0
N HIDDEN2:0
TRAIN:CANSASSCALED.DAT
OUTPUT FILE:CANSASOUT.TXT
LEARN:
SAVE WEIGHTS:CANSAS.WTS
EXECUTE:CANSASSCALED.DAT
QUIT:
```

Control file commands are listed on the Neural Form. One can also generate control files for classification in a manner similar to discriminant function analysis or hierarchical analysis in traditional multivariate statistics. Figure 17.5 below shows the dialogue form for specifying a classification control file. Default names have been entered for the name of two files created when the control file is "run". The "Confusion" file will contain the number of records (subjects) classified in each group. The neural net is "trained" to recognize the group classification on the basis of the "predictor" or classification variables. The confusion data is comparable to a contingency chi-square table in traditional statistics. A row will be generated for each group and a column will be generated for each predicted group (plus a column for unknowns) . In training the net, the data for each group is entered separately. Once the neuron weights are "learned", one can then classify unknown subjects. Often one analyzes the same data as used for training the net to see how well the network does in classifying the original data.

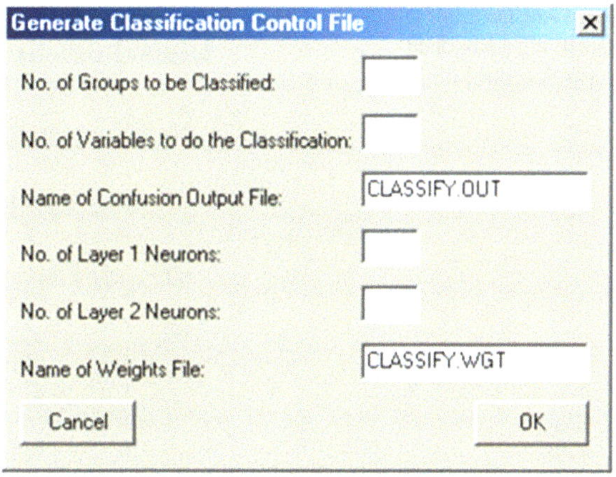

Fig. 17.5 The form for generating a classification control file

Figure below shows the generated control file for classifying subjects in three groups on the basis of two continuous variables. The continuous variables have been scaled to have a range from .1 to .9 as in the prediction problem previously discussed.

```
QUIT ERROR:0.1
QUIT RETRIES:5
CONFUSION THRESHOLD:50
NETWORK MODEL:LAYER
LAYER INIT:GENETIC
OUTPUT MODEL:CLASSIFY
N INPUTS:2
N OUTPUTS:3
N HIDDEN1:2
N HIDDEN2:0
CLASSIFY OUTPUT:1
TRAIN:GROUP1.DAT
CLASSIFY OUTPUT:2
TRAIN:GROUP2.DAT
CLASSIFY OUTPUT:3
TRAIN:GROUP3.DAT
LEARN:
SAVE WEIGHTS:CLASSIFY.WGT
RESET CONFUSION:
CLASSIFY:GROUP1.DAT
SHOW CONFUSION:
SAVE CONFUSION:CLASSIFY.OUT
RESET CONFUSION:
CLASSIFY:GROUP2.DAT
SHOW CONFUSION:
SAVE CONFUSION:CLASSIFY.OUT
RESET CONFUSION:
CLASSIFY:GROUP3.DAT
SHOW CONFUSION:
SAVE CONFUSION:CLASSIFY.OUT
RESET CONFUSION:
CLEAR TRAINING:
QUIT:
```

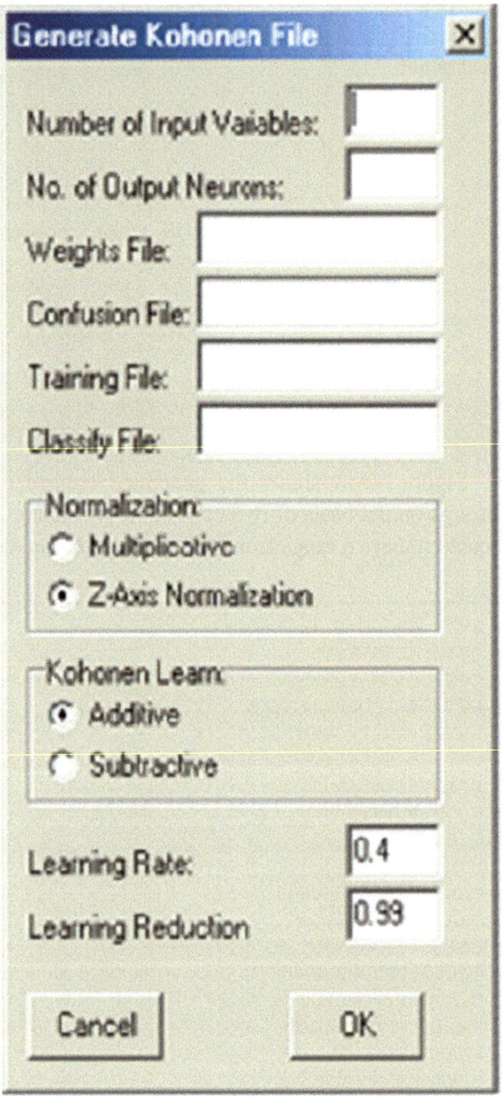

Fig. 17.6 Form for specifying a Kohonen network control file

In traditional multivariate statistics, hierarchical grouping analyses are sometimes performed in an attempt to identify "natural" groups on the basis of one or more continuous variables. One type of neural network called the "Kohonen" network may be utilized for a similar purpose. The user specifies the number of variables to analyze and the number of "output groups" that is expected. By repeated "runs" of the network with different numbers of output groups, one can examine the number of subjects classified into "self-organized" groups. Figure 17.6 above illustrates the dialogue box for specifying a Kohonen control file and program code below shows a sample control file for classifying data.

```
QUIT ERROR:0.1
QUIT RETRIES:5
CONFUSION THRESHOLD:50
KOHONEN NORMALIZATION MULTIPLICATIVE:
NETWORK MODEL:KOHONEN
KOHONEN INIT:RANDOM
OUTPUT MODEL:CLASSIFY
N INPUTS:3
N OUTPUTS:10
N HIDDEN1:0
N HIDDEN2:0
TRAIN:kohonen.dat
KOHONEN LEARN SUBTRACTIVE:
LEARN:
SAVE WEIGHTS:koh2.wts
RESET CONFUSION:
CLASSIFY:kohonen.dat
SHOW CONFUSION:
SAVE CONFUSION:confuse.txt
RESET CONFUSION:
CLASSIFY:kohonen.dat
SHOW CONFUSION:
SAVE CONFUSION:confuse.txt
CLEAR TRAINING:
QUIT:
```

Examples

Regression Analysis with One Predictor

A sample of 200 observations with two continuous variables were generated using the OpenStat simulation procedure for generating multivariate distributions. The data were generated to come from a population with a product–moment correlation of .60 and have means and standard deviations of 100 and 15 for each variable. The sample data generated had a correlation of 0.579 with means of 99.363, 99.267 and standard deviations of 15.675 and 16.988 respectively for the two variables.

To analyze this data with the neural network, we saved the generated data from OpenStat as a tab-separated variables file for importation into the Neural program. We used the import command in the Neural program to read the original tab file and then transformed the data into z scores. We did this in order to have scores we could later compare to the predicted scores obtained from the Neural program. We next transformed (scaled) these z scores to have a range between .1 and .9 a necessary step in order for the neurons of the network to have values with which it can work.

The control file for the analysis was created by selecting the option to generate a prediction control file into the grid of the program. The names of relevant files were then entered in the grid. The completed file is shown below:

```
QUIT ERROR:.1
QUIT RETRIES:3
NETWORK MODEL:LAYER
LAYER INIT:ANNEAL
OUTPUT MODEL:GENERAL
N INPUTS:1
N OUTPUTS:1
N HIDDEN1:0
N HIDDEN2:0
TRAIN:CORGENEDSCLD.DAT
OUTPUT FILE:CORGENED.TXT
LEARN:
SAVE WEIGHTS:CORGENED.WTS
EXECUTE:CORGENEDSCLD.DAT
QUIT:
```

Notice that there is one input and one output neuron defined. The Neural program will expect the output neuron values to follow the input neuron values when training the network. In this example, we want to train the network to predict the second value (Y) given the first value (X). In a basic statistics course we learn that the product–moment correlation is the linear relationship between an observed score (Y) and a predicted score Y' such that the squared difference between the observed "True" score Y and the observed predicted score (Y') is a minimum. The correlation between the predicted scores Y' and the observed scores Y should be the same as the correlation between X and Y. Of course, in traditional statistics this is because we are fitting the data to a straight line. If the data happen to fit a *curved line* better, then it is possible for the neural network to predict scores that are closer to the observed scores than that obtained using linear regression analysis. This is because the output of neurons is essentially non-linear, usually logistic in nature.

When we saved our control file and then clicked on the menu item to run the file, we obtained for following output:

Examples

```
NEURAL - Program to train and test neural networks
Written by William Miller

QUIT ERROR : 0.05
QUIT RETRIES : 5
NETWORK MODEL : LAYER
LAYER INIT : ANNEAL
OUTPUT MODEL : GENERAL
N INPUTS : 1
N OUTPUTS : 1
N HIDDEN1 : 0
N HIDDEN2 : 0
TRAIN : CORGENEDSCLD.DAT
SAVE WEIGHTS : CORGENED.WTS
There are no learned weights to save.
OUTPUT FILE : CORGENEDSCLD.TXT
LEARN :
Final error = 1.3720% of max possible
EXECUTE : CORGENED.DAT
QUIT :
```

You may notice that the value for the QUIT ERROR has been changed to 0.05 and the number of QUIT RETRIES changed to 5.

The .TXT file specified as the OUTPUT FILE now contains the 200 predicted scores obtained by the EXECUTE command. This command utilizes the weights obtained by the network (and now stored in CORGENED.WTS) to predict the output given new input values. We have elected to predict the same values as in the original training data sets X values and stored in a file labeled CORGENED.DAT which, of course, has also been transformed to z scores and scaled to values between .1 and .9 as were the original training values. These predicted values in the CORGENEDSCLD.TXT file were then re-transformed to z scores for comparison with the actual Y scores. The predicted and the transformed predicted scores were entered into the original (.TAB) data file and analyzed using the OpenStat package. The following results were obtained:

	CORRELATIONS		
	Y	YPREDICTED	ZPREDICTED
Y	1.0	0.580083	0.580083
YPREDICTED		1.0	1.0
ZPREDICTED			1.0

When X and Y were correlated following the initial generation of the data, the obtained value for the correlation of X with Y was 0.579. We conclude that the prediction with the neural network is, within a reasonable error, the same as that obtained with our traditional statistical procedure.

Regression Analysis with Multiple Predictors

Our next example examines the use of a neural network for prediction when there are multiple predictors. Our data comes from a file labeled "CANSAS.TAB" with which OpenStat users may be familiar. The file contains three body measurements and three measures of physical strength observed on 20 subjects. We have arbitrarily selected to predict the last performance measure with the five preceding measures.

The TAB file was imported into the Neural program grid and transformed to both z scores and scaled scores ranging from .1 to .9. Each transformation file was saved for later use.

We next generated a prediction control file and modified it to reflect the five input neurons and 1 output neuron. The control file is shown below:

```
QUIT ERROR:0.5
QUIT RETRIES:3
CONFUSION THRESHOLD:50
NETWORK MODEL:LAYER
LAYER INIT:ANNEAL
OUTPUT MODEL:GENERAL
N INPUTS:5
N OUTPUTS:1
N HIDDEN1:2
N HIDDEN2:0
TRAIN:CANSASSCALED.DAT
SAVE WEIGHTS:CANSAS.WTS
OUTPUT FILE:CANSASOUT.TXT
LEARN:
EXECUTE:CANSASSCALED.DAT
```

In order to compare the results with traditional multiple regression analysis, we needed to calculate the product–moment correlation between the values predicted by the Neural network using the same data as would be used to obtain the multiple correlation coefficient in traditional statistical analysis. We used the predicted scores from the CANSASOUT.TXT file and correlated them with the original dependent variable in the CANSAS.TAB file. The results of the classical multiple regression are shown first:

Examples

```
===========================================================
Block Entry Multiple Regression by Bill Miller
---------------- Trial Block 1 Variables Added --------------
Product-Moment Correlations Matrix with 20 cases.

Variables
              weight      waist      pulse      chins     situps
     weight    1.000      0.870     -0.366     -0.390    -0.493
      waist    0.870      1.000     -0.353     -0.552    -0.646
      pulse   -0.366     -0.353      1.000      0.151     0.225
      chins   -0.390     -0.552      0.151      1.000     0.696
     situps   -0.493     -0.646      0.225      0.696     1.000
      jumps   -0.226     -0.191      0.035      0.496     0.669

Variables
              jumps
     weight   -0.226
      waist   -0.191
      pulse    0.035
      chins    0.496
     situps    0.669
      jumps    1.000

Means with 20 valid cases.

Variables     weight      waist      pulse      chins     situps
             178.600     35.400     56.100      9.450    145.550

Variables     jumps
              70.300

Standard Deviations with 20 valid cases.

Variables     weight      waist      pulse      chins     situps
              24.691      3.202      7.210      5.286     62.567

Variables     jumps
              51.277

Dependent Variable: jumps
     R         R2         F     Prob.>F   DF1  DF2
   0.798     0.636      4.901    0.008    5    14
Adjusted R Squared = 0.507

Std. Error of Estimate =      36.020
```

```
Variable    Beta       B  Std.Error       t  Prob.>t    VIF    TOL
weight    -0.588  -1.221      0.704  -1.734    0.105  4.424  0.226
 waist     0.982  15.718      6.246   2.517    0.025  5.857  0.171
 pulse    -0.064  -0.453      1.236  -0.366    0.720  1.164  0.859
 chins     0.201   1.947      2.243   0.868    0.400  2.059  0.486
situps     0.888   0.728      0.205   3.546    0.003  2.413  0.414

Constant =    -366.967
Increase in R Squared =    0.636
F = 4.901 with probability =    0.008
Block 1 met entry requirements
```
==

Next, we show the correlations obtained between the values predicted by the Neural network and the original Y (jumps) variable:

==
```
Product-Moment Correlations Matrix with 20 cases.

Variables
               jumps     RawScaled
     jumps     1.000         0.826
 RawScaled     0.826         1.000

Means with 20 valid cases.
Variables      jumps     RawScaled
              70.300         0.256

Standard Deviations with 20 valid cases.
Variables      jumps     RawScaled
              51.277         0.152
```
==

The important thing to notice here is that the original multiple correlation coefficient was .798 using the traditional analysis method while the correlation of original scores to those predicted by the Neural network was .826. It appears the network captured some additional information that the linear model in multiple regression did not capture!

An additional analysis was performed using the following control file:

```
===============================================================
QUIT ERROR:0.5
QUIT RETRIES:3
NETWORK MODEL:LAYER
LAYER INIT:ANNEAL
OUTPUT MODEL:GENERAL
N INPUTS:5
N OUTPUTS:1
N HIDDEN1:2
N HIDDEN2:0
TRAIN:CANSASSCALED.DAT
SAVE WEIGHTS:CANSAS.WTS
OUTPUT FILE:CANSASOUT.TXT
LEARN:
EXECUTE:CANSASSCALED.DAT
===============================================================
```

Notice the addition of 2 neurons in a hidden layer. In this analysis, an even higher correlation was obtained between the original dependent score and the scores predicted by the Neural network:

The output for the above control file is shown below:

```
===============================================================
Variables
                   jumps      Raw Scaled   zscaled2hid
     jumps         1.000        0.826        0.919
  RawScaled        0.826        1.000        0.885
 scaled2hid        0.919        0.885        1.000

Means with    20 valid cases.

Variables         jumps       RawScaled    zscaled2hid
                 70.300         0.256        0.000

Standard Deviations with    20 valid cases.

Variables         jumps       RawScaled    zscaled2hid
                 51.277         0.152        1.000
===============================================================
```

The last variable, zscaled2hid, is the neural network predicted score using the 2 hidden layer neurons. The results also contain the results from the first analysis. Notice that we have gone from a multiple correlation coefficient of .798 to .919 with the neural network. It should be noted here that our "degrees of freedom" are quite low and we may be "over-fitting" the data by simply adding hidden level neurons.

Classification Analysis with Multiple Classification Predictors

In the realm of traditional multivariate statistical analyses, the discriminant function analysis method is used to identify raw or standardized weights of continuous variables that optimally separate groups of individuals in the "hyperspace" of discriminant space. Essentially, orthogonal axis of the original k-variable space are obtained. The number of axis is the smaller of the number of groups or the number of variables minus 1. Weights are then obtained that may be used to predict group membership based on the centroids (vector of means) of each group, the dispersion of each group and the prior probability of membership in each group.

With the Neural Program, we may create a Layer network for classifying objects based on the values of one or more input neurons. For our example, we have chosen to classify individuals that are members of one of three possible groups. We will classify them on the basis of 2 continuous variables. Our network will therefore have two input neurons, three output neurons and, we have added 2 neurons in a hidden layer. To train our network, we tell the network to classify objects for output neuron 1, then for output neuron 2 and finally for output neuron 3 that correspond to objects in groups 1, 2 and 3 respectively. This requires three data files with the objects from group 1 in one training file, the objects for group 2 in another file, etc.

The LEARN command will begin the network's training process for the three groups defined by the prior CLASSIFY OUTPUT and TRAIN filename commands. The obtained neural weights will be stored in the file name specified by the SAVE WEIGHTS command. Once the network has determined its weights, one can then utilize those weights to classify subjects of unknown membership into one of the groups. We have chosen to classify the same subjects in the groups that we used for the initial training. This is comparable to using the discriminant functions obtained in traditional statistics to classify the subjects on which the functions are based.

In traditional statistics, one will often create a "contingency table" with rows corresponding to the known group membership and the columns corresponding to the predicted group membership. If the functions can correctly classify all subjects in the groups, the diagonal of the table will contain the sample size of each group and the off-diagonal values will be zero. In other words, the table provides a count of objects that were correctly or incorrectly classified. Of course, it would be better to use a separate validation group drawn from the population which was NOT part of the training samples. In the case of the neural network, a file is created (or appended) with the count of predicted membership in each of the groups. An additional count column is also added to count objects which could not be correctly classified. This file is called the "CONFUSION" file. We reset the "confusion" table before each classification trial then CLASSIFY objects in a validation file. We show the confusion as well as save it in the confusion file. The SHOW CONFUSION will present the classifications in the output form while the SAVE CONFUSION filename command will cause the same output to be appended to the file.

```
================================================================
QUIT ERROR:0.1
QUIT RETRIES:5
CONFUSION THRESHOLD:50
NETWORK MODEL:LAYER
LAYER INIT:GENETIC
OUTPUT MODEL:CLASSIFY
N INPUTS:2
N OUTPUTS:3
N HIDDEN1:2
N HIDDEN2:0
CLASSIFY OUTPUT:1
TRAIN:DiscGrp1.DAT
CLASSIFY OUTPUT:2
TRAIN:DiscGrp2.DAT
CLASSIFY OUTPUT:3
TRAIN:DiscGrp3.DAT
LEARN:
SAVE WEIGHTS:Discrim.WGT
RESET CONFUSION:
CLASSIFY:DiscGrp1.DAT
SHOW CONFUSION:
SAVE CONFUSION:DISCRIM.OUT
RESET CONFUSION:
CLASSIFY:DiscGrp2.DAT
SHOW CONFUSION:
SAVE CONFUSION:DISCRIM.OUT
RESET CONFUSION:
CLASSIFY:DiscGrp3.DAT
SHOW CONFUSION:
SAVE CONFUSION:DISCRIM.OUT
RESET CONFUSION:
CLEAR TRAINING:
QUIT:
================================================================
```

The listing presented below shows a print out of the confusion file for the above run. Notice that one line was created each time a group of data were classified. Since we had submitted our classification tasks in the same order as the original grouping, the result is a table with counts of subject classifications in each of the known groups. In this example, all subjects were correctly classified.

```
================================================================
NEURAL - Program to train and test neural networks
Written by William Miller

QUIT ERROR : 0.1
QUIT RETRIES : 5
CONFUSION THRESHOLD : 50
NETWORK MODEL : LAYER
LAYER INIT : GENETIC
OUTPUT MODEL : CLASSIFY
N INPUTS : 2
N OUTPUTS : 3
N HIDDEN1 : 2
N HIDDEN2 : 0
CLASSIFY OUTPUT : 1
TRAIN : DISCGRP1.DAT
CLASSIFY OUTPUT : 2
TRAIN : DISCGRP2.DAT
CLASSIFY OUTPUT : 3
TRAIN : DISCGRP3.DAT
LEARN :
Final error = 0.0997% of max possible
SAVE WEIGHTS : DISCRIM.WGT
RESET CONFUSION :
CLASSIFY : DISCGRP1.DAT
SHOW CONFUSION :
Confusion: 5 0 0 0
SAVE CONFUSION : DISCRIM.OUT
RESET CONFUSION :
CLASSIFY : DISCGRP2.DAT
SHOW CONFUSION :
Confusion: 0 5 0 0
SAVE CONFUSION : DISCRIM.OUT
RESET CONFUSION :
CLASSIFY : DISCGRP3.DAT
SHOW CONFUSION :
Confusion: 0 0 5 0
SAVE CONFUSION : DISCRIM.OUT
RESET CONFUSION :
CLEAR TRAINING :
QUIT :
================================================================
```

Examples

When we classify each of the objects in the original three groups, we see that subjects in group 1 were all classified in the first group, all in group 2 classified into group 2, etc. In this case, training provided 100 % correct classification by the network of all our original objects. Of course, one would normally cross-validate a network with subjects not in the original training group. If you run a traditional discriminant analysis on this same data, you will see that the two methods are in complete agreement.

Pattern Recognition

A number of medical, industrial and military activities rely on recognizing certain patterns. For example, digital pictures of a heart may be scanned for abnormalities, and a manufacturer of automobile parts may use a digital scanned image to rotate and/ or flip a part on an assembly line for its next processing. The military may use a digitized scan of a sonar sounding to differentiate among whales, dauphins, sea turtles, schools of fish, torpedoes and submarines. In each of these applications, a sequence of binary "bits" (0 or 1) representing, say, horizontal rows of the digitized image are "mapped" to a specific object (itself represented perhaps by an integer value.)

As an example of pattern recognition, we will create digital "images" of the numbers 0, 1, 2, ..., 9. Each image will consist of a sequence of 25 bits (neural inputs of 0 or 1) and the image will be mapped to 10 output neurons which contain the number of images possible and corresponding to the digits 0 through 9 (0000 to 1001.) We will train a network by entering the image values randomly into a training set. We will then "test" the network by entering a data file with 20 images in sequence (10) and randomly placed (10). Examine the Confusion output to verify that (1) when we classify the original data there is one value for each digit and (2) when we enter 20 images we obtain 2 digits in each group.

Notice we have used a 5 by 5 grid to "digitize" a digit. For example, the number 8 is obtained from an image of:

$$
\begin{array}{ccccc}
0 & 1 & 1 & 1 & 0 \\
0 & 1 & 0 & 1 & 0 \\
0 & 0 & 1 & 0 & 0 \\
0 & 1 & 0 & 1 & 0 \\
0 & 1 & 1 & 1 & 0 \\
\end{array}
$$

and the number 2 is:

$$
\begin{array}{ccccc}
0 & 1 & 1 & 0 & 0 \\
1 & 0 & 0 & 1 & 0 \\
0 & 0 & 1 & 0 & 0 \\
0 & 1 & 0 & 0 & 0 \\
1 & 1 & 1 & 1 & 0 \\
\end{array}
$$

The values of 0 and 2 above are mapped to the output of 0000 and 0010 respectively.

The training file of the digitized images is shown below:
```
===============================================================
0 1 1 1 0 1 0 0 0 1 1 0 0 0 1 1 0 0 0 1 0 1 1 1 0
0 0 1 0 0 0 1 1 0 0 0 0 1 0 0 0 0 1 0 0 0 1 1 1 0
0 1 1 0 0 1 0 0 1 0 0 0 1 0 0 0 1 0 0 0 1 1 1 1 0
0 1 1 0 0 1 0 0 1 0 0 0 1 0 0 1 0 0 1 0 0 1 1 0 0
0 0 0 1 0 0 0 1 1 0 0 1 0 1 0 1 1 1 1 0 0 0 1 0
0 1 1 1 1 0 1 0 0 0 0 1 1 1 0 1 0 0 0 1 0 1 1 1 0
0 0 1 1 0 0 1 0 0 0 0 1 1 1 0 0 1 0 1 0 0 1 1 1 0
0 1 1 1 1 0 0 0 1 0 0 0 1 0 0 0 1 0 0 0 1 0 0 0 0
0 1 1 1 0 0 1 0 1 0 0 0 1 0 0 0 1 0 1 0 0 1 1 1 0
0 1 1 1 0 0 1 0 1 0 0 1 1 1 0 0 0 0 1 0 0 1 1 0 0
===============================================================
```

The listings below represent the Control File and Output of the training and testing of the neural network. Notice the model for the network and the command file entries.

```
===============================================================
QUIT ERROR:0.1
QUIT RETRIES:5
CONFUSION THRESHOLD:50
KOHONEN NORMALIZATION MULTIPLICATIVE:
NETWORK MODEL:KOHONEN
KOHONEN INIT:RANDOM
OUTPUT MODEL:CLASSIFY
N INPUTS:25
N OUTPUTS:10
N HIDDEN1:0
N HIDDEN2:0
TRAIN:scandigits.doc
KOHONEN LEARN ADDITIVE:
KOHONEN LEARNING RATE:0.4
KOHONEN LEARNING REDUCTION:0.99
LEARN:
SAVE WEIGHTS:scan.wts
RESET CONFUSION:
CLASSIFY:scandigits.doc
SHOW CONFUSION:
SAVE CONFUSION:scan.txt
RESET CONFUSION:
CLASSIFY:scantest.dat
SHOW CONFUSION:
SAVE CONFUSION:scan.txt
CLEAR TRAINING:
QUIT:
===============================================================
```

```
================================================================
NEURAL - Program to train and test neural networks
Written by William Miller

QUIT ERROR : 0.1
QUIT RETRIES : 5
CONFUSION THRESHOLD : 50
KOHONEN NORMALIZATION MULTIPLICATIVE :
NETWORK MODEL : KOHONEN
KOHONEN INIT : RANDOM
OUTPUT MODEL : CLASSIFY
N INPUTS : 25
N OUTPUTS : 10
N HIDDEN1 : 0
N HIDDEN2 : 0
TRAIN : SCANDIGITS.DOC
KOHONEN LEARN ADDITIVE :
KOHONEN LEARNING RATE : 0.4
KOHONEN LEARNING REDUCTION : 0.99
LEARN :
Final error = 0.0000% of max possible
SAVE WEIGHTS : SCAN.WTS
RESET CONFUSION :
CLASSIFY : SCANDIGITS.DOC
SHOW CONFUSION :
Confusion:    1   1   1   1   1   1   1   1   1   1   0
SAVE CONFUSION : SCAN.TXT
RESET CONFUSION :
CLASSIFY : SCANTEST.DAT
SHOW CONFUSION :
Confusion:    2   2   2   2   2   2   2   2   2   2   0
SAVE CONFUSION : SCAN.TXT
CLEAR TRAINING :
QUIT :
================================================================
```

Exploration of Natural Groups

Researchers often attempt to "tease" information or relationships out of a set of measurements without prior knowledge of those relationships. This "data-mining" might be simply to aggregate objects with similar profiles in order to examine other aspects of those objects that they may share. A variety of statistical methods for "grouping" objects on the basis of multiple continuous measures have been developed. The "Hierarchical Grouping" procedure is one of the more popular ones. The criteria for grouping may vary from procedure to procedure however. Many procedures examine the distance between each object and all other objects in the Euclidean space of the grouping variables. Of course, the distance is affected by the scale of each measurement. For that reason, one often transforms all measures to a common scale like the z score scale which has a mean of 0 and a standard deviation of 1.0. Still, this may ignore the different distribution shapes of the variables. Some grouping methods take this into account and measure the distance among objects using distribution characteristics. Most of the procedures "create" groups by first combining the two "closest" objects and replacing the two objects with a single group that is the average of the two objects in the group. The process is begun again, each time replacing the two objects with a group that combines the two objects. The user can typically print out the group membership at each iteration of the grouping process.

The Kohonen Neural Network provides an excellent basis for exploring natural groups which may exist among objects with multiple measures. One can train this network to classify objects into "M" number of groups based on values of "k" variables. One specifies an input neuron for each of the k variables and an output neuron for each group. Following the training one then uses the network to classify objects into the M groups. By varying the number of output neurons, one can utilize multiple networks to explore the objects classified into each group.

The Kohonen network model has a number of parameters that may be specified to control the operation of the training. One may use a multiplicative or a z method for normalization of the weights. You can initialize weights using random values or no random values. The learning method may be additive or subtractive. The learning rate and reduction parameters may each be specified. See Appendix A for further details on all parameters.

To demonstrate the use of the Kohonen net for classification, we will employ a file of data that may be analyzed by traditional hierarchical grouping as well as a neural network. The results of each will be explored.

The file to be analyzed is labeled "MANODISCRIM.TAB" with the contents shown below:

Y1	Y2	Group
3	7	1
4	7	1
5	8	1
5	9	1
6	10	1
4	5	2
4	6	2
5	7	2
6	7	2
6	8	2
5	5	3
6	5	3
6	6	3
7	7	3
7	8	3

Examples

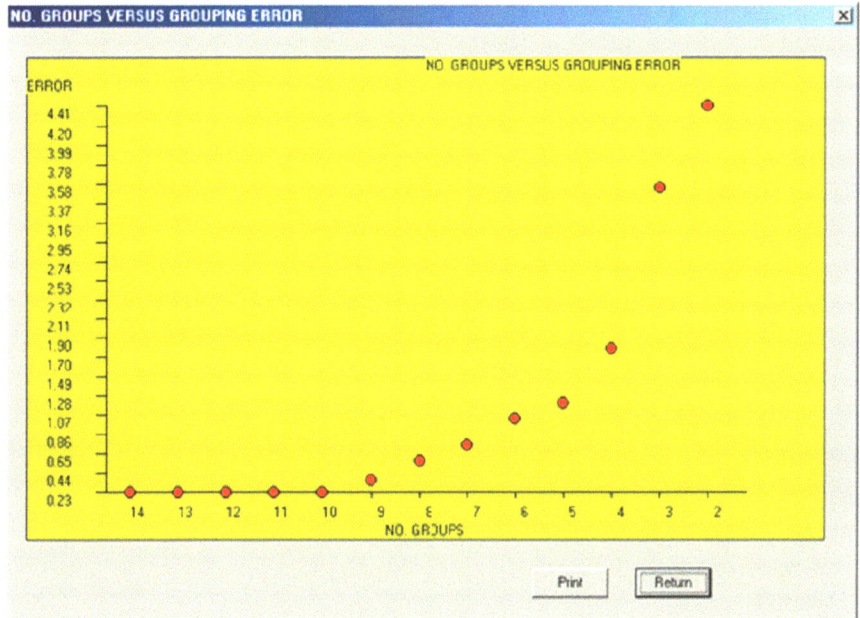

Fig. 17.7 Groups versus between group error

When we analyzed the above data using the Hierarchical grouping procedure of OPENSTAT we obtained the following groupings of data and error plot (Fig. 17.7):

```
================================================================
14 groups after combining group 2  (n := 1 ) and group 7  (n := 1) error = 0.233
13 groups after combining group 3  (n := 1 ) and group 4  (n := 1) error = 0.233
12 groups after combining group 9  (n := 1 ) and group 10 (n := 1) error = 0.233
11 groups after combining group 12 (n := 1 ) and group 13 (n := 1) error = 0.233
10 groups after combining group 14 (n := 1 ) and group 15 (n := 1) error = 0.233
9 groups after combining group 6  (n := 1 ) and group 11 (n := 1) error = 0.370
8 groups after combining group 2  (n := 2 ) and group 8  (n := 1) error = 0.571
7 groups after combining group 9  (n := 2 ) and group 14 (n := 2) error = 0.739
6 groups after combining group 1  (n := 1 ) and group 2  (n := 3) error = 1.025
```

```
Group 1 (n = 4)
  Object = 0
  Object = 1
  Object = 6
  Object = 7
Group 3 (n = 2)
  Object = 2
  Object = 3
Group 5 (n = 1)
  Object = 4
Group 6 (n = 2)
  Object = 5
  Object = 10
Group 9 (n = 4)
  Object = 8
  Object = 9
  Object = 13
  Object = 14
Group 12 (n = 2)
  Object = 11
  Object = 12

5 groups after combining group 3 (n = 2 ) and group 5 (n = 1)
error = 1.193
Group 1 (n = 4)
  Object = 0
  Object = 1
  Object = 6
  Object = 7
Group 3 (n = 3)
  Object = 2
  Object = 3
  Object = 4
Group 6 (n = 2)
  Object = 5
  Object = 10
Group 9 (n = 4)
  Object = 8
  Object = 9
  Object = 13
  Object = 14
Group 12 (n = 2)
  Object = 11
  Object = 12
```

Examples

```
4 groups after combining group 6 (n = 2 ) and group 12 (n = 2)
error = 1.780
Group 1 (n = 4)
  Object = 0
  Object = 1
  Object = 6
  Object = 7
Group 3 (n = 3)
  Object = 2
  Object = 3
  Object = 4
Group 6 (n = 4)
  Object = 5
  Object = 10
  Object = 11
  Object = 12
Group 9 (n = 4)
  Object = 8
  Object = 9
  Object = 13
  Object = 14

3 groups after combining group 3 (n = 3 ) and group 9 (n = 4)
error = 3.525
Group 1 (n = 4)
  Object = 0
  Object = 1
  Object = 6
  Object = 7
Group 3 (n = 7)
  Object = 2
  Object = 3
  Object = 4
  Object = 8
  Object = 9
  Object = 13
  Object = 14
Group 6 (n = 4)
  Object = 5
  Object = 10
  Object = 11
  Object = 12
```

```
2 groups after combining group 1 (n = 4 ) and group 6 (n = 4)
error = 4.411
Group 1 (n = 8)
  Object = 0
  Object = 1
  Object = 5
  Object = 6
  Object = 7
  Object = 10
  Object = 11
  Object = 12
Group 3 (n = 7)
  Object = 2
  Object = 3
  Object = 4
  Object = 8
  Object = 9
  Object = 13
  Object = 14
```
==

To complete a similar analysis with the neural network program we created the following control file and then modified it for two additional runs:
==
```
QUIT ERROR:0.1
QUIT RETRIES:5
CONFUSION THRESHOLD:50
KOHONEN NORMALIZATION Z:
NETWORK MODEL:KOHONEN
KOHONEN INIT:RANDOM
OUTPUT MODEL:CLASSIFY
N INPUTS:2
N OUTPUTS:6
N HIDDEN1:0
N HIDDEN2:0
TRAIN:HIER.DAT
KOHONEN LEARN ADDITIVE:
KOHONEN LEARNING RATE:0.4
KOHONEN LEARNING REDUCTION:0.99
LEARN:
SAVE WEIGHTS:HIER.WTS
RESET CONFUSION:
CLASSIFY:HIER.DAT
SHOW CONFUSION:
SAVE CONFUSION:HIER.TXT
RESET CONFUSION:
CLASSIFY:HIER1.DAT
SHOW CONFUSION:
SAVE CONFUSION:HIER.TXT
```

Examples

```
RESET CONFUSION:
CLASSIFY:HIER2.DAT
SHOW CONFUSION:
SAVE CONFUSION:HIER.TXT
RESET CONFUSION:
CLASSIFY:HIER3.DAT
SHOW CONFUSION:
SAVE CONFUSION:HIER.TXT
CLEAR TRAINING:
QUIT:
```
==

Control File for Exploration of Groups Using a Kohonen Neural Network for Six Groups

In the above file we specified six output neurons. This is our initial guess as to the number of "natural groups" in the data. The output from this run is shown below:

```
NEURAL - Program to train and test neural networks
Written by William Miller
QUIT ERROR : 0.1
QUIT RETRIES : 5
CONFUSION THRESHOLD : 50
KOHONEN NORMALIZATION Z :
NETWORK MODEL : KOHONEN
KOHONEN INIT : RANDOM
OUTPUT MODEL : CLASSIFY
N INPUTS : 2
N OUTPUTS : 6
N HIDDEN1 : 0
N HIDDEN2 : 0
TRAIN : HIER.DAT
KOHONEN LEARN ADDITIVE :
KOHONEN LEARNING RATE : 0.4
KOHONEN LEARNING REDUCTION : 0.99
LEARN :
Final error = 12.6482% of max possible
SAVE WEIGHTS : HIER.WTS
RESET CONFUSION :
CLASSIFY : HIER.DAT
SHOW CONFUSION :
Confusion: 3 1 3 3 3 2 0
SAVE CONFUSION : HIER.TXT
RESET CONFUSION :
CLASSIFY : HIER1.DAT
SHOW CONFUSION :
Confusion: 2 1 0 0 0 2 0
SAVE CONFUSION : HIER.TXT
RESET CONFUSION :
CLASSIFY : HIER2.DAT
```

```
SHOW CONFUSION :
Confusion:   1   0   1   2   1   0   0
SAVE CONFUSION : HIER.TXT
RESET CONFUSION :
CLASSIFY : HIER3.DAT
SHOW CONFUSION :
Confusion:   0   0   2   1   2   0   0
SAVE CONFUSION : HIER.TXT
CLEAR TRAINING :
QUIT :
```
==
Kohonen Network Output for Exploratory Grouping with Six Groups Estimated

You may compare the number of objects out of the total 15 that were classified in each of the groups (i.e. 3, 1, 3, 3, 3 ,2) and compare this with the number in six groups obtained with the Hierarchical Grouping procedure (4,2, 1,2,4,2). There is obviously some difference in the grouping. One can also see how the subjects who belong to groups 1, 2 or 3 are classified by each program.

For the second neural network analysis we modified the first control file to contain three output neurons, our next guess as to the number of "natural groups". The output obtained is as follows:
==
```
NEURAL - Program to train and test neural networks
Written by William Miller

QUIT ERROR : 0.1
QUIT RETRIES : 5
CONFUSION THRESHOLD : 50
KOHONEN NORMALIZATION Z :
NETWORK MODEL : KOHONEN
KOHONEN INIT : RANDOM
OUTPUT MODEL : CLASSIFY
N INPUTS : 2
N OUTPUTS : 3
N HIDDEN1 : 0
N HIDDEN2 : 0
TRAIN : HIER.DAT
KOHONEN LEARN ADDITIVE :
KOHONEN LEARNING RATE : 0.4
KOHONEN LEARNING REDUCTION : 0.99
LEARN :
Final error = 21.3618% of max possible
SAVE WEIGHTS : HIER.WTS
RESET CONFUSION :
CLASSIFY : HIER.DAT
SHOW CONFUSION :
Confusion:   4   6   5   0
SAVE CONFUSION : HIER.TXT
RESET CONFUSION :
CLASSIFY : HIER1.DAT
```

Examples

```
SHOW CONFUSION :
Confusion:    0    3    2    0
SAVE CONFUSION : HIER.TXT
RESET CONFUSION :
CLASSIFY : HIER2.DAT
SHOW CONFUSION :
Confusion:    1    3    1    0
SAVE CONFUSION : HIER.TXT
RESET CONFUSION :
CLASSIFY : HIER3.DAT
SHOW CONFUSION :
Confusion:    3    0    2    0
SAVE CONFUSION : HIER.TXT
CLEAR TRAINING :
QUIT :
```
==

Kohonen Network Output for Exploratory Grouping with Three Groups

Notice that number of subjects classified in each group are 4, 6 and 5 respectively. The Hierarchical Grouping procedure placed 4, 7 and 4 respectively. It should be pointed out that the output neurons do not necessarily follow the same order as the "true" groups, i.e. 1, 2 and 3. In fact, it appears in our last analysis that the 3rd neuron may be sensitive to subjects in group 1, and neuron 1 most sensitive to subjects in group 3. Neurons 1 and 2 seem about equally sensitive to members of both groups 1 and 2. To determine the prediction for each object (subject) we would classify each of the objects by themselves rather that read them by group.

We can construct contingency tables of actual versus predicted groups if we wish for either type of analysis. For example, the Hierarchical Grouping analysis would yield the following:

	PREDICTED GROUP		
ACTUAL GROUP	1	2	3
1	2	3	0
2	2	2	1
3	0	2	3

For the Kohonen Neural Network we would have:

	PREDICTED GROUP		
ACTUAL GROUP	1	2	3
1	3	0	2
2	1	3	1
3	0	3	2

Comparison of Grouping by Hierarchical Analysis and a Kohonen Neural Network

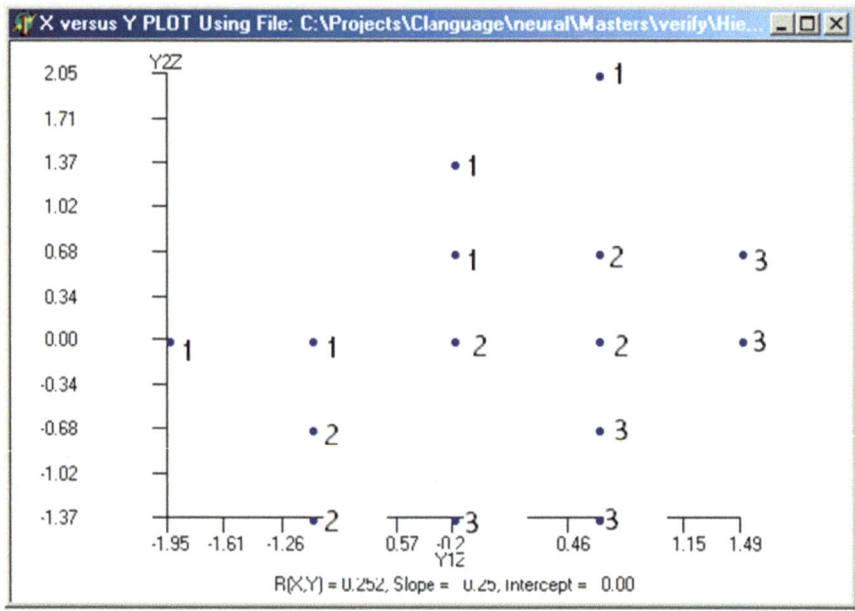

Fig. 17.8 Plot of subjects in three groups, each subject measured on two variables

Seven subjects in the original groups were predicted to be in the "natural" groups by the first method while eight subjects in the original groups were in "natural" groups by the second method. Of course, one does not typically know, a priori, what the "true" group memberships are. Thus, whether one uses traditional statistics or neural networks, one must still explore what seems to be common denominators among the grouped subjects. It is sometimes useful to actually plot the objects in the standardized score space to initially speculate on the number of "natural" groups. Above is a plot of the 15 scores of our original data (Fig. 17.8):

Group 1, 2, and 3 subjects are labeled with the values 1, 2 and 3. Notice that when you try to "split" the groups using Y1 or Y2 (horizontal or vertical) axis there is overlap and confusion regarding group membership. On the other hand, if you drew diagonal lines you can see how each of the three groups COULD be separated by considering both Y1 and Y2 concurrently. In Fact, that is just what the discriminant function analysis in traditional statistics does. Go back up and examine the results for our earlier example of discriminant analysis using a neural network. The data for that example is exactly the same as was analyzed with the present network!

Time Series Analysis

This example is based on the needs of grocery store retailers to predict customer purchases for items they stock. Over-stocking costs them shelf space while under-stocking might cost them sales. Ideally, the shelves are stocked with just enough

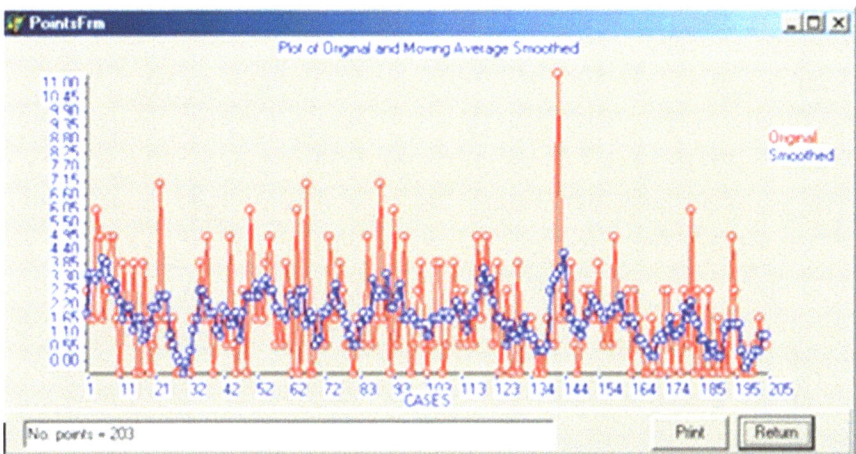

Fig. 17.9 Original daily sales of creamed chicken with smoothed averages (3 values in each average)

items to meet the demand for a day's purchases. It would be possible to use historical data to give us a reasonable estimate of the purchases to be made for a given item. Of course, the historical data would have to be for the same day of the week, same sales promotion for the item, same weather factors, same store location, same customer base, etc. to yield the "best" prediction of purchases for a given day. Most stores however do not have such historical data and often may have only one or two preceding week's data. In our example, we are assuming we have collected weekly data over a period of 28 weeks and wish to be able to predict customer purchases of Creamed Chicken Soup for a given day, in this example, Sunday. Our data consists of 28 records in a data file. Each record contains the number of cans of Creamed Chicken Soup sold on Sunday, Monday, Tuesday, Wednesday, Thursday, Friday, Saturday and (the next) Sunday. In other words, we have 8 consecutive day's sales in each record. We will attempt to predict the sales on the 8th day using the sales data from the previous seven days.

A variety of time-series analyses have been developed utilizing traditional statistical methods. Many are based on "auto-correlation" analyses. Users of the OpenStat package can perform a variety of analyses on the same data to attempt the best prediction. Shown below are two graphs obtained from the autocorrelation procedure. The data were the units of Creamed Chicken sold each day from Sunday through Saturday for 28 weeks. A lag of 6 (0 through 7) was utilized for the autocorrelation analysis and smoothing average was utilized to project for 2 additional data points (Figs. 17.9, 17.10):

Autoregressive methods along with smoothing average methods are sometimes used to project (estimate) subsequent data points in a series. If one examines the first figure above, one can observe some cyclic tendencies in the data. Fast Fourier smoothing or exponential smoothing might "flatten" these cyclic tendencies (which

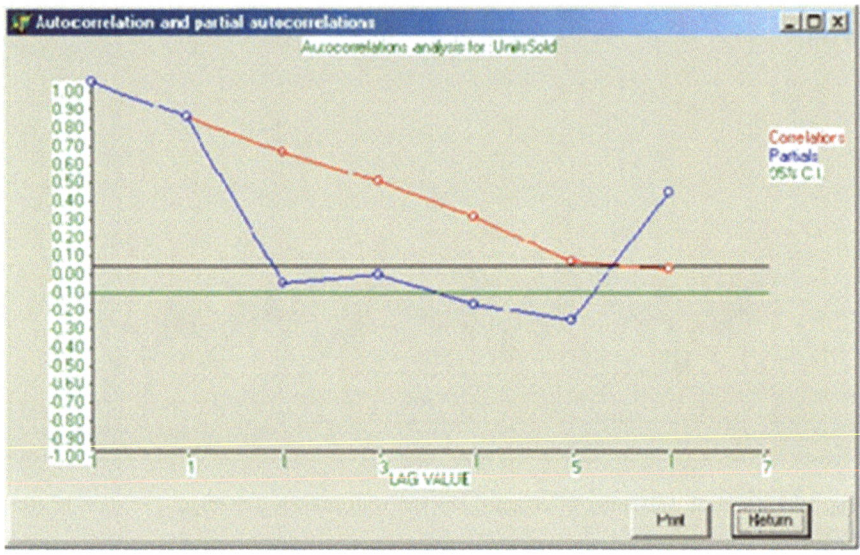

Fig. 17.10 Auto and partial correlations for lags from Sunday (lag 1 = Saturday, etc.)

appear to be a week long in duration.) Nearly all methods will result in an estimate for Sunday sales which reflect some "smoothing" of the data and estimate a new values that are, on the average, somewhat less than those actually observed.

The neural network involves identifying the series and building a network that will predict the next value. To do this, we recorded Sunday through Sunday sequences of sales for 28 weeks. In our Neural Program, the last variable is always the output neuron. If our desire had been to predict Monday sales, then the sequence recorded would have been Monday through the subsequent Monday. We transformed the number of sales for each day into z scores and then to values having a range of .1 to .9 as required for our network. The predicted values we obtain from executing the network weights are re-translated into z scores for comparison with the observed z score data for Sunday sales.

There are a variety of variables which one can modify when training the network. In the Feed-Forward network, you have several alternatives for estimating the neural weights. You also have alternatives in the use of hidden layers and the number of neurons in those layers. You also have choices regarding the minimum error and the number of times the network attempts to obtain the least-squares error (QUIT ERROR and QUIT RETRIES.) We "experimented"with five variations of a control file for training the neural network in the prediction of Sunday sales. Three of those control files are shown below:

Examples

```
============================================================
QUIT ERROR:0.01
QUIT RETRIES:5
NETWORK MODEL:LAYER
LAYER INIT:ANNEAL
OUTPUT MODEL:GENERAL
N INPUTS:7
N OUTPUTS:1
N HIDDEN1:3
N HIDDEN2:1
TRAIN:CRMCHKZSCLD.DAT
OUTPUT FILE:CRMCHICK1.OUT
LEARN:
SAVE WEIGHTS:CRMCHICK1.WTS
EXECUTE:CRMCHKZSCLD.DAT
QUIT:
============================================================
```

Control Form for a Time Series Analysis - First Run

Notice that the above control file uses the Anneal method of minimizing the least squares function obtained by the neural weights. In addition, two hidden layers of neurons were used with three and one neuron respectively in those layers. The output obtained from this run is shown in the following figure:

```
============================================================
NEURAL - Program to train and test neural networks
Written by William Miller

QUIT ERROR : 0.01
QUIT RETRIES : 5
NETWORK MODEL : LAYER
LAYER INIT : ANNEAL
OUTPUT MODEL : GENERAL
N INPUTS : 7
N OUTPUTS : 1
N HIDDEN1 : 3
N HIDDEN2 : 1
TRAIN : CRMCHKZSCLD.DAT
There are no learned weights to save.
OUTPUT FILE : CRMCHICK1.OUT
LEARN :
SAVE WEIGHTS : CRMCHICK1.WTS
Final error = 0.0825% of max possible
EXECUTE : CRMCHKZSCLD.DAT
QUIT :
============================================================
```

Time Series Analysis Output -First Run

Notice the final error reported in the output above and compare it with the next two examples.

```
================================================================
QUIT ERROR:0.01
QUIT RETRIES:5
NETWORK MODEL:LAYER
LAYER INIT:ANNEAL
OUTPUT MODEL:GENERAL
N INPUTS:7
N OUTPUTS:1
N HIDDEN1:0
N HIDDEN2:0
TRAIN:CRMCHKZSCLD.DAT
OUTPUT FILE:CRMCHICK3.TXT
LEARN:
SAVE WEIGHTS:CRMCHICK3.WTS
EXECUTE:CRMCHKZSCLD.DAT
QUIT:
================================================================
```

Control Form for a Time Series Analysis - Third Run

In this last example (run three), we have eliminated the neurons in the hidden layers that were present in our first example. The output is shown below. Note that the size of the final error is considerably larger than the previous analysis.

```
================================================================
NEURAL - Program to train and test neural networks
Written by William Miller

QUIT ERROR : 0.01
QUIT RETRIES : 5
CONFUSION THRESHOLD : 50
NETWORK MODEL : LAYER
LAYER INIT : ANNEAL
OUTPUT MODEL : GENERAL
N INPUTS : 7
N OUTPUTS : 1
N HIDDEN1 : 0
N HIDDEN2 : 0
TRAIN : CRMCHKZSCLD.DAT
OUTPUT FILE : CRMCHICK3.TXT
LEARN :
Final error = 4.5999% of max possible
SAVE WEIGHTS : CRMCHICK3.WTS
EXECUTE : CRMCHKZSCLD.DAT
QUIT :
================================================================
```

Time Series Analysis Output for Run Three

In our last experimental time series analysis we have utilized a different method for initializing the neural weights. We used the genetic method for simulating a population to evolve with weights that minimized the least squares criterion. We also used just one hidden layer containing two neurons in contrast to our first example which used two hidden layers. The output final error is more than the first example but less than our second example.

```
===============================================================
QUIT ERROR:0.01
QUIT RETRIES:5
CONFUSION THRESHOLD:50
NETWORK MODEL:LAYER
LAYER INIT:GENETIC
OUTPUT MODEL:GENERAL
N INPUTS:7
N OUTPUTS:1
N HIDDEN1:2
N HIDDEN2:0
TRAIN:CRMCHKZSCLD.DAT
OUTPUT FILE:CRMCHICK5.TXT
LEARN:
SAVE WEIGHTS:CRMCHICK5.WTS
EXECUTE:CRMCHKZSCLD.DAT
QUIT:
===============================================================
```

Control Form for a Time Series Analysis - Fifth Run

```
===============================================================
NEURAL - Program to train and test neural networks
Written by William Miller

QUIT ERROR : 0.01
QUIT RETRIES : 5
CONFUSION THRESHOLD : 50
NETWORK MODEL : LAYER
LAYER INIT : GENETIC
OUTPUT MODEL : GENERAL
N INPUTS : 7
N OUTPUTS : 1
N HIDDEN1 : 2
N HIDDEN2 : 0
TRAIN : CRMCHKZSCLD.DAT
OUTPUT FILE : CRMCHICK5.TXT
LEARN :
Final error = 0.2805% of max possible
SAVE WEIGHTS : CRMCHICK5.WTS
EXECUTE : CRMCHKZSCLD.DAT
QUIT :
===============================================================
```

Time Series Analysis Output for Run Five

For each of the above examples, we "z-score" translated the predicted outputs obtained through use of the six days of predictor data. We then copied these three sets of predicted scores into a data file containing our original Sunday Sales data and obtained the product–moment correlation among the four sets. The results are shown below:

```
===============================================================
Product-Moment Correlations Matrix with 28 cases.
Variables
                  VAR. 8        Pred8_1       Pred8_3       Pred8_5
     VAR. 8       1.000         0.993         0.480         0.976
     Pred8_1      0.993         1.000         0.484         0.970
     Pred8_3      0.480         0.484         1.000         0.501
     Pred8_5      0.976         0.970         0.501         1.000

Means with 28 valid cases.

Variables         VAR. 8        Pred8_1       Pred8_3       Pred8_5
                  0.000         0.020         -0.066        0.012

Standard Deviations with 28 valid cases.

Variables         VAR. 8        Pred8_1       Pred8_3       Pred8_5
                  1.000         1.013         0.952         1.016
===============================================================
```

Correlations Among Variable 8 (Sunday Sales) and Predicted Sales Obtained From The Neural Network for Runs 1, 3 and 5. Note: Sales Measures in Z Score Units.

Notice that the "best" predictions were obtained from our first control file in which we utilized two hidden layers of neurons. The last analysis performed nearly as well as the first with fewer neurons. It also "learned" much faster than the first example. It should be noted that we would normally re-scale our values again to translate them from z scores to "raw" scores using the mean and standard deviation of the Sunday sales data.

Bibliography

Afifi AA, Azen SP. Statistical analysis. A computer oriented approach. New York: Academic; 1972.
Anderberg MR. Cluster analysis for applications. New York: Academic; 1973.
Bennett S, Bowers D. An introduction to multivariate techniques for social and behavioral sciences. New York: Wiley; 1977.
Besterfield DH. Quality control. Englewood Ciffs: Prentice-Hall; 1986.
Bishop YM, Fienberg SE, Holland PW. Discrete multivariate analysis. Theory and practice. Cambridge, MA: The MIT Press; 1975.
Blommers PJ, Forsyth RA. Elementary statistical methods in psychology and education. 2nd ed. Boston: Houghton Mifflin Company; 1977.
Borg WR, Gall MD. Educational research. An introduction. 5th ed. New York: Longman; 1989.
Brierley, Phil. MLP neural network in C++. http://www.philbrierly.com, 2012
Brockwell PJ, Davis RA. Introduction to time series and forecasting. New York: Springer; 1996.
Bruning JL, Kintz BL. Computational handbook of statistics. 2nd ed. Glenview: Scott, Foresman and Company; 1977.
Campbell DT, Stanley JC. Experimental and quasi-experimental designs for research. Chicago: Rand McNally College; 1963.
Chapman DG, Schaufele RA. Elementary probability models and statistical inference. Waltham: Ginn-Blaisdell; 1970.
Cody RP, Smith JK. Applied statistics and the SAS programming language. 4th ed. Upper Saddle River: Prentice Hall; 1997.
Cohen J. Statistical power analysis for the behavioral sciences. 2nd ed. Hillsdale: Lawrence Erlbaum Associates; 1988.
Cohen J, Cohen P. Applied multiple regression/ correlation analysis for the behavioral sciences. Hillsdale: Lawrence Erlbaum Associates; 1975.
Comrey AL. A first course in factor analysis. New York: Academic; 1973.
Cook TD, Campbell DT. Quasi-experimentation. Design and analysis issues for field settings. Chicago: Rand McNally College; 1979.
Cooley WW, Lohnes PR. Multivariate data analysis. New York: Wiley; 1971.
Crocker L, Algina J. Introduction to classical and modern test theory. New York: Holt, Rinehart and Winston; 1986.
Diekhoff GM. Basic statistics for the social and behavioral sciences. Upper Sadle River: Prentice Hall; 1996.
Edwards AL. Techniques of attitude scale construction. New York: Appleton-Century-Crofts; 1957.
Efromovich S. Nonparametric curve estimation. Methods, theory, and applications. New York: Springer; 1999.

Ferrguson GA. Statistical analysis in psychology and education. 2nd ed. New York: McGraw-Hill Book Company; 1966.
Fienberg SE. The analysis of cross-classified categorical data. 2nd ed. Cambridge, MA: The MIT Press; 1980.
Fox J. Multiple and generalized nonparametric regression. Thousand Oaks: Sage; 2000.
Freund JE, Walpole RE. Mathematical statistics. 4th ed. Englewood Ciffs: Prentice-Hall; 1987.
Fruchter B. Introduction to factor analysis. Princeton: D. Van Nostrand Company; 1954.
Gay LR. Educational research. Competencies for analysis and application. 4th ed. New York: Macmillan; 1992.
Gentle JE, Kennedy Jr WJ. Statistical computing. New York: Marcel Dekker; 1980.
Glass GV, Stanley JC. Statistical methods in education and psychology. Englewood Ciffs: Prentice-Hall; 1970.
Gottman JM, Leiblum SR. How to do psychotherapy and how to evaluate it. A manual for beginners. New York: Holt, Rinehart and Winston; 1974.
Guertin WH, Bailey Jr JP, Bailey Jr JP. Introduction to modern factor analysis. Ann Arbor: Edwards Brothers; 1970.
Gulliksen H. Theory of mental tests. New York: Wiley; 1950.
Hambleton RK, Swaminathan H. Item response theory. Principles and applications. Boston: Kluwer-Nijhoff Publishing; 1985.
Hansen BL, Chare PM. Quality control and applications. Englewood Ciffs: Prentice-Hall; 1987.
Harman HH. Modern factor analysis. Chicago: The University of Chicago Press; 1960.
Hays WL. Statistics for psychologists. New York: Holt, Rinehart and Winston; 1963.
Heise DR. Causal analysis. New York: Wiley; 1975.
Hinkle DE, Wiersma W, Jurs SG. Applied statistics for the behavioral sciences. Boston: Houghton Mifflin Company; 1988.
Huntsberger DH, Billingsley P. Elements of statistical inference. 6th ed. Boston: Allyn and Bacon; 1987.
Kelly LG. Handbook of numerical methods and applications. Reading: Addison-Wesley; 1967.
Kerlinger FN, Pedhazur EJ. Multiple regression in behavioral research. New York: Holt, Rinehart and Winston; 1973.
Lieberman B, editor. Contemporary problems in statistics. A book of readings for the behavioral sciences. New York: Oxford University Press; 1971.
Lindgren BW, McElrath GW. Introduction to probability and statistics. 2nd ed. New York: Macmillan; 1966.
Marcoulides GA, Schumacker RE, editors. Advanced structural equation modeling. Issues and techniques. Mahwah: Lawrence Erlbaum Associates; 1996.
Masters T. Practical neural network recipes in C++. San Diego: Morgan Kaufmann; 1993.
McNeil K, Newman I, Kelly FJ. Testing research hypotheses with the general linear model. Carbondale: Southern Illinois University Press; 1996.
McNemar Q. Psychological statistics. 4th ed. New York: Wiley; 1969.
Minium EW. Statistical reasoning in psychology and education. 2nd ed. New York: Wiley; 1978.
Montgomery DC. Statistical quality control. New York: Wiley; 1985.
Mulaik SA. The foundations of factor analysis. New York: McGraw-Hill; 1972.
Myers JL. Fundamentals of experimental design. Boston: Allyn and Bacon; 1966.
Nunnally JC. Psychometric theory. New York: McGraw-Hill; 1967.
Olson CL. Essentials of statistics. Making sense of data. Boston: Allyn and Bacon; 1987.
Payne DA, editor. Curriculum evaluation. Commentaries on purpose, process, product. Lexington: D. C. Heath and Company; 1974.
Pedhazur EJ. Multiple regression in behavioral research. Explanation and prediction. 3rd ed. Fort Worth: Holt, Rinehart and Winston; 1997.
Press WH, Flannery BP, Teukolsky SA, Vetterling WT. Numerical recipes in C. The art of scientific computing. Cambridge: Cambridge University Press; 1988.
Ralston A, Wilf HS. Mathematical methods for digital computers. New York: Wiley; 1966.

Rao CR. Linear statistical inference and its applications. New York: Wiley; 1965.
Rao V, Rao H. C++ neural networks and fuzzy logic. 2nd ed. New York: MIS Press; 1995.
Rich E, Knight K. Artificial intelligence. New York: McGraw-Hill; 1983.
Rogers J. Object-oriented neural networks in C++. San Diego: Academic; 1997.
Roscoe JT, Research F. Statistics for the behavioral sciences. 2nd ed. New York: Holt, Rinehart and Winston; 1975.
Rummel RJ. Applied factor analysis. Evanston: Northwestern University Press; 1970.
Scheffe' H. The analysis of variance. New York: Wiley; 1959.
Schumacker RE, Lomax RG. A beginner's guide to structural equation modeling. Mahwah: Lawrence Erlbaum Associates; 1996.
Siegel S. Nonparametric statistics for the behavioral sciences. New York: McGraw-Hill Book Company; 1956.
Silverman EN, Brody LA. Statistics. A common sense approach. Boston: Prindle, Weber and Schmidt; 1973.
SPSS, Inc. SPSS-X user's guide. 3rd ed. Chicago: SPSS; 1988.
Steele SM. Contemporary approaches to program evaluation: implications for evaluating programs for disadvantaged adults. Syracuse: ERIC Clearinghouse on Adult Education (undated).
Stevens J. Applied multivariate statistics for the social sciences. 3rd ed. Mahwah: Lawrence Erlbaum Associates; 1996.
Stodala Q, Stordahl K. Basic educational tests and measurement. Chicago: Science Research Associates; 1967.
Thomson G. The factorial analysis of human ability. 5th ed. Boston: Houghton Mifflin Company; 1951.
Thorndike RL, editor. Educational measurement. 2nd ed. One Dupont Circle, Washington, DC: American Council on Education; 1971.
Thorndike RL. Applied psychometrics. Boston: Houghton Mifflin Company; 1982.
Veldman DJ. Fortran programming for the behavioral sciences. New York: Holt, Rinehart and Winston; 1967. p. 308–17.
Walker HM, Lev J. Statistical inference. New York: Henry Holt and Company; 1953.
Winer BJ. Statistical, principles in experimental design. New York: McGraw-Hill; 1962.
Worthen BR, Sanders JR. Educational evaluation: theory and practice. Belmont: Wadsworth Publishing Company; 1973.
Yamane T. Mathematics for economists. An elementary survey. Englewood Ciffs: Prentice-Hall; 1962.

Index

A
Adjustment of reliability for variance
    change, 307–308
Analyses menu, 19
Analysis of variance, 87–100
Analysis of variance-treatments by subjects
    design, 90–92
Analysis of variance using multiple regression
    methods, 132–136
Auto and partial autocorrelation, 78
Auto-correlation, 72–78
Average linkage hierarchical cluster
    analysis, 178–180
AxB log linear analysis, 214–216
AxBxC log linear analysis, 217–235
The AxS ANOVA, 92

B
Bartlett Chi-square test for homogeneity, 88
Bartlett test of sphericity, 204–205
Binary files, 7
Binary logistic regression, 145–146
Binary receiver operating
    characteristics, 66–69
Box plots, 35–38
Breakdown, 32–35
Breakdown procedure, 23–24
Bubble plot, 49–51

C
Canonical correlations, 141–145
Cluster analyses, 172–180
Cochran Q test, 249–250
Comma separated field files, 7
Common errors, 20–22

Compare observed to a theoretical
    distribution, 56–57
Comparison of two sample means, 85
Contingency chi-square, 237–239
Correlations in dependent samples, 65–66
Correspondence analysis, 206–210
Cox proportional hazards survival
    regression, 147–148
Creating a file, 7–9
Cross-tabulation, 30–32
CUSUM chart, 324–325

D
Data smoothing, 72
Defect (Non-conformity)
    c chart, 328–330
Defects per unit u chart, 330–332
Differential item functioning, 289–307
Discriminant function / MANOVA, 163–172
Distribution parameter estimates, 23
Distribution plots, 25

E
Eigenvalues and vectors, 356–357
Entering data, 10
Exploration of natural groups, 391–400

F
Factor analysis, 189–194
Files, 7–22
Fisher's exact test, 244
Fixed format files, 7
Frequencies, 27–30
Friedman two way ANOVA, 251–252

G
The general linear model, 141
Generate test data, 312–314
The GradeBook, 363
Guttman scalogram analysis, 285–286

H
Hartley Fmax test, 88
Help, 11
Hierarchical cluster analysis, 172–177
Hoyt reliability, 275–277

I
Installing OpenStat, 3
Item analysis, 269–275
Item banking, 367–371

K
Kaplan-Meier survival test, 256–264
Kendall's coefficient of
    concordance, 245–246
Kendall's tau and partial tau, 255–256
K-means clustering analysis, 177–178
The Kolmogorov-Smirnov test, 265–267
Kruskal-wallis one-way
    ANOVA, 246–248
Kuder-richardson #21 reliability, 277–278

L
Latin and Greco-Latin square designs,
    105–126
Linear programming, 333–335
Log linear screening, 210–235

M
Mann-Whitney U test, 241–243
Matrix files, 7
Matrix operations, 351–361
Median polish analysis, 203–204
Microsoft excel, 15
Multiple groups x *versus* y plot, 57–59

N
Nested factors analysis of variance, 99–100
Neural networks, 373–406
Non-linear regression, 158–161
Normality tests, 46–47

O
Observed and theoretical distributions, 44
One sample tests, 79–82
One, two or three way ANOVA, 87–100
The options form, 9
Options menu, 9

P
Partial and semi_partial correlations, 70–71
Partial auto-correlation, 78
Path analysis, 181–189
Pattern recognition, 389–391
p chart, 326–328
Pie chart, 41–42
Polynomial regression smoothing, 74
Polytomous DIF analysis, 308–312
Probability of a binomial event, 252–253
Product moment correlation, 61–62
Proportion differences, 82–84

Q
QQ and PP plots, 45

R
Random selection, 17
Range chart, 320–322
Rasch one parameter item analysis, 279–285
Resistant line, 47–49
Runs test, 253–254

S
Saving a file, 10–11
S control chart, 322–324
Select a specified range of cases, 16
Select cases, 15, 16
Select if, 17
Sign test, 250–251
Simple linear regression, 63–64
Simulation menu, 20
Single sample proportion test, 81
Single sample variance test, 82
Smooth data, 51–52
Sort, 14
Space separated field files, 7
Spearman-Brown reliability prophecy, 315
Spearman rank correlation, 240–241
2-Stage least-squares regression, 153–157
Stem and leaf plot, 43–44
String labels, 21

Index

Successive interval scaling, 287–289
Sums of squares by regression, 136–140
SVDInverse, 352–353

T
Tab separated field f, 7
Testing equality of correlations, 65
Text files, 7
Three factor nested ANOVA, 101
Three variable rotation, 38–39
Time series analysis, 400–406
T-test, 85–87
Two factor repeated measures
	analysis, 95–99
Two within subjects ANOVA, 129–132

U
Using MatMan, 337–338

V
Variables definition, 8
The variables equation option, 13
The variables menu, 12–14
Variable transformation, 12

W
2 or 3 way fixed ANOVA with 1 case
	per cell, 126–129
Weighted composite test reliability, 278–279
Weighted least-squares regression, 148–153
Wilcoxon matched-pairs signed
	ranks test, 248–249

X
XBAR chart, 317–332
X *versus* multiple Y plot, 52–55
X *versus* Y plots, 39–41

MIX
Papier aus verantwortungsvollen Quellen
Paper from responsible sources
FSC® C105338

If you have any concerns about our products,
you can contact us on
ProductSafety@springernature.com

In case Publisher is established outside the EU,
the EU authorized representative is:
**Springer Nature Customer Service Center GmbH
Europaplatz 3, 69115 Heidelberg, Germany**

Printed by Libri Plureos GmbH
in Hamburg, Germany